Mereology and the Sciences

SYNTHESE LIBRARY

STUDIES IN EPISTEMOLOGY,
LOGIC, METHODOLOGY, AND PHILOSOPHY OF SCIENCE

VOLUME 371

For further volumes:
http://www.springer.com/series/6607

Claudio Calosi • Pierluigi Graziani
Editors

Mereology and the Sciences

Parts and Wholes in the Contemporary Scientific Context

Springer

Editors
Claudio Calosi
Department of Basic Sciences and Foundations
University of Urbino
Urbino
Italy

Pierluigi Graziani
Department of Basic Sciences and Foundations
University of Urbino
Urbino
Italy

ISBN 978-3-319-05355-4 ISBN 978-3-319-05356-1 (eBook)
DOI 10.1007/978-3-319-05356-1
Springer Cham Heidelberg New York Dordrecht London

Library of Congress Control Number: 2014939928

Printed on acid-free paper

Springer is part of Springer Science+Business Media (www.springer.com)

Preface and Acknowledgements

The first idea behind this book came to us at the end of 2011 when we taught a seminar on Mereology and Set Theory at the University of Urbino. During several discussions it became clear that a comprehensive treatment of possible applications of mereology to contemporary sciences was missing. The present work is our modest attempt, as far as it is possible, to fill this gap. It grew from being a proposal for a special issue of a journal into a fully-fledged book.

We want to thank all the contributors to the volume who accepted enthusiastically our invitation and had the patience to satisfy our requests. Each one of them has written a significant and important contribution.

We are particularly grateful to Vincenzo Fano, Yuri Balashov and Achille Varzi for having supported this project right from the very start. A special mention goes to the referees for Springer who read the entire manuscript more than once. They provided invaluable suggestions and detailed criticisms on each and every part of the book: without them it would have been substantially worse.

We would also like to thank Christie Lue at Springer. She was always kind and helpful.

We have discussed many of the materials that are now in the book with both colleagues and friends. They know who they are so there is no need to mention them by name. We want to thank them all.

Contents

Notes on Contributors

Yuri Balashov is a Professor of Philosophy at the University of Georgia. He has published extensively in philosophy of science and metaphysics in leading philosophical journals. He is the author of *Persistence and Spacetime* and a co-editor of *Einstein Studies in Russia* and *Philosophy of Science: Contemporary Readings*.

Dines Bjørner taught at Technical University of Denmark in Copenhagen before retiring in 2007. He was also a computer scientist working for IBM and responsible for establishing the United Nations University International Institute for Software Technology in Macau. He is an expert in domain engineering, requirements engineering and formal methods. He is the author of the influential three-volume monograph *Software Engineering*.

Claudio Calosi is currently a post-doc at University of Urbino, Department of Basic Sciences and Foundations. He has been a Visiting Research Scholar at University of Pittsburgh, University of California at Irvine and New York University. His main research interests are philosophy of science and analytic metaphysics and he has published extensively on both in journals such as *Synthese, Dialectica*, and *Foundations of Physics* to mention a few.

Cristina Coppola studied mathematics at University of Salento and University of Naples. She is currently a research fellow in mathematics at University of Salento. Her primary research interests are mathematics and education on the one hand and foundations of geometry on the other. She is the author of numerous scientific articles published in journals, books and conference proceedings.

Vincenzo Fano is an Associate Professor of Logic and Philosophy of Science at University of Urbino. His works focuses particularly on the foundations of physics and philosophy of psychology. He is the author of five monographs and has authored more than a hundred papers in journals such as *Foundations of Physics* and *Studia Liebniziana*.

Antony Galton is Reader in Knowledge Representation at the University of Exeter, where he is based in the College of Engineering, Mathematics and Physical

Sciences. He works in the area of logic, ontology, and knowledge representation, with a particular focus on space and time. His published books include *The Logic of Aspect* (1984) and *Qualitative Spatial Change* (2000).

Giangiacomo Gerla is a Full Professor of Mathematics at University of Salento. His research focuses, in particular, on fuzzy logic, fuzzy set-theory, multi-valued logic and point-free geometry. He is the author of numerous scientific articles in mathematics and logic. He is also the author of numerous monographs including *Fuzzy Logic. Mathematical Tools for Approximate Reasoning*.

Cody Gilmore is an Associate Professor of Philosophy at the University of California, Davis. He is interested in persistence, death, time travel, universals and propositions. He has published papers on these topics in *Oxford Studies in Metaphysics, Philosophical Perspectives, Philosophical Studies, Australasian Journal of Philosophy, Philosophy Compass, The Stanford Encyclopedia of Philosophy, The Oxford Handbook of Philosophy of Death*, and elsewhere.

Pierluigi Graziani received his Ph.D. from University of Rome La Sapienza and is currently a post-doc and Assistant Professor of Logic at University of Urbino, Department of Basic Sciences and Foundations. His works spans from logic to history and philosophy of mathematics. He has been editor of different volumes and published in different distinguished journals.

Rafał Gruszczyński is an Assistant Professor at the Department of Logic, Nicolaus Copernicus University in Poland. His scientific work focuses on mereology and point-free geometry. He published in *Bulletin of Symbolic Logic*, *Fundamenta Informaticae* and *Bulletin of the Section of Logic*. He is an assistant editor of a quarterly *Logic and Logical Philosophy* (ERIH list) and a supervisor of *Logical Society*, an organization at Nicolaus Copernicus University whose mission is to deepen students' knowledge of advanced topics in logic.

Rom Harré studied chemical engineering but went on to teach mathematics for several years until turning to logic and philosophy of science. He was University Lecturer in Philosophy Science at Oxford until joining Georgetown University in the psychology department where he has been involved in the development of discursive psychology. His most recent books include *Pavlov's Dogs and Schrödinger's Cat* and *Psychology for the Third Millennium*.

Paul Hovda is an Associate Professor of Philosophy and Humanities at Reed College. He specializes in Logic and Metaphysics and has published several papers in distinguished journals including *Journal of Philosophical Logic* and *Philosophy and Phenomenological Research*.

Ludger Jansen is a Lecturer at the Department of Philosophy of the University of Münster, Germany. His research focuses on the metaphysics of science, including biomedical and social ontology. Together with Barry Smith he edited the first introduction to applied ontology in the German language (*Biomedizinische Ontologie*, Zürich 2008).

Jean-Pierre Llored was first trained as a chemist (chemical engineer and then Professor agrégé of Chemistry). He resumed his studies in history and philosophy of science (Master Degree) and obtained two Ph.D.s in epistemology, history of science and technology (Ecole Polytechnique, France) and in philosophy (ULB, Belgium). His research fields are mereology, quantum chemistry, nanotechnologies, and the concept of emergence. Deputy Editor of the journal *Foundations of Chemistry*, he also edited the volume *The Philosophy of Chemistry: Practices, Methodologies, and Concepts* (Cambridge Scholars Publishing, 2013) and is editing the forthcoming volume *La chimie, cette inconnue?* (Hermann, France, 2014).

Andrzej Pietruszczack is a Professor of Logic at the Department of Logic, Nicolaus Copernicus University in Poland. He is an author of two books on mereology: *Metamereology* and *Foundations of Theory of Parthood* (both in Polish) and over 50 papers devoted to mereology, classical and non-classical logic and set theory published, among others, in *Bulletin of Symbolic Logic*, *Studia Logica*, *Fundamenta Informaticae* and *Notre Dame Journal of Formal Logic*. He is a co-founder and an editor-in-chief of a quarterly *Logic and Logical Philosophy* (ERIH list).

Lech Polkowski studied mathematics at University of Warsaw with a particular emphasis on topology. His results are quoted in monographs on set-theoretic topology and topological dimension theory. He received his habilitation in Mathematical Foundations of Computer Science at Warsaw University of Technology in 1994. Since 2000 he has been a Full Professor there, in Polish-Japanese Institute of IT, and in the University of Warmia and Mazury in Olsztyn, Poland. He has given essential contributions in topology, robotics, and computer science. He is the creator of the so-called rough mereology, a fuzzified variant of standard mereology. On this very topic he has authored in 2011 the monograph *Approximate Reasoning by Parts. An Introduction to Rough Mereology*.

Stefan Schulz is a Professor of Medical Informatics at the Institute of Medical Informatics, Statistics and Documentation of the Medical University of Graz, Austria. His research encompasses electronic health records, medical language processing, biomedical terminologies, and the application of formal ontologies for biomedical knowledge representation.

Gino Tarozzi has studied physics and philosophy at University of Bologna. He is currently Full Professor of Logic and Philosophy of Science at University of Urbino. His main research interest is the foundations of quantum mechanics, a topic in which he has given highly significant contributions.

Achille C. Varzi is a Professor of Philosophy at Columbia University. He works mainly in Logic and Metaphysics. He has written extensively in areas as diverse as formal semantics, spatial logic, formal ontology and analytic metaphysics including mereology and mereotopology. He is an editor of the Journal of Philosophy and the author (together with Roberto Casati) of *Holes and other Superficialities* and *Parts and Places*.

Parts, Wholes and Contemporary Sciences

Claudio Calosi and Pierluigi Graziani

Talk about parts, wholes and parthood relations is pervasive. Consider these examples, taken from Winston et al. (1987)[1]:

(1 *a*)	A handle is part of a cup
(2 *a*)	Phonology is part of linguistics
(3 *b*)	A tree is part of a forest
(4 *b*)	This ship is part of a fleet
(5 *c*)	This slice is part of a pie
(6 *c*)	This hunk is part of my clay
(7 *d*)	A martini is partly alcohol
(8 *d*)	Water is partly hydrogen
(9 *e*)	Bidding is part of playing bridge
(10 *e*)	Ovulation is part of the menstrual cycle
(11 *f*)	The Everglades are part of Florida
(12 *f*)	An oasis is part of a desert

Where (*a*–*f*) exemplify six different types of mereological relations, namely (*a*) component-integral object, (*b*) member-collection, (*c*) portion-mass, (*d*) stuff-object, (*e*) feature-activity and (*f*) place-area. It is maybe possible to resist the claim that all of them exemplify truly mereological relations. Yet they provide substantive evidence that we seem to employ parts and wholes in a wide ranging number of contexts and circumstances.

It is not surprising then that reflections about parts and wholes were somehow part (pun intended) of philosophy since its very early days, back to the Pre-Socratics.

[1]Winston et al. (1987). See also Varzi (2014).

C. Calosi • P. Graziani
Department of Basic Sciences and Foundations, Urbino University, Urbino, Italy
e-mail: claudio.calosi@uniurb.it; pierluigi.graziani@uniurb.it

Parmenides and Zeno argued there exists only one object, the Universe. This claim can be understood to mean that the Universe is a mereological atom, i.e. an entity that does not have any parts (or better proper parts). Democritus, possibly to answer some of the Eleatic arguments, maintained that everything was composed of atoms. This translates easily into the claim that every composite object has some atomic parts.[2]

Already in Plato's *Thaetetus*, *Parmenides* and *Timaeus* we find sophisticated analysis of parts and wholes. Aristotle tackles these questions in his *De Partibus Animalium*, *De Generatione et Corruptione* and *Metereology*. Furthermore *Physics* (I.2, 185b11-14) and *Metaphysics* (Z.17 1041b11-33) contain arguably two of the most influential passages about parts and wholes in the entire philosophical literature. Hellenistic philosophy saw Epicurus embracing a form of Democritean atomism and Chrysippus discussing some variant of the mereological paradox which became known as the Growing Paradox,[3] the puzzle of how a thing can retain its identity in face of drastic mereological variations.

Boethius *De Divisione* was one of the cornerstone of philosophical reflection about parts and wholes and its legacy was widespread and influential throughout the middle ages.[4] It can be seen in thinkers as different as Peter Abelard (1079–1142), Albertus Magnus (1206–1280), Bonaventure (1221–1274), Thomas Aquinas (1225–1274), and Duns Scotus (1256–1308). To give just two examples of the depth of the mereological insights that can be found in medieval writings we could point at Ockham's *Questiones Variae* (q. vi art. ii) and Buridan's unedited *Questions on the Physics of Aristotle* (Book 1. Q.10).

Modern philosophy was no ignorant of mereology as witnessed in the works of Gottfried Leibniz (in particular *Monadology*) and Immanuel Kant, especially in his early writings. It is however only with Franz Brentano and his pupils, most notably Edmund Husserl, that a fully fledged theory about parthood relations was developed. In fact many consider his third *Logical Investigation* (1901) as the birthmark of modern mereology.

The first complete formal theory of parthood is however credited to the Polish mathematician Stanislaw Leśniewski in his (1916) *Foundations of General Theory of Sets*. His work was originally published in polish, which significantly restricted the number of people in the philosophical and mathematical community that had access to it. It was Leonard and Goodman's (1941) The *Calculus of Individuals and its Uses* that made the formal theory of parthood available for a wider audience. These two works can be considered the foundations for what is now known as classical mereology.

---

[2]Note that we are not claiming that these philosophers were engaged in thinking about mereology *per se*, but rather that their main metaphysical tenets have an implicit mereological import.

[3]That is closely related to what Unger calls the Problem of the Many. For the Growing Paradox see Sedley (1982). Unger's problem of the many can be found in his Unger (1980). We firstly found discussions of these issues in Normore (2006).

[4]The interested reader can start from Arlig (2011).

These works actually shared another feature. They were originally intended to be a *nominalistic* alternative to standard set theory. Already in 1935 Alfred Tarski[5] had proved that the parthood relation axiomatized by classical mereology forms a complete boolean algebra with no zero element. However this alternative foundational project turned out to be not viable.[6]

Another strand in thinking about parthood relations, that was independent of nominalistic commitments, was developed by Bertrand Russell, especially in his *Our Knowledge of the External World*, and Alfred Whitehead in numerous of his writings. In particular Whitehead's analysis, building also on the work of Theodore de Laguna, offered important insights on the relation between the mereological notion of parthood and the topological notion of connection. This approach did not receive a systematic development before Grzegorczyk's works in the 1960s.

After a somehow silent period, mereology became around 1980 a substantive part (again, pun intended) of another broad philosophical enterprise, namely *formal ontology*.[7] Formal ontology attempts at laying down the bare formal structure of all there is, whatever there is. On this account, regardless of what entities should be admitted in our domains of quantification, all there is must exhibit some general structure and obey some general laws, and the task of ontology would be to describe such structures and laws. It can be probably traced back to Aristotle's theory of being *qua* being, through Husserl theory of objects *as such*, to recent developments in analytic metaphysics. More often than not candidates for formal ontological relations include *identity*, *parthood* and *dependence*. Usually this way of understanding the aim and the scope of ontology is contrasted with what is called material ontology. On this account, made popular by Willard Quine, the task of ontology is instead that of providing a structured inventory of the world. Thus it is concerned with the *nature* of what there is, in sharp contrast with a formal approach that is allegedly domain independent.

This turn of events has made mereology a somewhat familiar formal tool in the hands of ontologists and metaphysicians but has distanced it from the contemporary scientific context. This is because scientific theories, at least when interpreted with even a mild realistic attitude, are closer to a material approach. However talk of part and wholes is no less pervasive in that very context and seems deeply entrenched in our scientific endeavor. Just consider for example the following questions:

(13*g*) Is a quantum system in an entangled state a composite system at all?
(14*g*) Are quarks constituents of baryons?
(15*g*) Is spacetime discrete?
(16*h*) Are singletons atoms and subsets parts of sets?
(17*h*) Does an affine space contain the same parts of a metric space defined over it?

---

[5]See Tarski (1935).

[6]For an analysis of various limits of mereology see Libardi (1994), Tsai (2009, 2011, 2013), and references therein.

[7]Our understanding is in line with Varzi (2010).

(18*h*) Are boundaries the mereological remainders distinguishing closed and open regions?
(19*i*) In what sense the same nucleotides can form different DNA molecules?
(20*i*) Do 1,2 dichloroethene (1,2 DCE) have the same mereological structure represented by the chemical formula $C_2H_2Cl_2$?
(21*l*) Is missing information a part of a database?
(22*l*) Is there a preferred mereological analysis that maximize robot efficiency when moving in a particularly structured environment?

Questions (13)–(22) refer either directly or indirectly to mereological issues. Also, they are typical examples of questions that are addressed by our best scientific theories broadly constructed. We have in fact taken the (*g*)-questions from physics, the (*h*)-questions from mathematics, the (*i*)-questions from natural sciences, biology and chemistry in particular, and finally the (*l*)-questions from computer science and engineering. It is not by chance that the application of mereology to these scientific theories is exactly what is explored in this volume.

This is because, for some of the reasons (and undoubtedly many others) we have outlined in this introduction, a thorough study of parts and wholes within the contemporary scientific context is missing. The aim of the present volume is to fill this gap. It gathers contributions from well renown scholars that are widely recognized as leading contributors to the field and young researchers alike.

Let us then spend a few words about its structure and contents, so as to show clearly how it does indeed fill the gap.

The book is divided in four main parts: (I) Physics, (II) Mathematics, (III) Natural Sciences, (IV) Computer Science and Engineering, exactly those disciplines from which we took questions (13*g*)–(22*l*).

Even if some of the papers contain introductory sections, it was simply not possible to provide an even incomplete overview of the scientific theories and practices that the papers deal with. We have then added a brief introduction to each part that attempts to single out the most relevant philosophical results of each paper. This way the reader could read the most technical materials through a privileged lens and perspective, so to say. Or so we hope. Also, we have suggested background readings that would greatly help the reader that is interested yet unfamiliar with most of the details of the scientific theories that are addressed within each part. These readings include works at different levels of technical sophistication and different levels of philosophical analysis. We are confident that each reader can find the materials that are best suited for both her interests and her knowledge.

Let us just briefly mention here that the series of *Handbook for the Philosophy of Science* published by North Holland contains several volumes that cover the scientific disciplines addressed in the present book. They include Philosophy of Physics, Philosophy of Mathematics, Philosophy of Logic, Philosophy of Chemistry, Philosophy of Biology, Philosophy of Medicine and Philosophy of Technology and Engineering Science.Within each handbook there are scholarly articles, historical surveys and recent developments in the respective fields. Naturally there are many other Handbooks, for example the ones published by Oxford University Press or Cambridge University Press. We just mentioned the series published by North Holland for it is the most complete when it comes to disciplines that are covered in this book.

It should be clear that the present volume fills indeed the gap between philosophical reflections on parts and wholes and the contemporary scientific context. Or better, it begins to fill this gap. We do not pretend this volume to be exhaustive. Rather we would like it to be a first step, a first brick in a much larger bridge (we do mean that the brick is part of the bridge: fourth pun!).

The volume also shows how fruitful an interaction between philosophical considerations and scientific applications can be. On the one hand our understanding of parts and wholes cannot but be significantly improved by a careful study of how such notions work within our best scientific theories. On the other hand, scientific theories usually do not wear their own interpretations and ontological commitments on their sleeves. A difficult, interpretative, philosophical work is required. And rigorous tools such as different formal theories of parthood can play a significant role in elucidating, widening, revealing and clarifying some of the issues that such an interpretative work touches upon. The present volume intends to cover different application of mereological frameworks to different scientific theories. Given the richness and the variety of these scientific domains, differences in methodology, addressed topics, languages, and also in writing styles should be expected. For these (and other) reasons this book is addressed to a wide range of scholars coming from different disciplines and backgrounds and should be important for different communities such as scientists (chemists, physicists, computer scientists), engineers, and philosophers (philosophers of sciences, analytic metaphysicians, ontologists). We believe that, despite an initial difficulty, all these different readers will be rewarded for the effort that is put in reading the entire volume. It encourages an interdisciplinary dialogue that we think is a crucial ingredient in any true and fruitful advancement of knowledge. Our world is varied and complex. We cannot hope for a deep understanding of its complexity if we simply stick to the investigations and results of various disciplines without any attempt to integrate them. Philosophy began with dialogue between different forms of knowledge and descriptions of the world, given by natural sciences, mathematics, physics, even poetry. It is to this kind of dialogue that we should turn back to. And this volume can be read as a part (pun intended) of this dialogue.

We have made sure that the volume covers different approaches to mereological thinking. We have hinted at some of them in this introduction. Some contributions follow the path of what might be called a whiteheadian conception (Graziani and Fano, Coppola and Gerla), whereas some other contributions can be considered the heirs of the Leśniewski-Tarski approach (Gruszczyński and Pietruszczack, Polkowski). Some of them are more reminiscent of the Leonard and Goodman's work (Calosi and Tarozzi, Gilmore), and others even advocate a different mereological perspective (Llored and Harrè).

The volume ends with a brief appendix on different formal theories of parthood that can be found in the literature.[8] It is a significant addition to the volume that it was written by Achille Varzi, one of the leading figures in the field. Also we have included (i) a table that sums up the different symbols used throughout the contributions to help the reader orient herself within different formal notations, and (ii) a selected general bibliography on formal mereology.

As we have already pointed out, we do not consider this volume to offer an exhaustive treatment of parts and wholes within the contemporary scientific context. Rather our hope is that it can lay down the foundations for an excitingly new and not so well trodden territory of interdisciplinary work. And maybe also suggests new possibilities for future researches and researchers.

References

Arlig, A. (2011). Medieval mereology. In Edward N. Zalta (Ed.), *The Stanford encyclopedia of philosophy* (Spring 2014 ed.). http://plato.stanford.edu/archives/spr2014/entries/mereology-medieval/.

Graziani, P. (2014). Proposta di Nuovi Simboli per la Mereologia Formale. In E. Casetta & V. Giardino (Eds.), *Mettere a Fuoco il Mondo*. Special Issue of Isonomia Epistemologica. Available at http://isonomia.uniurb.it/epistemologica.

Libardi, M. (1994). Applications and limits of mereology. From the theory of parts to the theory of wholes. *Axiomathes, 5*(1), 13–54.

Normore, C. (2006). Ockham's metaphysics of parts. *The Journal of Philosophy, CIII*, 737–754.

Sedley, D. (1982). The stoic criterion of identity. *Phronesis, XXVII*, 255–275.

Tarski, A. (1935). Zur Grundlegung der Booleschen Algebra. I. *Fundamenta Mathematicae*, 24, 177–198; Eng. trans. by Woodger, J. H. (1956). On the foundations of the Boolean algebra. In A. Tarski (Ed.), *Logics, semantics, metamathematics, papers from 1923 to 1938* (pp. 320–341). Oxford: Clarendon.

Tsai, H. (2009). Decidability of mereological theories. *Logic and Logical Philosophy, 18*, 45–63.

Tsai, H. (2011). More on the decidability of mereological theories. *Logic and Logical Philosophy 20*, 251–265.

Tsai, H. (2013). Decidability of general extensional mereology. *Studia Logica, 101*(3), 619–636.

Unger, P. (1980). The problem of the many. *Midwest Studies in Philosophy, V*, 411–467.

Varzi, A. C. (2010). On the boundary between material and formal ontology. In B. Smith, R. Mizoguchi, & S. Nakagawa (Eds.), *Interdisciplinary ontology* (Vol. 3, pp. 3–8). Tokio: Keio University Press.

Varzi, A. C. (2014). Mereology. In E. N. Zalta (Ed.), *The Stanford encyclopedia of philosophy* (Spring 2014 ed.). http://plato.stanford.edu/archives/spr2014/entries/mereology.

Winston, M., Chaffin, R., & Herrmann D. (1987) A Taxonomy of part-whole relations. *Cognitive Science, 11*, 417–444.

[8]This renders the volume self sufficient, as far as it is possible. The reader has in fact in one volume an introductory appendix, different applications, list of symbols and selected bibliography.

List of Symbols

Unfortunately there is no standard notation for different mereological notions. Thus the following provides the reader with a minimal list of different mereological symbols found throughout the literature on parts and wholes.[1]

Parthood: x is part of y
$x < y, x \prec y, x \sqsubseteq y, P(x, y)$

Proper Parthood: x is a proper part of y
$x << y, x \prec\prec y, x \sqsubset y, PP(x, y)$

Proper Extension: x is a proper extension of y
$x >> y, x \succ\succ y$

Overlap: x overlaps y
$x \circ y, O(x, y)$

Disjointness: x is disjoint from y
$x \wr y, D(x, y)$

Underlap: x underlaps y
$U(x, y)$

Equality: x equals y
$x \equiv y, EQ(x, y)$

[1]An interesting evaluation of different symbolic notations for classical mereology and a proposal for a new one can be found in Graziani (2014).

Binary Sum: mereological sum [2] of x and y
$Sum(x, y)$, $x + y$

Binary Product: mereological product of x and y
$Prod(x, y)$, $x \times y$

Infinitary Sum: mereological sum of the $\varphi(x)$
$Sum(\varphi(x))$, $S(\varphi(x))$

[2] The second notation is often used in the context of an extensional mereology, when the sum in question is *unique*. The same goes for the second notation for the mereological product below.

Part I
Physics

Introduction to Part I: Mereology and Physics

Physics has stumbled across mereological notions since its early days. Democritean physics is probably the first known explicit example of atomism. On the other hand Aristotelian physics only features gunky objects, i.e. objects that decompose infinitely into proper parts. This debate is far from being settled. Dirac and Feynman, two of the founding fathers of modern quantum physics, both argued that quantum mechanics vindicated atomism, whereas recent works in philosophy of physics, such as Arntzenius (2003),[3] present serious challenges to this view. Furthermore almost every physical theory seem to quantify over composite objects. As such all these theories engage in mereological thinking.

Not only physical theories employ mereological notions. Metaphysical intimations coming from physics seem to suggest or even call for a revision of some of those notions and their behavior. The papers in this section are perfect examples of the fruitful interaction between philosophical reflection on mereology and foundational issues in physics.

Relativity theory, with its replacement of three-dimensional space and one-dimensional time with a more fundamental four-dimensional entity known as spacetime, seems at first sight to be an inhospitable setting for a metaphysics of material objects that maintain that they persist through time by being wholly present at each time of their existence, thus being multilocated in spacetime. This metaphysics is known as Three-dimensionalism. Rather, the argument goes, Relativity favors the rival conception, according to which material objects persist through time by having a temporal part at each time of their existence, thus being singly located in spacetime. This rival conception is called Four-dimensionalism. It is immediately clear the mereological import of such debate, for the alternative

---

[3] Arntzenius (2003).

metaphysics of persistence ascribe different mereological structures to persisting objects. The first two papers deal with these issues.

In the first one (*Building Enduring Objects Out of Spacetime*) Gilmore explores different strategies to build three-dimensional objects out of spacetime regions that are untouched by the relativistic threats. He actually starts off from arguments drawn from General Relativity and Quantum Field Theory and he argues that *coincidentalism*, the view that material objects are constituted by, yet not identical with, spacetime regions, could do the trick. This entails the rejection of two principles of classical mereology, namely *extensionality* and *unrestricted composition*.

In the second one (*Relativistic Parts and Places*) Balashov gives us reasons to abandon a crucial premise of the relativistic argument against Three-dimensionalism, namely that three-dimensional objects are *multilocated at each and every slice* of their spatiotemporal career. He claims that before even asking at which slices of that career relativistic three-dimensional objects are located at we should ask which of those slices are *eligible* to be their locations and that Relativity does not sanction the claim that all of them are indeed eligible. Once again mereological consequences are noteworthy for this entails the rejection of the *unrestricted composition principle*.

The third paper (*Parthood and Composition in Quantum Mechanics*) by Calosi and Tarozzi switches the focus to quantum mechanics. That theory has been credited to have shown that the reductive claim according to which a *composite object is nothing over and above its parts* is false. The paper can be seen as a thorough investigation of whether that claim is warranted and in what sense it is. It is argued that whereas there could be a sense in which this is true, namely in that composite systems have properties that are not immediately reducible to properties of the component parts, in a stricter mereological sense, namely the failure of *extensionality of composition*, this is not true. It is actually shown that quantum systems are a model of extensional mereology.

Finally the paper by Fano and Graziani (*Continuity of Motion in Whitehead's Geometrical Space*) explores a neglected conception in the foundations of spacetime theories, namely the conception of gunk, point-free spaces inaugurated by De Laguna and Whitehead, thus taking us back to some of the issues we opened this brief overview with. Despite the epistemological merits of the proposal they argue that this would have rather unwelcome consequences for the description of motion that is provided by most of our physical theories, even simple ones such as classical mechanics. The tension, they claim, is generated by the following facts: (i) classical mechanics crucially adopts the notion of a *point-particle* in its description of motion; (ii) *sets of (constructed) points* in these Whiteheadian spaces turn out to be *non-connected*; (iii) *connectedness* is a *necessary* condition for *continuity*.

Needless to say there are several good introductions to the physical theories explored in the papers. We limit ourselves to just a few which we believe are both simple (yet rigorous) enough for non-physicists and are not dismissive of more foundational and philosophical issues. Extensive bibliographies are contained in each volume. A classic in Special Relativity is Rindler (1991). A good introduction to the mathematically more demanding General Relativity is Schutz (1985). Two very accessible, yet profound and rigorous introductions to Quantum Mechanics that are particularly suited for philosophers are Albert (1992) and Hughes (1992). A classic, yet more sophisticated textbook is Beltrametti and Cassinelli (1981). Since Gilmore's contribution also mentions Quantum Field Theory we recommend the philosophically inclined Kuhlmann et al. (2002). An advanced must-read in classical mechanics is Goldstein et al. (2001).

References

Albert, D. (1992). *Quantum mechanics and experience*. Cambridge: Harvard University Press.
Arntzenius, F. (2003). Is quantum mechanics pointless? *Philosophy of Science, 70*(5), 1447–1457.
Beltrametti, E., & Cassinelli, G. (1981). *The logic of quantum mechanics*. Reading: Addison-Wesley.
Goldstein, H., Poole, C. P., & Safko, J. L. (2001). *Classical mechanics* (3rd ed.). Reading: Addison Wesley.
Hughes, R. I. G. (1992). *The structure and the interpretation of quantum mechanics*. Cambridge: Harvard University Press.
Kuhlmann, M., Lyre, H., & Wayne, A. (Eds.). (2002). *Ontological aspects of quantum field theory*. Singapore: World Scientific.
Rindler, W. (1991). *Introduction to special relativity* (2nd ed.). Oxford: Clarendon Press.
Schutz, B. (1985). *A first course in general relativity*. Cambridge: Cambridge University Press.

Chapter 1
Building Enduring Objects Out of Spacetime

Cody Gilmore

Endurantism, the view that material objects are wholly present at each moment of their careers, is under threat from supersubstantivalism, the view that material objects are identical to spacetime regions. I discuss three compromise positions. They are alike in that they all take material objects to be composed of spacetime points or regions without being identical to any such point or region. They differ in whether they permit multilocation and in whether they generate cases of mereologically coincident entities.

1.1 Introduction

Let me start with a rough characterization of two main views about persistence:

Endurantism. At least some material objects persist through time; and every material object is temporally unextended and wholly present at each instant at which it exists at all. Moreover, it is not the case that every material object has a different instantaneous temporal part[1] at each different instant at which it exists.

---

[1] The standard definition of 'instantaneous temporal part' runs as follows: 'x is an instantaneous temporal part of y at t' means '(i) t is an instant, (ii) x is a part of y at t, (iii) x overlaps-at-t every part-at-t of y, (iii) x is present at t, and (iv) x is not present at any other instant'. (This is based on Sider 2001, 59.) For other definitions, see Gibson and Pooley (2006, 163), Parsons (2007), and Balashov (2010, 73). The key point is that, in order for a thing y to count as a temporal part of a thing x, y must be a part of x and y must be *spatially co-located* with x at any moment at which y is present.

C. Gilmore (✉)
Department of Philosophy, UC Davis, Davis, CA, USA
e-mail: csgilmore@ucdavis.edu

C. Calosi and P. Graziani (eds.), *Mereology and the Sciences*, Synthese Library 371,
DOI 10.1007/978-3-319-05356-1_1,

Perdurantism. At least some material objects persist through time; every material object has a different instantaneous temporal part at each different instant at which it exists. Material objects that do persist are temporally extended and are at most partially present (not wholly present) at any one instant.

I will introduce more carefully formulated views later on (from Gilmore 2006), but these are adequate for present purposes. Endurantism fits comfortably with presentism and certain other A-theories of time.[2] It also fits together fairly well with a certain brand of B-theoretic eternalism. What I have in mind here is a view like Newton's, according to which substantival space and substantival time are two separate and fundamental entities, and spacetime, if there is such a thing at all, is merely a construct of some sort. (Perhaps spacetime points are identified with ordered <point of space, instant of time> pairs.) Call this view about space and time 'separatist substantivalism'; it should be understood as incorporating eternalism and the B-theory.

But eternalist, B-theoretic endurantism begins to run into trouble as soon as we shift to (i) *relationism* about time or to (ii) a *spacetime* framework, be it substantivalist or relationist. Start with (i). Given eternalism and the B-theory, endurantists face pressure to invoke times or spacetime regions to handle the problem of change.

Suppose that Bob changes from being bent (an hour ago) to being straight (now). If perdurantism is true and Bob has temporal parts, then we can say that it was one temporal part of Bob that was bent and it is a different temporal part of Bob that is straight. If the A-theory is true and there is a metaphysically privileged time, then we can say that Bob himself is straight, not bent (though he was bent). Without temporal parts or a privileged present, however, the most natural account of change is to 'relativize to times': say that Bob is bent at one time (or spacetime region) and straight at another.[3] The idea is that Bob's shapes are really relations: he bears the bent-at relation to one time (or region) and the straight-at relation to another. If, as the relationist claims, there are no such things as times or regions, then this account fails, and it is unclear what else endurantist can put in its place.[4] I will assume, then, that if endurantism is going to find a home in an eternalist, B-theoretic world, such a world will need to include substantival times or spacetime regions. Now consider (ii). Is eternalistic, B-theoretic endurantism tenable in the spacetime framework? By 'the spacetime framework', I mean, roughly, the view that the spatiotemporal is more fundamental than the purely spatial or the purely temporal. Given the spacetime framework, we have a choice between spacetime relationism

[2] A-theories of time all say that there is a time that is present in some absolute, not-merely-indexical sense. That is, they say that there is a 'metaphysically privileged' present time. The B-theory of time denies this. Presentism is an A-theory of time according to which there are no non-present entities (such as, presumably, pre-Socratic philosophers and Martian outposts). Eternalism is the view that the past, present, and future all exist equally. See Sider (2001) and Markosian (2010) for more on these views.

[3] See Haslanger (2003) for an overview of these issues.

[4] See Hawthorne and Sider (2006) for a sophisticated discussion of this issue.

and spacetime substantivalism. Spacetime relationism, according to which there are objects and/or events standing in spatiotemporal relations but there are no spacetime points or regions, is inhospitable to endurantism for reasons that I have just sketched. So we can focus on the substantivalist version of the spacetime framework, which I state as follows:

Spacetime Substantivalism. Spacetime is more fundamental than space or time. There are such things as concrete, substantival spacetime points and/or regions. If there are such things as points or regions of space, these are merely spacetime regions of certain sorts ('columns'). Likewise, if there are such things as instants or intervals of time, these are merely spacetime regions of certain other sorts ('rows').

The view is neutral as to whether spacetime is relativistic. The question we now face is this: how is endurantism affected by the transition from *separatist* substantivalism to *spacetime* substantivalism, be it pre-relativistic or relativistic? (As with separatist substantivalism, I will understand spacetime substantivalism as incorporating eternalism and the B-theory.) Sider (2001) and Schaffer (2009) both argue that endurantism is harmed by this transition. Their argument runs through two claims:

(1) If spacetime substantivalism is true, then so is supersubstantivalism, the view that each material object just is a spacetime region. (They appeal to considerations of parsimony and, in Schaffer's case, fit with physics; more on this in Sect. 1.2.)
(2) If supersubstantivalism is true then perdurantism, not endurantism, is true. (Persisting spacetime regions perdure; they don't endure.)

No analogous argument is available given separatist substantivalism. In particular, separatists have no analogue of premise (1). For they have no locations with which material objects can be plausibly identified.[5] However, as soon as one makes the shift from separatist substantivalism to spacetime substantivalism, one gains the option of identifying material objects with locations (spacetime regions), and with that option available, parsimony (among other things) counts heavily in favor of taking it.

The argument carries real weight. In light of it, there's no denying that the transition from separatist substantivalism to spacetime substantivalism does some harm to endurantism. Still, it's worth asking: if one insists on combining spacetime

[5] They can't identify an object with its location in space, since objects often occupy different regions of space at different times, but no region of space occupies different regions at different times. And of course they can't identify an object with its location in *time* – say, the interval that is the object's total timespan. There are many reasons for this, but one of them is that, again, an object typically occupies different regions of space at different times, but no interval of time does this. Finally, they can't identify a material object with a spacetime region, since they either reject spacetime regions altogether or treat them as set-theoretic constructs; and presumably material objects are not set-theoretic constructs. The shift from space and time to spacetime solves these problems. No region of space is in different places at different times, but there are regions of spacetime that are. And spacetime substantivalists are free to deny that *spacetime* regions are set-theoretic constructs.

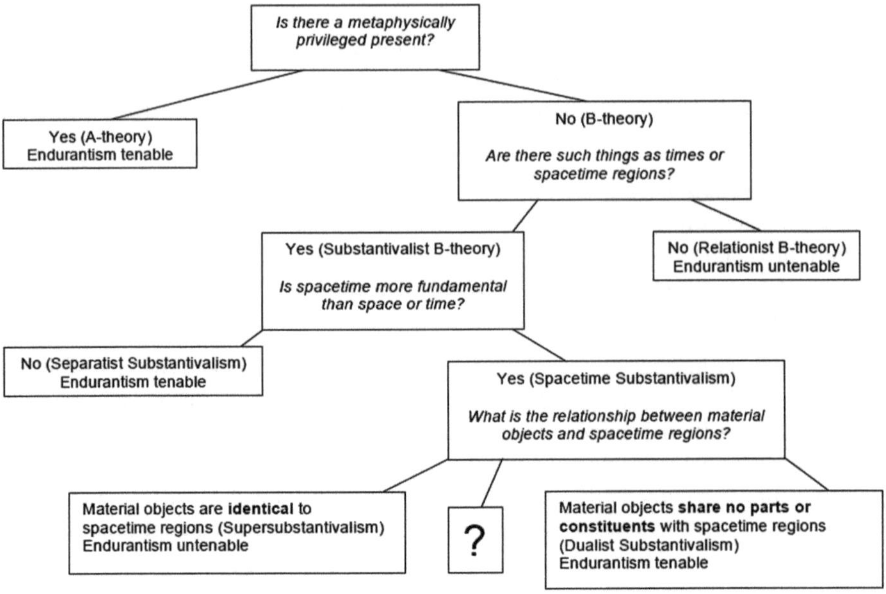

Fig. 1.1 Time and persistence

substantivalism with endurantism, how should one do it? Let me be more specific. Suppose that, on the basis of considerations given in support of (1), one rejects *dualistic substantivalism*, the view that material objects *occupy* spacetime regions but are never identical with any region and indeed never even share any parts or constituents with any region. In that case, how should one combine spacetime substantivalism and endurantism?

In this chapter I explore several such combinations, some of them new, and I chart pros and cons of each. Though I take no stance on which, if any, of these packages is true, I suggest that some are promising and worthy of further attention. (See Fig. 1.1 for a map of the terrain covered so far. A 'close up' on spacetime substantivalism – and its species – appears toward the end of the chapter.)

1.2 From Substantivalism to Supersubstantivalism to Perdurantism

In this section I give a quick sketch of the considerations in support of (1) and (2).

(1) Given spacetime substantivalism, there seem to be two main options concerning the status of material objects. First, one can be a dualist substantivalist, in the sense described above. This has been the standard default position for virtually all spacetime-friendly endurantists and even for some perdurantists (Hudson 2001, 2005). Second, one can say that each material object is identical

to some spacetime region – specifically, the object's path, the region that exactly contains the object's complete career or life-history. This is *supersubstantivalism.*[6] (As I noted above, supersubtantivalism becomes a tenable option only given substantivalism about *spacetime*. Substantivalists who take space and time to be separate and fundamental entities have no locations with which material objects can be plausibly identified.) Of these two views – dualist substantivalism and supersubstantivalism – considerations of parsimony favor the latter. Dualist substantivalism is unparsimonious with respect to *ontology*, since it embraces (i) sui generis, substantival spacetime points and/or regions and (ii) sui generis material objects that occupy spacetime but that are not in any way constructed from the same basic ingredients as spacetime. And dualist substantivalism is unparsimonious with respect to *ideology*, since its proponents will presumably need some primitive, fundamental occupation predicate to state the facts about how material objects relate to spacetime regions. Supersubstantivalism economizes on ontology, since it avoids sui generis material objects, and it economizes on ideology, since it has no need for a primitive, fundamental occupation predicate. According to the supersubstantivalist, for a material object to *occupy* a region is just for the material object to *be* that region.

Jonathan Schaffer offers a number of further considerations that he takes to favor supersubstantivalism over dualist substantivalism. Two of his arguments are worth quoting at length:

> The argument from General Relativity: General Relativistic models are Triples $<M, g, T>$ where M is a four-dimensional continuously differentiable point manifold, g is a metric-field tensor, and T is a stress-energy tensor (with both g and T defined at every point of M, and with g and T coupled by Einstein's field equations). *There are no material occupants in* $<M, g, T>$ *triples*. That is, the distribution of matter in General Relativity is not given via a list of material objects in occupation relations to regions. Rather the distribution is given by the stress-energy tensor, which is a field, and thus naturally interpreted as a *property of the spacetime* ... Thus Earman suggests identifying M with the spacetime manifold, and treating g and T as properties of M: 'Indeed, modern field theory is not implausibly read as saying the physical world is fully described by giving the values of various fields, whether scalar, vector, or tensor, which fields are attributes of the space-time manifold M' (Earman 1989, p. 115; Schaffer 2009, 142, italics original).

> The argument from Quantum Field Theory: Quantum Field Theory, like General Relativity, is a theory of fields (which again are naturally interpreted as states of the spacetime) rather than material occupants. ... Thus in quantum field theory, 'particles' turn out to be excitation properties of spacetime itself, as d'Espagnat explains: 'Within [quantum field theory] particles are admittedly given the status of mere properties, ...but they are properties of something. This something is nothing other than space or space-time, which, being locally structured (variable curvature), have indeed enough 'flexibility' to possess

[6]As I will understand it, supersubstantivalism is neutral as to which regions count as material objects. (Every region? Every 'matter-filled' region? Every maximal continuous matter-filled region?) And then there is the further question of what counts as being 'matter-filled'. Presumably this will need to be spelled out in field-theoretic terms, but even so the answer is hardly straightforward. Again, supersubstantivalists are free to disagree amongst themselves on these questions. They are united only in claiming that all material objects are regions.

infinitely many 'properties' or particular local configurations' (Schaffer 2009, 142–3, italics original).

At the very least, Schaffer makes a convincing case to the effect that *many leading authorities* in physics and the philosophy of physics believe that sui generis material objects play no role in General Relativity or Quantum Field Theory and, further, that the existence of such material objects may be positively in tension with these theories. (See Schaffer's paper for many further quotations and references.) On the assumption that substantivalists must choose between dualistic substantivalism and supersubstantivalism, then, the case for (1) is strong.

(2) Why think that supersubstantivalists ought to be perdurantists, not endurantists? The answer, roughly, is that spacetime perdures. More carefully: if spacetime region r is the path of *persisting* object, then – barring some highly exotic view about spacetime[7] – r perdures; in particular, r is temporally extended and has (proper) temporal parts. So, if o is *identical* to r, then o perdures too. All persisting material objects perdure, according to supersubstantivalism.

In sum: for those metaphysicians who are seeking to develop a viable form of endurantism that harmonizes with physics, there is reason to hope that endurantism can be freed from a commitment to dualistic substantivalism. Not only does dualistic

[7]Here are four such views. (i) **Extended Simple Regions**. Spacetime might be composed of spatially and temporally extended but mereologically simple 'grains'. (See Braddon-Mitchell and Miller (2006) and Dainton (2010) for discussion of related views.) Such a grain might count as persisting (since it's temporally extended), but it wouldn't have any proper temporal parts, and so might not count as perduring. (ii) **Spatially Gunky Spacetime**. Spacetime might be 'spatially gunky' and altogether lacking in proper temporal parts: suppose that every spacetime region is complex, spatially extended, and of infinite temporal extent in both temporal directions, so that each region is eternal and composed of spatially smaller regions. These regions would count as persisting but not as perduring (and even opponents of extended simples can believe in them). (iii) **Restricted Composition on Spacetime Points**. Suppose that all spacetime regions are composed of spatially-and-temporally-unextended, mereologically simple spacetime points, and that some spacetime points compose something iff they are arranged 'complete path of a living organism'-wise. Then, since no living organism has a spacetime point or another living organism as a proper temporal part (let's assume), it's plausible that no temporally extended region has any proper temporal parts. (The pluralities of simples that would compose the temporal parts of such regions, if they composed anything, do not in fact compose anything.) In that case there could be regions that persist but do not perdure. (iv) **Mereologically Coinciding Regions Without Strong Supplementation**. Suppose that all spacetime regions are composed of spatially-and-temporally-unextended simple spacetime points and that every plurality of points composes a region. But suppose further that there is at least one plurality of points, the ps, that compose two different regions, r1 and r2, such that: (a) r1 and r2 are both spatially and temporally extended, (b) r1 has a full distribution of proper temporal parts, and (c) r2 does not have any proper temporal parts. Thus the relationship between r1 and r2 is like the relationship between a statue and a lump that are composed of the same simples but that differ in that the head of the statue is a part of the statue but not of the lump. (Strong Supplementation – the principle that if x is not a part of y, then x has some part that fails to overlap y – is violated in such cases.) In such a case r2 would persist but not perdure. (Eagle (2010) floats a view that sanctions mereologically coincident spacetime regions but does not suggest that they might different with respect to having temporal parts.)

substantivalism fail the parsimony test, but experts tell us that it's in tension with our best physical theories.

1.3 First Compromise: The Path Constitution View

Fortunately, there is room to maneuver here. For even if we accept spacetime substantivalism and reject full-blown dualism about regions and their material occupants, we need not embrace supersubstantivalism.

1.3.1 Outlining the View

Instead of taking material objects to be *identical* with the regions that are their paths, one might take them merely to *coincide mereologically* with those regions. The idea would be that the relationship between a material object and its spacetime path is the same as the relationship often taken to hold between a statue and the lump of clay that constitutes it: mereological coincidence without identity. (Say that x *mereologically coincides with* y if and only if x and y overlap – share parts with – exactly the same things.) As far as I am aware, this view was first entertained in print by John Hawthorne:

> One might take the further step of not treating occupation as fundamental. The statue and lump are mereologically coincident. Perhaps they are also mereologically coincident with a spatiotemporal region. Occupation can then be defined in terms of mereological relations to regions. And just as we typically picture the statue as inheriting certain properties – weight and so on – from the lump by mereological coincidence, we can here think of various objects as inheriting various magnitudes associated with fields by mereological coincidence with spacetime regions which in turn are the fundamental bearers of field values (Hawthorne 2006, 118, n. 18).

Following Schaffer, let's use the term 'monistic substantivalism' for the view that each material object is either identical to or mereologically coincident with some spacetime region. Monistic substantivalism comes in two main versions: the *identity* version, a.k.a. *super*substantivalism, which holds that each material object just is a region, and the *constitution* version, which holds that at least some material objects are not identical to any region, but that each of them coincides mereologically with a region.

The constitution view achieves some measure of ontological parsimony, since it treats material objects not as *sui generis* entities but as things that, informally speaking, are composed of the same basic ingredients as spacetime regions themselves, and it is parsimonious with respect to ideology, since it allows us to define 'occupies' as 'coincides with', rather than treating it as a fundamental primitive. Further, as Schaffer notes, it harmonizes with General Relativity and Quantum Field Theory:

> The constitution ... [version] of monism can claim parsimony, and can claim fit with General Relativity and Quantum Field Theory, insofar as these issues only concern the fundamental ontology. The constitution views preserve the fundamental ontology of a spacetime bearing fields (Schaffer 2009, 143–144).

Suppose, then, that we opt for the constitution view.

How would this help endurantism?[8] It's not at all clear that it would, since it's tempting to think that if x mereologically coincides with y, and y perdures, then x perdures too. But one possibility is this. In opting for the constitution view, we open up logical space for the doctrine that a given plurality of spacetime points, the ps, compose (at least) two things: (i) a region, r, which is temporally extended and has a full distribution of instantaneous and non-instantaneous temporal parts, *the ts*, and (ii) a material object, o, which is temporally extended and co-located with r, but which does not have any of the ts as parts and indeed does not have any proper temporal parts at all. (Presumably o and r differ with regard to their de re modal profile as well, so that o but not r could have had, say, a shorter temporal duration.) The core idea here is that the relationship between o and r is like the relationship between a statue and a lump of clay that are both composed of the same simples but that do not have exactly the same parts: e.g., the statue, but not the lump, has the *head* of the statue as a part (Lowe 2003). (As pointed out in note 7, this requires rejecting Strong Supplementation, the principle that says that if x is not a part of y, then x has a part that fails to overlap y.) At this point it will be convenient to fill in some details that have so far been implicit:

The Path Constitution View

Absolutism: There is only one fundamental parthood relation, it is a two-place relation (expressed by 'x is a part of y'), and it does not hold relative to times, locations, sortals, or anything else.

Plenitude for Regions: Each set of spacetime (points and/or[9]) regions has at least one spacetime region as a fusion.[10]

Path Coincidentalism: Each material object coincides with a spacetime region (its path), but no material object is identical to any spacetime region.

No Fundamental Occupation: The predicate 'x occupies y' is not fundamental; it is defined in mereological terms, as 'x coincides with y', or perhaps 'y is a region, and x coincides with y'.

---

[8]Hawthorne (2006) and Schaffer both seem to think that the constitution version is friendlier to certain forms of endurantism than is the identity version, although neither goes into much detail on this point. Hawthorne focuses mostly on forms of endurantism (framed in terms of grounding or metaphysical dependence) that will not concern us here. Schaffer's reason for taking the constitution-version to be endurance-friendly is not clear to me. He writes that 'the constitution view ... does not entail four-dimensionalism .... Presumably the constituted object could have different persistence conditions than its constituting matter [a spacetime region]' (Schaffer 2009, 137).

[9]Henceforth points (if there are any) count as regions.

[10]This view would fail if (i) some sets of regions had no fusion at all, in which case a form of restricted composition would be true or (ii) some set of regions had more than one spacetime region as a fusion, as discussed in note 7.

Regions Have Temporal Parts: Each persisting spacetime region has proper temporal parts.

Objects Lack Temporal Parts: There are material objects, but none of them has proper temporal parts.

Parts of Objects: A material object x is a part of a material object y only if some region that x occupies is a part of some region that y occupies.

The Path Constitution View (PCV) takes no stand on which spacetime regions constitute material objects. (Every region? Every region at which certain fields have an everywhere positive value?) Nor does it take a stand on *how many* material objects are constituted by a given region that constitutes at least one material object. (One? Two? Continuum-many?)

We've already mentioned the main virtues of the PCV: parsimony, fit with GTR and QFT, and – for those with endurantist sympathies – avoidance of temporal parts of material objects.

One potential drawback of PCV – for those who are attracted to a certain brand of endurantism – is that it treats persisting material objects as temporally extended and singly located in spacetime. Second, and relatedly, PCV denies that any fundamental parthood relation ever holds between, say, an oxygen atom with a one billion year-long career and a human being with a 90-year-long career. For it often happens that the path of such an atom *overlaps* the path of a human being, but it never happens that the path of such an atom is a *part* of the path of a human being. I elaborate on these issues below. Toward the end of the paper I will mention a pair of problems that afflict all three of the compromise positions to be discussed in this paper.

1.3.2 Problems for the Path Constitution View

In stating these problems it will be convenient to work with precise definitions of three notions: the notion of being *weakly located* at a region, the notion of an object's *path*, and the notion of *persisting*. Our definitions will invoke (i) a primitive predicate for parthood (which we take to be reflexive and transitive) and (ii) a predicate for occupation. Informally, to say that x occupies r is to say that x has (or has-at-r) exactly the same shape and size as r and stands (or stands-at-r) in all the same spatiotemporal relations to things as does r. But of course the friend of PCV does not take 'occupies' as primitive; rather she defines it in terms of mereological coincidence as specified earlier.

Now for the definition of 'is weakly located at'. Intuitively, to say that x is weakly located at r is to say that r is 'not completely free of' x (Parsons 2007); thus Russia is weakly located in Europe, in Asia, in Siberia, and in the Milky Way, but not in the Andromeda Galaxy. (Pretend that Russia is a material object and the rest are all spacetime regions). Our official definition will be this: 'x is weakly located at r' means '$\exists r^*$[x occupies r^*& r^* overlaps r]'. In words: 'x occupies something that overlaps r', where 'overlaps' means 'shares a part with'.

As for the notion of an object's path: intuitively, my path is the spacetime region that I exactly sweep out over the course of my career. Although it is natural to speak as though each object has at most one path, we will not build this into our definition. We will say: 'r is a path of x' means '$\forall r^*$ [r overlaps $r^* \leftrightarrow x$ is weakly located at r^*]', that is, 'r overlaps all and only those entities at which x is weakly located'. It follows from this definition (together with the reflexivity of parthood) that if both r and r* are paths of x, then r and r* coincide. So, although we won't assume that no object has more than one path, we are committed to the view that no object has two paths that fail to coincide with each other.

Finally, we can say that 'x *persists*' means '$\exists r \exists r_1 \exists r_2$ [r is a path of x & r_1 is a part of r& r_2 is a part of r& r_1 absolutely earlier than r_2]'. In other words, to persist is to have a path some parts of which are absolutely earlier than others. So much for definitions.

Now, just as a matter of usage, when one says that a thing 'endures', one can mean at least two things. First, one can mean that the thing persists but does not have temporal parts. Call this *mereological endurance.* Second, one can mean that the thing persists and occupies many different spacetime regions, each of them instantaneous or spacelike. Call this *locational endurance.* There is a corresponding ambiguity in the term 'perdure'. When one says that a thing perdures, one can mean that it persists and has (a sufficiently full distribution of) temporal parts, or that it persists and occupies only its path (or paths, if it has more than one). Call the former *mereological perdurance* and the latter *locational perdurance.* (See Gilmore 2006 for more on this.)

Some philosophers seem to think that material objects endure both mereologically and locationally, while others seem to think that they perdure both mereologically and locationally. But there is logical space for mixed views. One might take material objects to mereologically endure but locationally perdure, or to mereologically perdure but locationally endure. See Fig. 1.2 (from Gilmore 2008, 1230) for an illustration of these options.

As we have seen, PCV accommodates mereological endurance. Since there is logical space to say that two entities coincide with having exactly the same parts, there is logical space to say that Obama lacks temporal parts but coincides with a spacetime region that has temporal parts.

Problem 1: PCV rules out locational endurantism. However, PCV does not accommodate locational endurance. Given the definition of 'occupies' in No Fundamental Occupation, we get the result that any two regions occupied by Obama coincide with each other. But, together with our other definitions, this entails that Obama occupies only his path(s), that is, that he locationally *perdures.*

Loosely stated, the problem is this. The locational endurantist wants to say that (i) although Obama's path is temporally extended, each of the regions that Obama occupies (each of his 'locations') is temporally unextended, and that (ii) there are a great many pairs of these locations that do not even overlap, much less coincide. But given the definition of 'occupies' built into PCV, we cannot say that. Instead, we have to say that Obama occupies only those regions with which he coincides. And he can coincide with two different regions only if they coincide with each other. So he

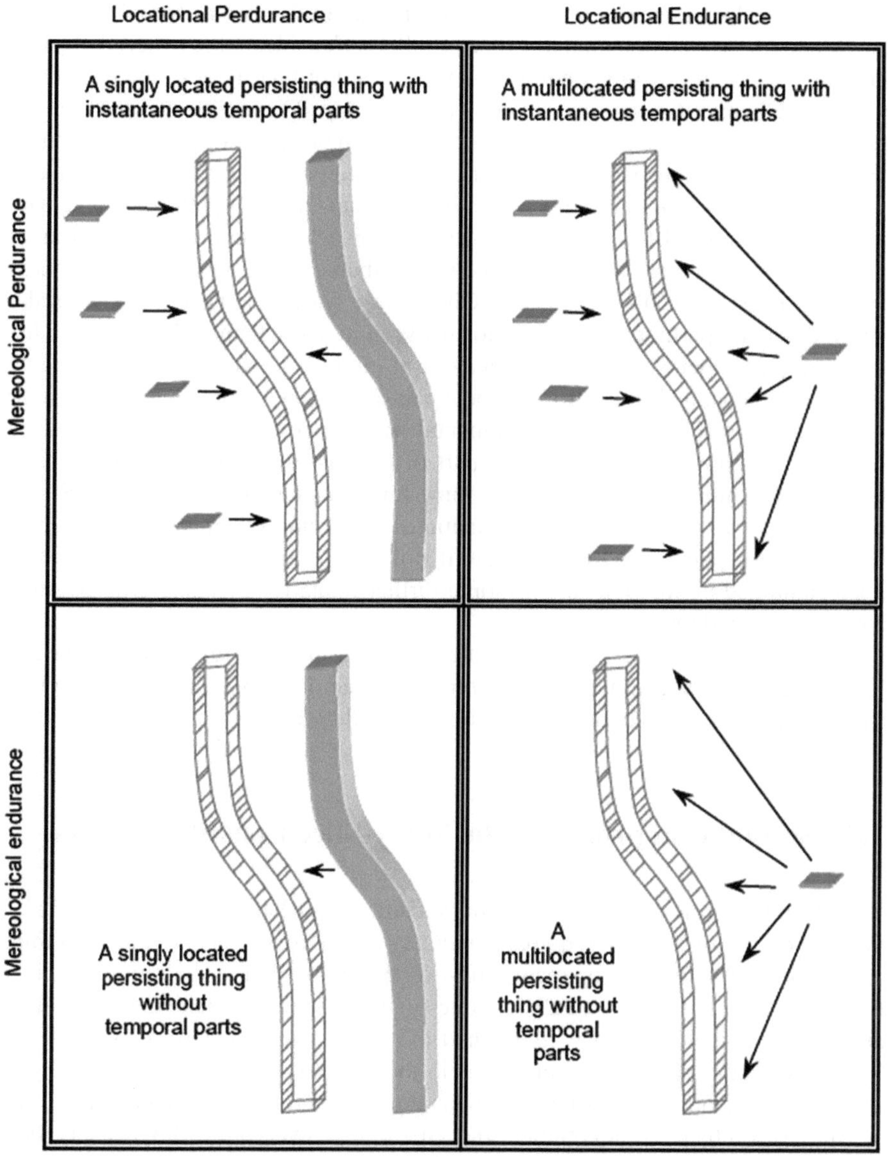

Fig. 1.2 Four forms of persistence

can *occupy* two different regions only if they coincide with each other. He cannot occupy two *non-coinciding* regions, not to mention two *non-overlapping* regions. So, for what it's worth, locational endurantists will need to reject PCV. This is the first potential drawback mentioned above.

Problem 2: Gain and loss of parts (in a fundamental sense of 'part'). Now let me turn to the second potential drawback for PCV. Consider some material object m that satisfies the following conditions: (i) we would ordinarily describe m as being a *part* of Obama *at some time*, (ii) m's path overlaps Obama's path, and (iii) m's path is not part of Obama's path, perhaps because m pre-dates or post-dates Obama, or perhaps because m is for some period of time spatially outside of Obama. In particular, (iv) some parts of m's path fail to overlap Obama's path, and some parts of Obama's path fail to overlap m's path. The object m might be an electron, an oxygen atom, or a tooth that was pulled when Obama was a boy. For concreteness, let's supposes it's a DNA molecule. Given these assumptions, PCV tells us that no fundamental parthood relation holds between m and Obama. Granted, if m had temporal parts, then some *temporal part* of m might be a part, in the fundamental sense, of Obama; and m itself might be a part of Obama in some *non-fundamental sense*; but m itself is not in any fundamental sense a part of Obama.[11]

Intuitively, however, m itself is a part of Obama, in some fundamental sense of 'part'. Put more carefully: there is some fundamental parthood relation R such that, if R is two-place, then R is instantiated by m and Obama in that order (or by the ordered pair <m, Obama>), and if R is a three-or-more-place relation, then it's instantiated by m, Obama, and some further relata (or by some ordered -tuple containing m and Obama).

In short, people have DNA molecules as parts, in some fundamental sense of 'part'. We should accommodate this point if we can do so without paying too high a price. PCV doesn't accommodate it. So we should look elsewhere.

1.4 Second Compromise: The Many-Slice Constitution View

Why does PCV rule out locational endurantism? In nutshell, it's because PCV says that (i) *occupying* a region requires *coinciding* with that region and that (ii) a thing can't coincide with each of many non-overlapping regions. The commitment to (ii) arises from the fact that PCV assumes that parthood is reflexive and transitive and that 'x coincides with y' is defined as '∀z [z overlaps x iff z overlaps y]'. These are highly plausible assumptions in the context of the claim, made explicit in Absolutism, that the relevant fundamental parthood relation is two-place.

But Absolutism is negotiable. Indeed, almost everyone who accepts both endurantism and B-theoretic eternalism *already* rejects Absolutism for independent

[11]To see this, note first that, given PCV together with our set-up, no region occupied by m is a part of any region occupied by Obama. But then, by Parts of Objects, we get the result that m is not a part of Obama. So the fundamental parthood relation expressed by 'is a part of' doesn't hold between m and Obama. And according to Absolutism, this is the *only* fundamental parthood relation.

reasons.[12] The idea goes roughly as follows. Objects gain and lose parts over time. A certain DNA molecule, m, is a part of Obama at one time but not at another. If the present were metaphysically privileged, we might be able to capture this fact in terms of tense operators and a two-place parthood predicate: $\neg Part(m, obama) \& WAS[Part(m, obama)]$. If things had temporal parts, we could try to capture the fact in terms of a non-fundamental, time-relative parthood predicate, defined in terms of the notion of a temporal part and ultimately in terms of a fundamental two-place parthood predicate (Sider 2001). But without temporal parts or a privileged present, the most natural option is to say that the fundamental parthood relation holding between material objects is a more-than-two-place relation.

It bears repeating that this is an *independent* motivation for dropping Absolutism. Making room for monistic substantivalism has typically been the farthest thing from endurantists' minds. And yet they – or at least the B-theoretic eternalists among them – have already rejected Absolutism almost universally.

But it turns out that once we drop Absolutism, we can articulate a natural notion of coincidence (or 'coincidence-at') in terms of which we can say that a given object coincides (at different times or locations) with different regions that do not overlap (at any time or location) each other. This lets us say that Obama occupies – and coincides with – each in a series of temporally unextended spacetime regions, just as a wave coincides (at different times) with each in a series of wave-shaped portions of water. Thus by dropping Absolutism, we open up a way to combine locational endurantism with monistic substantivalism.

As before, we will need to reject (the appropriately restated version of) Strong Supplementation if we are to avoid the result that persisting material objects have temporal parts. For we will assume that each material object *mereologically coincides* with each in a series of instantaneous slices of the object's path. If it turned out that the material object had these slices as *parts*, they would count as *temporal* parts of the object. So we will need to say that, in some cases, an object x mereologically coincides with an object y but does not have y as a part. This conflicts with Strong Supplementation.

1.4.1 Outlining the View

I suspect that this basic strategy can be implemented in a variety of ways, depending upon what Absolutism is replaced with. One tempting suggestion is to replace it with

[12]Many have argued that the fundamental parthood relation for material objects is a three-place relation expressed by 'x is a part of y at z', with two slots for material objects and one slot for a time (Thomson 1983; Van Inwagen 1990; Koslicki 2008) or a region of space or spacetime (Rea 1998; Hudson 2001; McDaniel 2004; Donnelly 2010). As far as I am aware, the only self-described B-theoretic endurantist who accepts Absolutism is Parsons (2000, 2007).

(3P) The fundamental parthood relation for material objects is a three-place relation expressed by ‘x is a part of y at z’, with one slot for the part, one slot for the whole, and a third slot for a time, region of space, or region of spacetime.

On the basis of considerations that do not concern monistic substantivalism, I have argued (Gilmore 2009) that 3P is inferior to

(4P) The fundamental parthood relation for material objects is a four-place relation expressed by ‘x at y is a part of z at w’, with one slot for the part, one slot for a *location* of the part (e.g., a spacetime region), one slot for the whole, and one slot for a *location* of the whole (e.g., a spacetime region).[13]

So I will make use of 4P in what follows. For all I know, 3P and 4P would both serve equally well for task at hand in this chapter. I am opting for 4P only because I take it to be preferable on grounds that will not concern us here.

Now, to help firm up the reader’s grasp of my proposed four-place parthood relation, let me set out some principles that plausibly govern it. I’ll get all these principles (and some associated definitions) out on the table quickly, then I’ll supply some examples, in diagram form, that should help to clarify the principles. So please bear with me. First, the *Location Location Principle*:

(*LLP*) $\forall x \forall y \forall z \forall w[P(x, y, z, w) \rightarrow [L(x, y) \& L(z, w)]]$
If x at y is a part of z at w, then x occupies y and z occupies w.

This just makes explicit the assumption that the second and fourth slots are reserved for locations of the part and whole, respectively. Second, an analogue of the reflexivity of parthood:

$(R_{4P})\ \forall x \forall y[L(x, y) \rightarrow P(x, y, x, y)]$
If x occupies y, then x at y is a part of x at y.

We can’t say ‘for all x and all y, x at y is a part of x at y’ since, together with *LLP*, this would entail that everything occupies everything, which is obviously false. R_{4P} is the most natural alternative. Third, an analogue of the transitivity of parthood:

$$(T_{4P})\ \forall x_1 \forall y_1 \forall x_2 \forall y_2 \forall x_3 \forall y_3\ [[P(x_1, y_1, x_2, y_2) \& P(x_2, y_2, x_3, y_3)] \rightarrow P(x_1, y_1, x_3, y_3)]$$

It will also be useful to define predicates for overlapping and coincidence:

$(D_O)\ O(x_1, y_1, x_2, y_2) =df\ \exists x_3 \exists y_3\ [P(x_3, y_3, x_1, y_1) \& P(x_3, y_3, x_2, y_2)]$
‘x_1 at y_1 overlaps x_2 at y_2’ means ‘some x_3 at some y_3, is a part both of x_1 at y_1 and of x_2 at y_2’

$$(D_C)\ CO(x_1, y_1, x_2, y_2) =df\ [L(x_1, y_1) \vee L(x_2, y_2)] \& \forall x_3 \forall y_3\ [O(x_3, y_3, x_1, y_1) \leftrightarrow O(x_3, y_3, x_2, y_2)]$$

[13] Kleinschmidt (2011) independently proposes 4P and some of the same 4P-appropriate mereological principles to be given here. But she eventually rejects 4P.

'x_1 at y_1 coincides with x_2 at y_2' means 'either x_1 occupies y_1 or x_2 occupies y_2, and for any x_3 and y_3, x_3 at y_3 overlaps x_1 at y_1 if and only if x_3 at y_3 overlaps x_2 at y_2'

The first clause in D_C is needed to avoid the result that Obama, at a region r_1 on the moon, coincides with Putin, at a region r_2 on Jupiter. (Since Obama does not occupy r_1, nothing (at any location) is a part of him there, and so nothing (at any location) overlaps him there. Similarly for Putin and r_2. It follows that exactly the same things, at exactly the same locations, overlap Obama at r_1 as overlap Putin at r_2.) With the first clause in place, however, we can show (given R_{4P}, T_{4P}, and D_O) that if o_1 at r_1 coincides with o_2 at r_2, then o_1 occupies r_1 *and* o_2 occupies r_2. In slogan form: you can't coincide with things at regions at which you don't occupy. We will also want to define a predicate for fusion. To do this, we can think of fusion as a three-place relation that holds between a *thing*, a *set*, and a *location* of the thing, where the set in question is a set of ordered <thing, location of that thing> pairs:

$$(D_F)\ F(y, s, y^*) =_{df} \exists z(z \in s) \& \forall z [z \in s \rightarrow \exists w \exists w^* [z = <w, w^*> \& P(w, w^*, y, y^*)]] \& \forall z \forall z^* [P(z, z^*, y, y^*) \rightarrow \exists u \exists w \exists w^* [u \in s \& u = <w, w^*> \& O(w, w^*, z, z^*)]]$$

In words, y fuses s at y* just in case: (i) s is a non-empty set, (ii) each member of s is an ordered pair whose first member at its second member is a part of y at y*, and (iii) for any z and any z*, if z at z* is a part of y at y*, then there is some ordered pair in s whose first member at its second member overlaps z at z*. D_F does not have the result that material objects have sets or ordered pairs as parts. When a thing fuses a set of ordered pairs, it has the *first members* of those ordered pairs as parts, not the pairs themselves, and not the set of them.

Now for a pair of diagrams to illustrate these concepts. Figure 1.3 depicts a case in which two different composite objects (f and g) fuse the same simples (a, b, and c) and hence count as coinciding. These composite objects also occupy the same region. Figure 1.3 may also be useful in that it illustrates cases of overlapping and cases in which our reflexivity and transitivity principles (R_{4P} and T_{4P}) apply. The *raison d' étre* of three-place or four-place parthood is the need to accommodate cases in which an object is *multilocated* (occupies two or more non-coinciding spacetime regions) and exhibits *mereological variation* from one location to another (has parts at one of its locations that it does not have at another). Multilocation is missing from Fig. 1.3. So it will be useful to consider another case.

In the case depicted by Fig. 1.4, m is a composite object that occupies two different regions: rm_1 and rm_2. Further, m has different parts at different locations: at rm_1, m has d but not a as a part, and at rm_2, m has a but not d as a part.

We can think of m as being analogous to an enduring human being who is composed of different parts at different times at which it exists or at different spacetime regions that it occupies. At the earlier region rm_1, m is composed of a, b, and c (at certain locations of these objects) and at the later region rm_2, m is composed of d, b, and c (at certain later locations of these objects).

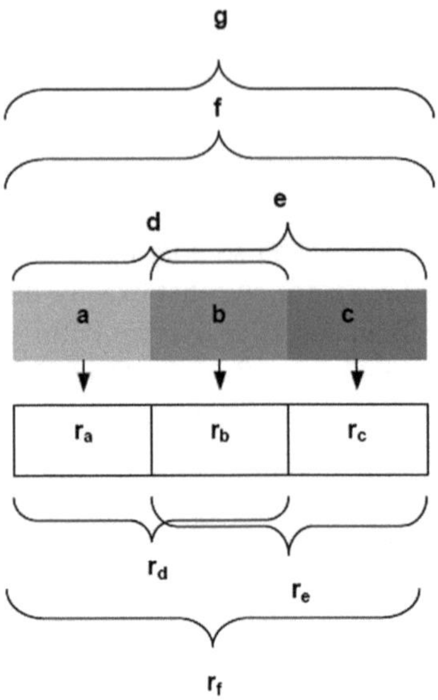

Some facts about the case

1. a – g are material objects
2. $r_a - r_f$ are spacetime regions
3. $g \neq f$
4. a occupies r_a, . . . , f occupies r_f, and g occupies r_f
5. a at r_a is a part of a at r_a (by 4 and R_{4P})
6. b at r_b is a part of d at r_d
7. b at r_b is a part of e at r_e
8. d at r_d overlaps e at r_e (by 6, 7, and D_O)
9. e at r_e is a part of f at r_f
10. b at r_b is a part of f at r_f (by 7, 9, and T_{4P})
11. f fuses the set $\{<a, r_a>, <b, r_b>, <c, r_c>\}$ at r_f, as does g
12. f fuses the set $\{<d, r_d>, <e, r_e>\}$ at r_f , as does g
13. f at r_f coincides with g at r_f

Fig. 1.3 Four-place parthood

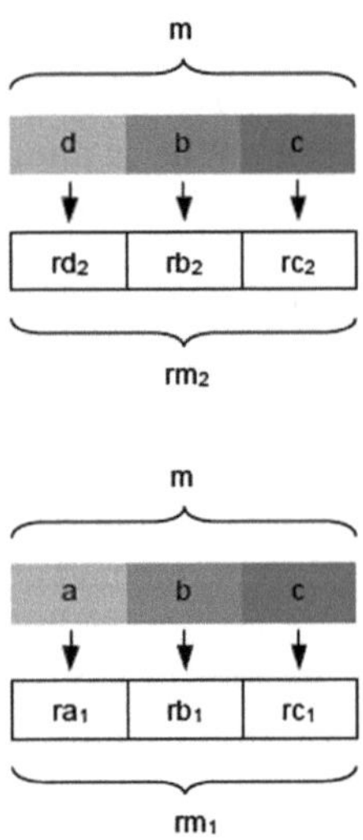

Some facts about the case

1. a – m are material objects
2. $ra_1 - rm_2$ are instantaneous spacetime regions
3. rm_1 and each of its parts is absolutely earlier than rm_2 and each of its parts; rm_1 and rm_2 do not overlap
4. m occupies exactly two regions, rm_1 and rm_2; m does *not* occupy any proper sub- or superregions of these
5. b and c each occupies exactly two regions: b occupies rb_1 and rb_2, c occupies rc_1 and rc_2
6. a at ra_1 is a part of m at rm_1
7. it is not the case that: a at ra_1 is a part of m at rm_2
8. $\sim\exists r$[a at r is a part of m at rm_2], in words: a is not, anywhere, a part of m at rm_2.
9. d at rd_2 is a part of m at rm_2.
10. $\sim\exists r$[d at r is a part of m at rm_1], in words: d is not, anywhere, a part of m at rm_1
11. m fuses $\{<a, ra_1>, <b, rb_1>, <c, rc_1>\}$ at rm_1
12. m does *not* fuse $\{<d, rd_2>, <b, rb_2>, <c, rc_2>\}$ at rm_1
13. m fuses $\{<d, rd_2>, <b, rb_2>, <c, rc_2>\}$ at rm_2
14. m does *not* fuse $\{<a, ra_1>, <b, rb_1>, <c, rc_1>\}$ at rm_2

Fig. 1.4 Four-place parthood and multilocation

So far I have been suppressing a pair of important questions. First, is 'occupies' ('L') defined, and if so how? Second, how does the sub-region relation that holds between regions relate to the parthood relation that holds between material objects? I answer the first question affirmatively and give the following definition:

(D_L) $L(x, y) =_{df} \exists z \exists w\, [P(x, y, z, w) \vee P(z, w, x, y)]$
'x occupies y' means 'either x at y is part of some z at some w, or some z at some w has x at y as a part'

As a slogan: to occupy a location is to *be* a part of something there or to *have*, there, something as a part. As for the second question, since we are emphasizing ideological parsimony in this chapter, we will operate under the assumption that there is just one fundamental parthood relation, and that it holds both between material objects and between regions (among other things, perhaps). Thus, if region r_1 is, intuitively, a *subregion* or *part-simpliciter* of r_2, then we should say that r_1 at r_1 is a part of r_2 at r_2. If we like, we can go further and define a two-place predicate for parthood *simpliciter* (which then comes out as a non-fundamental relation):

$$(D_{PS})\ P^2(x_1, x_2) =_{df} \exists y_1 \exists y_2\, [P(x_1, y_1, x_2, y_2) \& \forall y_3 \forall y_4 \\ [[L(x_1, y_3) \& L(x_2, y_4)] \rightarrow P(x_1, y_3, x_2, y_4)]]$$

In words, 'x_1 is a part-simpliciter of x_2' means 'x_1, somewhere, is a part of x_2, somewhere, and for any locations y_3 and y_4 of x_1 and x_2, respectively, x_1 at y_3 is a part of x_2 at y_4'. Presumably regions are singly located (occupying themselves only); at the very least it seems certain that no region occupies two non-coinciding regions. This makes it plausible that if region r_1 is, at r_1, a part of region r_2 at r_2, then r_1 is a part-simpliciter of r_2.

With the understanding that *regions*, no less than *material objects*, can fill the 'part' slot and 'whole' slot in our four-place parthood relation, we are now in a position to set out a principle that links (i) facts about the mereological relationship between a pair of objects with (ii) facts about the mereological relationship between the *locations* of those objects. I will call it *Withinness*:

(W_{4P}) $\forall x_1 \forall y_1 \forall x_2 \forall y_2\, [(x_1, y_1, x_2, y_2) \rightarrow P(y_1, y_1, y_2, y_2)]$
If x_1 at y_1 is a part of x_2 at y_2, then y_1 at y_1 is a part of y_2 at y_2.

Intuitively, this says that if Obama's right arm, at region r_{ra}, is a part of Obama, at region r_o, then the 'arm-region' is a part of the 'Obamba-region' – in four-place terms, r_{ra} at r_{ra} is a part of r_o at r_o. There are a number of further principles that we could set out that plausibly link 'mereological' facts to 'locational' facts, but W_{4P} is enough for now. (Strictly speaking, these facts are all mereological, since we're working with just one non-logical primitive: our four-place parthood predicate.)

So far, everything that I have said concerning our four-place parthood relation should seem at least as plausible to the *dualist* substantivalist as to the *monist* substantivalist. Indeed, most of what I have said about that relation here just summarizes what I have said elsewhere (Gilmore 2009; forthcoming a), working under dualist presuppositions.

But now that we have this framework in place, we can simply *drop the tacit dualism*, and everything else should remain intact. In particular, we can say that each material object *mereologically coincides* with each spacetime region that it occupies:

($Monism_{4P}$) For any material object o and spacetime region r, if o occupies r, then: o at r coincides with r at r.

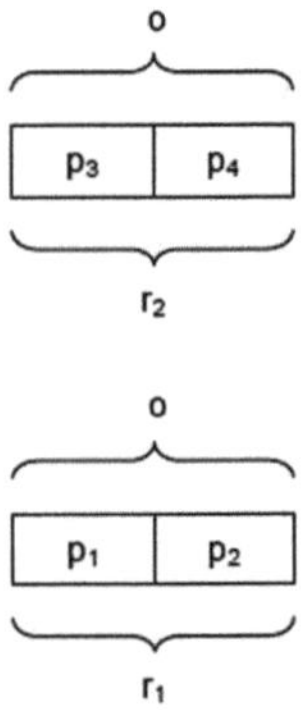

Some facts about the case

1. o is a material object
2. $p_1 - p_4$ are mereologically simple spacetime regions ('points')
3. r_1 and r_2 are mereologically complex spacetime regions
4. Each spacetime region occupies exactly one spacetime region, itself
5. o occupies exactly two spacetime regions, r_1 and r_2 (hence $o \neq r_1$ and $o \neq r_2$)
6. o fuses $\{\langle p_1, p_1\rangle, \langle p_2, p_2\rangle\}$ at r_1, as does r_1
7. o fuses $\{\langle p_3, p_3\rangle, \langle p_4, p_4\rangle\}$ at r_2, as does r_2
8. $\sim\exists r\exists r^*[r_1$ at r overlaps r_2 at $r^*]$
9. o at r_1 coincides with r_1 at r_1
10. o at r_2 coincides with r_2 at r_2
11. r_1 does not coincide^2 with r_2

Fig. 1.5 Monism$_{4P}$ and Locational Endurantism

According to this view, if Obama occupies the three-dimensional, instantaneous spacetime region r_1, then he, at r_1, mereologically coincides with r_1, at r_1; and in that case the relationship between him and r_1 is like the relationship between the material objects f and g in Fig. 1.1: coincidence without identity. Moreover, he can bear this relation of coincidence to many different regions that do not bear it to each other. Perhaps he also occupies the distinct region r_2. Then, given *Monism*$_{4P}$, it follows that Obama, at r_2, coincides with r_2, at r_2. But it does not follow that r_1 and r_2 even *overlap*, much less that they *coincide*: we remain free to say that $\neg\exists r\exists r*$ (r_1 at r overlaps r_2 at $r*$) and hence that $\neg\exists r\exists r*$ (r_1 at r coincides with r_2 at $r*$). If we define a two-place coincidence predicate as follows,

(D_{C2}) $CO^2(x_1, x_2) =_{df} \exists y_1\exists y_2 CO(x_1, y_1, x_2, y_2)$
'x_2 coincides^2 with x_2' means 'x_1, at some y_1, coincides with x_2, at some y_2'

then we can say, speaking quite strictly, that Obama coincides^2 with regions that do not coincide^2 with each other (since the two-place relation so defined is symmetric but not transitive). And of course, what goes for Obama also goes for any material objects that he ever has as parts, such as arms, legs, cells, or DNA molecules.

The analogy with waves is worth repeating. Just as a wave mereologically coincides, at different times, with different portions of water (where many pairs of these portions do not even overlap with each other), an enduring material object coincides^2 with many different instantaneous regions (many pairs of which do not overlap each other).

It may help to consider Fig. 1.5, which illustrates this combination of Monism$_{4P}$ and locational endurantism.

We should think of this as a simplified, highly unrealistic situation in which an object, o, has a complete spacetime path that is composed of just two instantaneous regions, the earlier region r_1 and the later region r_2.

The most important thing to note about the case is that it satisfies, in a precise way, both locational endurantism and monist substantivalism. The one material object in the case, o, occupies just the instantaneous, non-overlapping regions r_1 and

r_2; o does not occupy its temporally extended path, which, for simplicity, we have left out of the diagram altogether. Hence o locationally endures. Further, monist substantivalism is respected since, informally put, everything in the diagram is ultimately composed of spacetime regions. The regions themselves are of course composed of regions. But so is the one material object in the situation, o. It is composed of different spacetime points at different locations. At its location r_1, it is composed of the points p_1 and p_2. At its later location r_2, it is composed of p_3 and p_4.

Here is a general statement of the Many-Slice Constitution View (MSCV) that makes explicit some further details.

The Many-Slice Constitution View

One Parthood: There is only one fundamental parthood relation, and it holds both among material objects and among spacetime regions.

4P: The fundamental parthood relation for material objects is a four-place relation expressed by 'x at y is a part of z at w'.

Plenitude for Regions$_{4P}$: Each set of spacetime (points and/or) regions has at least one spacetime region as a fusion. In 4P-appropriate terms: for any non-empty set s, if each member of s is a spacetime (point or) region, then $\exists r\,[F(r, \{x : \exists y\,[y \in s\ \&\ x = \langle y, y\rangle]\}, r)$ & r is a spacetime region.

Locational Endurantism: Some material objects persist (have temporally extended paths), but no material object occupies any non-instantaneous (or non-spacelike) region.

Constitution+Monism$_{4P}$: For any material object o and spacetime region r: (i) $o \neq r$ and (ii) if o occupies r, then: o at r coincides with r at r.

No Fundamental Occupation$_{4P}$: There is no fundamental occupation relation; the predicate 'occupies' is defined in terms of a four-place parthood predicate, as specified in D_L.

Regions Have Temporal Parts: Each persisting spacetime region has proper temporal parts.

Objects Lack Temporal Parts: There are material objects, but none of them has proper temporal parts.

W_{4P}: If x at y is a part of z at w, then y at y is a part of w at w

Like PCV, MSCV leaves a number of questions open. It leaves open the question of which spacetime regions are 'material-object-paths', and which instantaneous slices of those paths are occupied by material objects. Further, it leaves open the question of whether a given instantaneous slice ever coincides with more than one material object.

1.4.2 How MSCV Avoids the Problems Facing PCV

How does the Many-Slice Constitution View help with the two problems that we raised for the Path Constitution View?

Problem 1: Ruling out locational endurantism. The first problem (from an endurantist vantage point) was that PCV was committed to locational perdurantism and ruled out locational endurantism. Obviously MSCV solves that problem. It says that material objects occupy only instantaneous (or spacelike) regions.

Problem 2: Gain and loss of parts. The second problem was that PCV led to the result that a certain DNA molecule, m, is not a part of Obama, in any fundamental sense of 'part'. The underlying reason for this was that, in the case we considered, m's path extended outside of Obama's path, and according to PCV, the only regions that m and Obama occupy are their paths. Given the 'Withinness' principle, this forces us to say that m is not a part of Obama, in any fundamental sense of 'part'.

MSCV avoids this problem by maintaining that m has many locations, each of them an instantaneous slice of its path. Similarly for Obama. Given this, it will be natural for the friend of MSCV to say that many of m's locations are parts of some location or other of Obama, and that m is a part of Obama, in some fundamental sense of 'part'. To be more precise, the friend of MSCV will find it natural to say that there are spacetime region r_m and r_o such that: (i) m occupies r_m, (ii) Obama occupies r_o, (iii) r_m at r_m is a part of r_o at r_o (presumably r_m is a part-simpliciter of r_o, in the sense defined earlier), and (iv) m at r_m is a part of Obama at r_o.

Given $4P$, clause (iv) amounts to the claim that there is a fundamental parthood relation that holds between m and Obama (and two regions). Crucially, all this is perfectly consistent with the fact that m's path extends outside of Obama's path.

1.4.3 A Problem for the Many-Slice Constitution View and the Path Constitution View

There is one core feature of both PCV and MSCV that many endurantists will see as a drawback[14]: the commitment to mereological-coincidence-without-identity.

Some philosophers apparently reject coinciding entities on something like 'purely mereological' grounds. They see the ban on these entities as (a) intuitively compelling on its own, or as (b) being justified by an analogy between composition and identity, or as (c) following from intuitively compelling principles concerning the behavior of parthood (reflexivity, strong supplementation, and anti-symmetry).

Others reject coinciding entities on the basis of the *grounding problem* (Bennett 2004). If coinciding objects x and y are not identical, presumably they differ with respect to certain properties – modal or historical ones, for example. But what could ground these differences, given that x and y coincide and hence are so similar physically? As applied to the thesis of object-region coincidence, the grounding

[14]B-theoretic endurantists who are on record in opposition to mereological coincidence without identity include Van Inwagen (1990), Burke (1994), Olson (1997), Rea (1998, 2000), Hershenov (2005), McGrath (2007), and Koslicki (2008). The argument for locational endurantism given in Gilmore (2007) depends upon the impossibility of mereological coincidence without identity.

problem runs as follows. If region r and object o mereologically coincide (at region r), then they will, presumably, be quite similar (at that region). They will be alike with respect to size and shape. As Hawthorne and Schaffer note, it's plausible that they will be alike with respect to the values of the various fields associated with r. Given these similarities, it may seem o and r should not differ in any way whatsoever.

Some philosophers will be moved by none of these considerations. But those who are moved will want to find an alternative to PCV and MSCV. (I have a special interest in finding such an alternative. For I have given an argument in favor of locational endurantism that relies on the principle that it is impossible for two different objects to mereologically coincide (in Gilmore 2007)).

1.5 Third Compromise: Regions-as-Pluralities Multilocationism

Faced with a threat of mereological coincidence between two entities, a natural response is to keep one and eliminate the other. Our third compromise keeps the material objects and eliminates the regions with which they were said to coincide.

Informally, the idea is as follows. There are spatially and temporally unextended, mereologically simple spacetime *points*, and there are *sets* of these points, but there are no mereological sums/fusions of these points: there are no mereologically complex spacetime regions. When we would ordinarily speak of a certain complex region r as being composed of some simple points, the ps, we should instead just speak of the ps plurally (Hudson 2005, 17). Thus, when we would ordinarily say, concerning some material object o, that o occupies r, we should instead say that o occupies the ps, where 'occupies' is treated as a predicate that is non-distributive with respect to its second argument place. And when the 4Per would ordinarily say that o at r is a part of some other material object, o*, at some other complex region, *r** (composed of the p*s), we should instead say that o, at the ps, is a part of o*, at the p*s, where '... at ... is a part of ... at ...' is treated as a predicate that is non-distributive with respect to its second and fourth argument places (at least).

The *monist substantivalist* component of the new view is that a material object o occupies some points, the ps, only if o is (in a sense to be specified) *composed* of the ps. The *locational endurantist* component of the view is that a persisting material object occupies many different pluralities of points, each of them temporally unextended. Putting these pieces together, we can say that there are many different non-overlapping, temporally unextended pluralities of points, the p_1s, the p_2s, and so on, such that: Obama occupies p_1s and is 'temporarily' composed of them, Obama occupies the p_2s and is 'temporarily' composed of them, and so on. Thus we retain locational endurantism and monist substantivalism but, having eschewed talk of complex regions, we avoid the commitment to mereologically coinciding entities.

To state this view more precisely, we will need to introduce a new set of definitions, properly restated in plural terms. First, a pair of principles governing the four-place parthood relation that we are taking to be fundamental:

$$(\mathrm{R}_{4Pplural})\ \forall x_1 \forall yy_1 \left[\exists x_2 \exists yy_2 \left[P^{pl}(x_1, yy_1, x_2, yy_2) \vee P^{pl}(x_2, yy_2, x_1, yy_1)\right] \rightarrow P^{pl}(x_1, yy_1, x_1, yy_1)\right]$$

If x_1 at yy_1 is part of some x_2 at some yy_2 or has some x_2 at some yy_2 as a part, then x_1 at yy_1 is a part of x_1 at yy_1.

$$(\mathrm{T}_{4Pplural})\ \forall x_1 \forall yy_1 \forall x_2 \forall yy_2 \forall x_3 \forall yy_3 \left[\left[P^{pl}(x_1, yy_1, x_2, yy_2) \& P^{pl}(x_2, yy_2, x_3, yy_3)\right] \rightarrow P^{pl}(x_1, yy_1, x_3, yy_3)\right]$$

If x_1 at yy_1 is a part of x_2 at yy_2 and x_2 at yy_2 is part of x_3 at yy_3, then x_1 at yy_1 is a part of x_3 at yy_3.

These are just plural analogues of R_{4P} and T_{4P}. I take them to be self-explanatory. Now for the definitions. I use the symbol '$\prec$' for the predicate 'is one of'.[15].

$$(\mathrm{PD1})\ L^{pl}(x_1, yy_1) =_{df} \exists x_2 \exists yy_2 \left[P^{pl}(x_1, yy_1, x_2, yy_2) \vee P^{pl}(x_2, yy_2, x_1, yy_1)\right]$$

'x_1 occupies yy_1' means 'for some x_2 and some yy_2, either x_1 at yy_1 is a part of x_2 at yy_2 or x_2 at yy_2 is a part of x_1 at yy_1.

$$(\mathrm{PD2})\ PATH^{pl}(x, zz) =_{df} \forall z \left[z \prec zz \leftrightarrow \exists yy \left[L^{pl}(x, yy) \& z \prec yy\right]\right]$$

'x has zz as a path' means 'for any z: z is one of zz if and only if there are yy such that: (i) x occupies yy and (ii) z is one of yy'.

$$(\mathrm{PD3})\ ACHR(zz) =_{df} \forall x \forall y\ [[x \prec zz \& y \prec zz \& x \neq y] \rightarrow SPCLK(x, y)]$$

'zz are achronal' means 'for any x and any y, if x is one of zz and y is one of zz and $x \neq y$, then x is spacelike separated from y'.

$$(\mathrm{PD4})\ PERS(x) =_{df} \exists yy \left[PATH^{pl}(x, yy) \& \neg ACHR(yy)\right]$$

'x persists' means 'there are some yy such that: (i) x has yy as a path and (ii) yy are not achronal'

$$(\mathrm{PD5})\ L\text{-}END^{pl}(x) =_{df} PERS(x) \& \forall yy \left[L^{pl}(x, yy) \rightarrow ACHR(yy)\right]$$

'x locationally endures' means '(i) x persists and (ii) for any yy, if x occupies yy, then yy are achronal'.

Locational endurantism can then be stated as the view that at least one material object persists and all persisting material objects locationally endure. To state the remaining components of Regions-as-Pluralities Multilocationism, it will help to have definitions of plural versions of 'overlaps' and 'fuses'.

$$(\mathrm{PD6})\ O^{pl}(x_1, yy_1, x_2, yy_2) =_{df} \exists x_3 \exists yy_3 \left[P^{pl}(x_3, yy_3, x_1, yy_1) \& P^{pl}(x_3, yy_3, x_2, yy_2)\right]$$

[15]See Linnebo (2012) on 'is one of' and plural quantification. My notation follows his.

(PD7) $F^{pl}(x, s, zz) =_{df} \exists y(y \in s) \& \forall y_1 [y_1 \in s \rightarrow \exists u_1 \exists ww_1 [[y_1 = <u_1, \{w : w \prec ww_1\}>] \& P^{pl}(u_1, ww_1, x, zz)]] \& \forall u_1 \forall ww_1 [P^{pl}(u_1, ww_1, x, zz) \rightarrow \exists y_2 \exists u_2 \exists ww_2 [y_2 \in s \& y_2 = <u_2, \{w : w \prec ww_2\}> \& O^{pl}(u_2, ww_2, u_1, ww_1)]]$

The last definition requires some unpacking. As before, we are taking fusion to be a three-place relation, but now we take it to hold between a thing, a non-empty set of ordered <thing, set of things> pairs, and a plurality (of points). We say that a thing x fuses set s at plurality zz if and only if: (i) s is a non-empty set of ordered pairs, (ii) each of these pairs is such that its first member, at the *members* of its second member, is a part of x, at zz, and (iii) for any object u_1 and plurality ww_1, if u_1 at ww_1 is a part of x at zz, then some member $<u_2, \{w : w \prec ww_2\}>$ of s is such that u_1 at ww_1 overlaps u_2 at ww_2.

With all these terms in hand, we can give an official statement of our new view as follows.

Regions-as-Pluralities Multilocationism (RPM)

One Parthood: There is only one fundamental parthood relation, and it holds both among material objects and among spacetime regions.

$4P_{plural}$: The fundamental parthood relation for material objects is a four-place relation expressed by 'x at yy is a part of z at ww', with one slot for the part, one slot for some things (e.g., spacetime points) that are collectively occupied by the part, one slot for the whole, and one slot for some things (e.g., spacetime points) that are collectively occupied by the whole.

Compositional Nihilism about Regions: There are no mereologically complex spacetime regions: if r is a spacetime region and x at yy is a part of r at zz, then $x = r$ and, for any w, if w is one of yy or w is one of zz, then $w = r$. In other words, the only cases of parthood holding among spacetime regions are cases in which a region r, at itself, is a part of r, at itself.

Locational Endurantism$_{plural}$: Some material objects persist, and all persisting material objects locationally endure, in the sense defined by PD5.

Objects Fuse Spacetime Points: For any material object o and spacetime points, zz, if o occupies zz, then o fuses $\{x : \exists y [y \prec zz \& x = <y, \{y\}>]\}$ at zz.

Unique Fusion$_{4Pplural}$: For any x, y, s, and zz: if x fuses s at zz and y fuses s at zz, then $x = y$.

No Fundamental Occupation$_{4P}$: There is no fundamental occupation relation; the predicate 'occupies' is defined in terms of 'x at yy is a part of z at ww', as specified in PD1.

Objects Lack Temporal Parts: There are material objects, but none of them has proper temporal parts.

Withinness$_{4Pplural}$: If x at yy is a part of z at ww, then yy are among ww, i.e., $\forall u [u \prec yy \rightarrow u \prec ww]$

Figure 1.6 depicts a simplified situation in which both RPM and monistic substantivalism are satisfied.

We can think of RPM as a theory of restricted composition, á la Van Inwagen (1990), but applied to spacetime points rather than to 'simple, enduring,

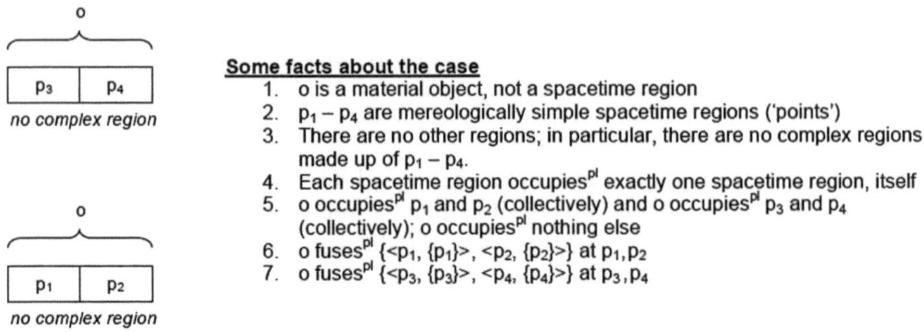

Fig. 1.6 Regions-as-pluralities Multilocationism

fundamental particles' such as electrons and quarks. Simplified somewhat, the theory makes the following claims:

(i) for any spacetime points, pp, if there are more than one of pp, then pp compose something if and only if:

  (a) they are achronal, and
  (b) they are arranged R-wise [where R is unspecified so far],

  and

(ii) if there are more than one of pp and they do compose some entity o, then:

  (a) o is a material object, not a spacetime region,
  (b) o occupies pp,
  (c) o is the only entity that pp compose,

  and

(iii) there are some persisting material objects, each such object occupies more than one plurality of spacetime points, and each such object is composed ('temporarily') of each of the pluralities of spacetime points that it occupies.

The analogy with van Inwagen's position is more than incidental. In (i) above, one could replace 'R' with 'living organism' and the result would be a version of van Inwagen's position that is consistent with monistic substantivalism. (Of course, RPM is flexible enough to accommodate other views about composition as well.) According to the 'van Inwagen-ized' version of RPM, there are mereologically simple spacetime points, living organisms (which are composed of certain pluralities of these points and which are multilocated in spacetime), and no other concrete entities.[16]

[16]Similarly, the dominant-kinds view developed in a dualist-substantivalist context by Burke (1994) and Rea (2000) can be developed in monist-substantivalist context with an appropriate replacement for 'R' in (i). Likewise for virtually any uniqueness-friendly endurantist theory of persistence-and-composition. See Markosian (2008) for more on restricted composition.

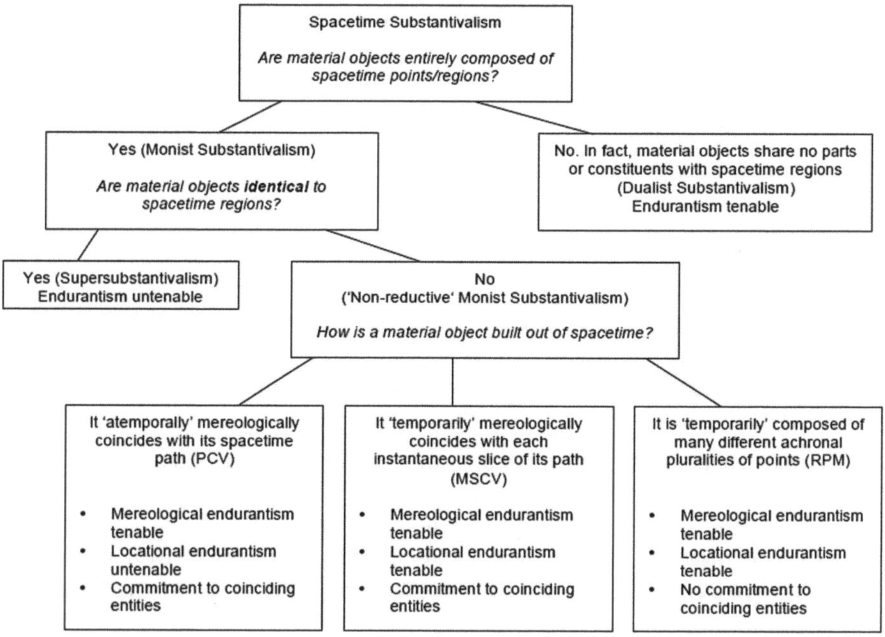

Fig. 1.7 Spacetime Substantivalism and Endurantism

Admittedly, the van Inwagen-ized version of RPM faces worries about arbitrariness and anthropocentrism. Why privilege living organisms over artifacts or non-living natural formations such as molecules or planets? And it depends upon an assumption which some will doubt: viz., that spacetime ultimately bottoms out in simple spacetime points, rather than being gunky. But these criticisms apply equally to van Inwagen's actual view, a form of dualist substantivalism. (van Inwagen assumes that matter bottoms out in simple particles rather than being gunky.)

Similarly, *any* version of RPM will need to make some maneuver that guarantees that no two material objects ever coincide even temporarily, and any such maneuver will have drawbacks. When I start with a lump-shaped piece of clay and mold it into a statue, it seems that I end up the following thing(s) in my hands: a piece of clay that has existed for several hours at least, and a statue that has existed only for a few minutes. This generates a Leibniz-law argument for the conclusion that the statue and the piece of clay are not identical, despite coinciding mereologically. One might deny the existence of statues and/or lumps (Van Inwagen 1990; Merricks 2001), one might say that a lump ceases to exist when molded into a statue (Burke 1994; Rea 2000), or might find something else to say. Presumably these moves all have their costs. But the crucial point is that RPM is not to blame. Coincidence-deniers already had to make these moves in the context of dualist substantivalism, before RPM came on the scene. (Figure 1.7 summarizes the terrain covered in Sects. 1.2–1.5.)

1.6 Two Problems for All Three Compromises

So far I have suppressed a pair of problems that afflict all three compromise positions. To these I now turn.

1.6.1 New Chalk

Judith Jarvis Thomson has offered the following argument against perdurantism:

> [If perdurantism is true, then] as I hold the bit of chalk in my hand, new stuff, new chalk keeps constantly coming into existence ex nihilo. That strikes me as obviously false. Thomson (1983, 213).

This argument is often dismissed as question-begging or otherwise not worth taking seriously. Whether or not such a dismissive attitude is justified, one thing does seem clear: of those philosophers who take perdurantism to be a 'live option' prior to encountering Thomson's argument, few will be convinced by that argument when they are exposed to it. Still, I suspect that, for better or for worse, Thomson's central premise,

> (T) it is not the case that, as I hold a bit of chalk in my hand, new stuff, new chalk keeps constantly popping into existence,

is a deeply held (perhaps basic) commitment for a significant number of endurantists. And if (T) is incompatible with perdurantism, it is also in tension with PCV, MSCV, and RPM. Let me take these in turn.

PCV and new chalk. There is logical space to say that this bit of chalk mereologically coincides with its spacetime path, that its path has temporal parts, but that the bit of chalk itself does not. Still, on this view, there are entities (instantaneous temporal parts of the path) that are very much *like* temporal parts of the bit of chalk. Each of them is composed entirely of simples (spacetime points) none of which were present at previous moments. Presumably each of these temporal parts of the bit of chalk have many or all of the same intrinsic physical properties as the bit of chalk (at the relevant times) – so much so that it would be accurate to say that they are 'chalky'. And they keep constantly popping into existence as I hold the bit of chalk in my hand. This conflicts with (T) if perdurantism does. (Strictly speaking, it is not obvious that if perdurantism is true, then new stuff or new chalk keeps popping into existence. This depends on subtle questions about the semantics of mass expressions like 'stuff' and 'chalk'. We cannot address these questions here.)

MSCV and the new chalk. The only major difference between MSCV and PCV is that MSCV takes the fundamental parthood relation, and the mereological coincidence relation defined in terms of it, to be relativized to regions. MSCV agrees with PCV that the bit of chalk has a spacetime path, that this path has instantaneous temporal parts, that these parts are 'chalky', and that they keep constantly popping into existence as I hold the bit of chalk in my hand.

RPM and the new chalk. RPM denies the existence of mereologically complex spacetime regions. So it denies that the bit of chalk has a spacetime path that has instantaneous temporal parts. But it still conflicts with the spirit of (T). Roughly put, RPM tells us that as I hold the bit of chalk in my hand at time t1, the bit of chalk is composed of some simple spacetime points, the p1s, and a few seconds later, at t2, the bit of chalk is composed of some other simple spacetime points, the p2s, where none of the p1s is identical to any of the p2s. Indeed, according to RPM (and PCV and MSCV too), there is a clear sense in which, between any two instants (in the same frame of reference), the bit of chalk undergoes *complete mereological turnover* at the fundamental level, the level of simples. There may not be any new *complex entities* popping into existence, but at each moment, there is an entirely new *plurality of simples* that pop into existence (and that compose the bit of chalk). Does this force us to say that *new stuff* or *new chalk* pops into existence? Again, this depends on questions about the semantics of mass expressions that I cannot address here. But either way, RPM surely conflicts with the spirit of (T), if perdurantism does.

Of course, the endurantist who embraces *dualist* substantivalism does not face the problem about new chalk. Since he denies that material objects share any parts or constituents with spacetime, he is free to say that the bit of chalk undergoes *no mereological variation at all*. For example, he is free to say that it is composed of some simple, fundamental particles, the ps, at t1, and that it is composed of these very same simple particles at t2.

1.6.2 Spatially Point-Like Enduring Objects

Interestingly, the existence of spatially point-like material objects would wreak havoc on all three compromise positions.[17] We can take them in turn once again.

PCV and spatially point-like material objects. Let e be a spatially point-like persisting material object, with a (one-dimensional) timelike curve as a spacetime path. Since's e's path is a timelike curve, the instantaneous temporal parts of that path are simple spacetime points. So, since e mereologically coincides with this path, these points are parts of e too. (Mereologically coincident objects may be able to differ with respect to their *complex* parts, but, given that parthood is reflexive, they

[17] And if spacetime is composed of spatially extended simple 'grains', the existence of 'spatially-grain-like' material objects would be equally problematic for all three compromise positions, for parallel reasons.

cannot differ with respect to their *simple* parts.[18]) But if the given points are parts of e, they are instantaneous temporal parts of e. So e has *instantaneous temporal parts*: it mereologically perdures. Thus PCV loses its appeal for the endurantist.

MSCV and spatially point-like material objects. The Many-Slice Constitution View does not say that e mereologically coincides with its path. Rather, it says that e mereologically coincides, in the '4P-appropriate way', with each instantaneous slice of that path. As I noted above, each of those slices is just a simple spacetime point. So, according to MSCV, e mereologically coincides with each in a series of simple points, without being identical to any of those points. But this is problematic. As I have noted elsewhere (forthcoming), the claim that

(3) e mereologically coincides with a simple to which it is not identical

is inconsistent with the reflexivity of parthood together with a plausible supplementation principle, 'quasi-supplementation':

(*QS*) if x is a part of y and x is not identical to y, then y has parts z and z* that do not overlap each other.

(Similarly, the '4P-appropriate' version of (3) is inconsistent with the 4P-appropriate versions of reflexivity and *QS*, taken together.) The bottom line is that if *QS* and the reflexivity of parthood are both necessary truths, then, while it may be possible for two different *complex* objects to coincide with one another, it is not possible for a *simple* object to coincide with any other object. This gives the friend of MSCV a reason to hope that there are no spatially point-like material objects.

RPM and spatially point-like material objects. The problem is essentially the same for RPM. Roughly put, RPM says that e is composed, at each moment of its career, of a different achronal plurality of simple spacetime points. In the case of a spatially extended material object, each of the given pluralities would include more than one point. But in the case of a spatially point-like object such as e, each of the given 'pluralities' includes just one thing, a simple point. So, at each moment of its career, e is composed of some simple (non-persisting) spacetime point – with which e is not identical. This is inconsistent with the conjunction of the relevant versions of the reflexivity of parthood and *QS*.

---

[18]We will assume that (i) x and y mereologically coincide, (ii) z is simple, and (iii) z is a part of x; we will show that z is a part of y, too. By (i) and the definition of 'mereologically coincide', it follows that (v) x and y overlap exactly the same things. By the reflexivity of parthood, (vi) z is a part of itself. Together with (iii) and the definition of 'overlaps', this entails that (vii) x overlaps z. Together with (v), this entails that y overlaps z. So, by the definition of 'overlap', (viii) there is a thing, call it w, that is a part of z and a part of y. But by the definition of 'simple', the only part of z is z itself. So w=z, and hence z is a part of y.

1.7 Conclusion

Of those who work on the metaphysics of persistence, most seem to assume that only perdurantists can build material objects out of spacetime. But the situation is not so straightforward.

If one is willing to embrace coinciding entities and reject Strong Supplementation, one can say that material objects lack temporal parts even though they coincide with temporally extended regions that have temporal parts. (This is the Path Constitution View, PCV.) Indeed, one can get this far while confining oneself to the *perdurantist*'s attractively simple fundamental ideology – a primitive two-place predicate for parthood *simpliciter*.

If one is willing to go a bit farther and help oneself to a slightly more exotic piece of fundamental ideology (a primitive, more-than-two-place predicate for a 'location-relative' parthood relation), and if one is still willing to embrace coinciding objects and reject Strong Supplementation, then one can say that material objects both (i) lack temporal parts, in the manner of 'mereological endurantism', and (ii) are multilocated in spacetime, in the manner of 'locational endurantism'. (This is the Many-Slice Constitution View, MSCV.) It is worth repeating, however, that virtually all eternalistic, B-theoretic endurantists already help themselves to a fundamental relativized parthood predicate, even in the context of dualist substantivalism.

Finally, if one is willing to eliminate complex spacetime regions (in favor of sets or pluralities of points) and treat the fundamental parthood predicate as being not merely 'location-relative' but also non-distributive, one can (i) reject temporal parts, (ii) retain locational endurantism, and (iii) avoid coinciding entities. (This is Regions-as-Pluralities Multilocationism, RPM.)

However, all three views come with further costs. They are all subject to Thomson-esque worries about 'new chalk' constantly popping into existence. And none of them fits well with the view that there are spatially point-like material objects. But for the would-be endurantist who is impressed by the case against dualist substantivalism, all this may be a price worth paying.

References

Balashov, Y. (2010). *Persistence and spacetime*. Oxford: Oxford University Press.

Bennett, K. (2004). Spatiotemporal coincidence and the grounding problem. *Philosophical Studies, 118*(3), 339–371.

Braddon-Mitchell, D., & Miller, K. (2006). The physics of extended simples. *Analysis, 66*, 222–226.

Burke, M. (1994). Preserving the principle of one object to a place: a novel account of the relations among objects, sorts, sortals and persistence conditions. *Philosophy and Phenomenological Research, 54*, 591–624.

Dainton, B. (2010) *Time and space* (2nd ed.). Montreal: McGill-Queen's.

Donnelly, M. (2010). Parthood and multi-location. In D. Zimmerman (Ed.), *Oxford studies in metaphysics* (Vol. 5, pp. 203–243).

Eagle, A. (2010). Perdurance and location. In D. Zimmerman (Ed.), *Oxford studies in metaphysics* (Vol. 5, pp. 53–94). Oxford: Oxford University Press.
Earman, J. (1989). *World enough and spacetime*. Cambridge: MIT.
Gibson, I. & Pooley, O. (2006). Relativistic persistence. In J. Hawthorne (Ed.), *Philosophical Perspectives, 20*, 157–198.
Gilmore, C. (forthcoming). Quasi-supplementation, plenitudinous coincidentalism, and gunk. In R. Garcia (Ed.), *Substance: New essays*. Philosophia Verlag.
Gilmore, C. (2006). Where in the relativistic world are we? In J. Hawthorne (Ed.), *Philosophical perspectives: metaphysics* (Vol. 20, pp. 199–236). Oxford: Blackwell.
Gilmore, C. (2007). Time travel, coinciding objects, and persistence. In D. Zimmerman (Ed.), *Oxford studies in metaphysics* (Vol. 3, pp. 177–198). Oxford: Oxford University Press.
Gilmore, C. (2008). Persistence and location in relativistic spacetime. *Philosophy Compass, 3/6*, 1224–1254.
Gilmore, C. (2009). Why parthood might be a four place relation, and how it behaves if it is. In L. Honnefelder, E. Runggaldier, & B. Schick (Eds.), *Unity and time in metaphysics* (pp. 83–133). Berlin: de Gruyter.
Haslanger, S. (2003). Persistence through time. In M. J. Loux & D. W. Zimmerman (Eds.), *The Oxford handbook of metaphysics*, (pp. 315–354). Oxford: Oxford University Press.
Hawthorne, J. (2006). *Metaphysical essays*. Oxford: Oxford University Press.
Hawthorne, J., & Sider, T. (2006). Locations. In J. Hawthorne (Ed.), *Metaphysical essays*, (pp. 85–109). Ithaca: Oxford University Press.
Hershenov, D. (2005). Do dead bodies pose a problem for biological approaches to personal identity? *Mind, 114*, 31–59.
Hudson, H. (2001). *A materialist metaphysics of the human person*. Ithaca: Cornell University Press.
Hudson, H. (2005) *The metaphysics of hyperspace*. Oxford: Oxford Univerty Press.
Kleinschmidt, S. (2011). Multilocation and mereology. *Philosophical Perspectives, 25*(1), 253–276. J. Hawthorne (Ed.).
Koslicki, K. (2008). *The structure of objects*. Oxford: Oxford University Press.
Linnebo, Ø. (2012). Plural quantification. In E. N. Zalta (Ed.), *The Stanford encyclopedia of philosophy* (Spring 2013 ed.). http://plato.stanford.edu/archives/spr2013/entries/plural-quant/
Lowe, E. J. (2003). Substantial change and spatiotemporal coincidence. *Ratio, 16*, 140–160.
Markosian, N. (2008). Restricted composition. In T. Sider, J. Hawthorne, & D. W. Zimmerman (Eds.), *Contemporary debates in metaphysics*, (pp. 341–363). London: Blackwell.
Markosian, N. (2010). Time. In E. N. Zalta (Ed.), *The Stanford encyclopedia of philosophy* (Spring 2014 ed.). http://pluto.stanford.edu/archives/spr2014/entries/time/
McDaniel, K. (2004). Modal realism with overlap. *Australasian Journal of Philosophy, 82*, 137–152.
McGrath, M. (2007). Four-dimensionalism and the puzzles of coincidence. In D. Zimmerman (Ed.), *Oxford studies in metaphysics* (Vol. 3, pp. 143–176). Oxford: Oxford University Press.
Merricks, T. (2001). *Objects and persons*. Oxford: Oxford University Press.
Olson, E. (1997). *The human animal: Personal identity without psychology*. Oxford: Oxford University Press.
Parsons, J. (2000). Must a four-dimensionalist believe in temporal parts? *The Monist, 83*, 399–418.
Parsons, J. (2007). Theories of location. In D. W. Zimmerman (Ed.), *Oxford studies in metaphysics*, (Vol. 3, pp. 201–232). Oxford: Oxford University Press.
Rea, M. (1998). In defense of mereological universalism. *Philosophy and Phenomenological Research, 58*(2), 347–360.
Rea, M. (2000). Constitution and kind membership. *Philosophical Studies, 97*, 169–193.
Schaffer, J. (2009). Spacetime the one substance. *Philosophical Studies, 145*, 131–148.
Sider, T. (2001). *Four-dimensionalism: An ontology of persistence and time*. Oxford: Clarendon.
Thomson, J. J. (1983). Parthood and identity over time. *Journal of Philosophy, 80*, 201–220.
Van Inwagen, P. (1990). *Material beings*. Ithaca: Cornell University Press.

Chapter 2
Relativistic Parts and Places: A Note on Corner Slices and Shrinking Chairs

Yuri Balashov

2.1 Introduction

Worries about parthood and location continue to stimulate the debate about persistence over time. It is now widely recognized that physical considerations are highly relevant to this debate. Recent work investigating the impact of relativity theory on the ontology of persistence has revealed, not surprisingly, many unexpected dimensions and subtle nuances of this impact. There now appears to be a broad consensus that no interesting metaphysical view of persistence (endurance, perdurance, or exdurance) is decisively refuted by relativistic considerations. There is little consensus as to how and to what extent various such views are supported by them. One should proceed on a case by case basis.

In this paper I review some recent developments focused on an especially intriguing aspect of relativistic persistence. My goal is not so much to adjudicate a mini-dispute in this area as to use it as a case study to draw some lessons about the broader metaphysical implications of the transition from the classical to the relativistic worldview. Some relativistic phenomena (e.g., relativity of simultaneity and time dilation) have no classical analogs and force us to revise the very fundamentals of common-sense ontology (e.g., reject presentism). Others – those that do most of the work in the arguments discussed below – have more familiar classical limits and, as a result, less dramatic metaphysical consequences.

Y. Balashov (✉)
Department of Philosophy, University of Georgia, Athens, GA, USA
e-mail: yuri@uga.edu

C. Calosi and P. Graziani (eds.), *Mereology and the Sciences*, Synthese Library 371,
DOI 10.1007/978-3-319-05356-1_2,

2.2 Enduring and Perduring Objetcs in Classical Spacetime

We need to start by situating the major views of persistence in relativistic spacetime. This, by itself, requires taking a stance on a number of controversial issues. The approach sketched below is therefore rather opinionated. Fortunately, except for one aspect of it,[1] this will not bias my discussion of the arguments of interest to me and, at the same time, will allow to avoid orthogonal engagements. Since the arguments in question focus on two rival modes of persistence, *endurance* and *perdurance*, and abstract from *exdurance* (also known as *stage theory*) I will set the latter aside in my discussion too. Finally, I will restrict the discussion to *special* relativity. To smoothen the transition to it, let us begin with the familiar context of classical spacetime.

Classically, a material object *o endures* iff it persists by being multilocated, in its entirety, at many instantaneous 'time-slices' of its path in spacetime. 'Multilocated' here means multiple *exact location*[2]; 'in its entirety' means *wholly* but not *solely*[3]; and 'path' is a 4D (four-dimensional) region of spacetime 'swept' by *o* during its life career.[4] Enduring objects are 3D (three-dimensional) entities (i) extended in space but not in time, (ii) having spatial but not temporal parts (on which more below), and (iii) persisting by being wholly present at all moments of time at which they exist (Fig. 2.1).

Classical perdurance can, for our purposes, be taken as involving the denial of all the above. A material object *o perdures* iff it persists by being singly located only at its path. Perduring objects are 4D entities (i) extended in time as well as space, (ii) having temporal as well as spatial parts, and (iii) exactly located only at their 4D paths (Fig. 2.2).

A bit more precisely, one could start with a three-place relation of parthood '*p* is a part of *o* at a region $R_\perp$' relativized to a temporally unextended region of spacetime $R_\perp$. The regions of interest are, of course, instantaneous 'time-slices' ('*t*-slices') of objects' paths, which can be indexed by moments of time (in the classical context) or by moments of time in frames (in the relativistic context), allowing one to simplify the notation and speak of 'parts at times' (or 'parts at frame-relative times') and thus anchor the technical language of persistence in familiar notions of common

[1]Noted in Sect. 2.4, note 22.

[2]Intuitively, a material object *o* can be said to be exactly located at a spacetime region R iff *o* and R have exactly the same shape, size, and position. Exact location can be taken as an unanalyzed and intuitively clear primitive (as is done, e.g., in Hudson (2001), Bittner and Donnelly (2004), Gilmore (2006), and Balashov (2008, 2010)) or as a defined notion (see, e.g., Parsons 2007 and Gilmore 2008). The choice affects other commitments. Below we abstract from this issue and adopt the first approach.

[3]Roughly, *o* is *wholly* located at R iff no part of *o* is missing from R; while *o* is *solely* located at R iff no region disjoint from R contains any part of *o*. An enduring object is (typically) wholly and exactly located at multiple regions of spacetime without being solely located at any of them.

[4]For now; we will need to make the notion of path more precise later.

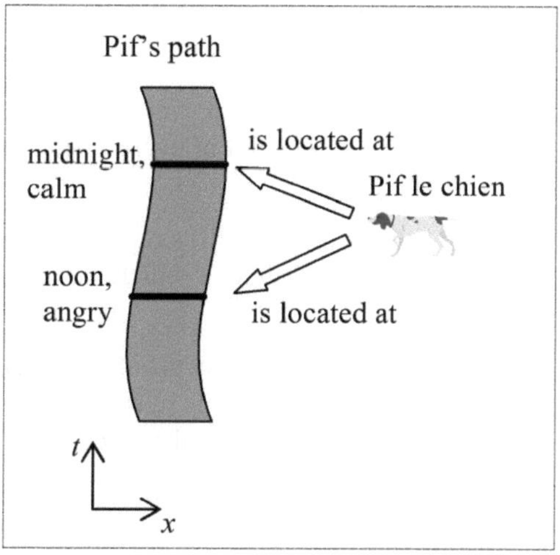

Fig. 2.1 Endurance in classical spacetime

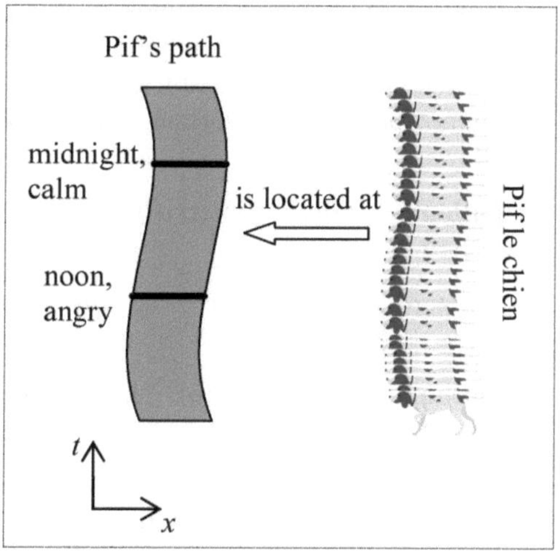

Fig. 2.2 Perdurance in classical spacetime

language. Where p, o and a t-slice of o's path, $\boldsymbol{o}_{\perp t}$, stand in such a relativized parthood relation we shall say that p is a *spatial part* (*s-part*) *of* o *at* t:

Definition 1. $p_{\perp}$ is a *spatial part* (*s-part*) of o at $t =_{df} p_{\perp}$ is a part of o at $\boldsymbol{o}_{\perp t}$.

Temporal parthood can then be defined as follows (Sider 2001, 59):

Definition 2. $p_{\parallel}$ is a *temporal part* (*t-part*) of o at $t =_{df}$ (i) $p_{\parallel}$ is located at $\boldsymbol{o}_{\perp t}$ but only at $\boldsymbol{o}_{\perp t}$, (ii) $p_{\parallel}$ is a part of o at $\boldsymbol{o}_{\perp t}$, and (iii) $p_{\parallel}$ overlaps at $\boldsymbol{o}_{\perp t}$ everything that is a part of o at $\boldsymbol{o}_{\perp t}$.

The subscripts '$\perp$' and '$\parallel$' indicate that the relevant dimensions are, respectively, 'orthogonal' or 'parallel' to the direction of time.

Given this background, classical endurance and perdurance amount to the following:

Definition 3. o *endures* in classical spacetime $=_{df}$ (i) o's path is temporally extended, (ii) o is located at every t-slice of its path, (iii) o is located only at t-slices of its path.

(i) ensures that o persists; (ii) says that an enduring object is 'wholly present' at all moments of classical time at which it exists; (iii) precludes o from being extended in time.

Definition 4. o *perdures* in classical spacetime $=_{df}$ (i) o's path is temporally extended, (ii) o is located only at its path, (iii) the object located at any t-slice of o's path is a proper t-part of o at that slice.

(ii) indicates that o is temporally extended and is as long as its path, while (iii) guarantees that o has a distinct proper temporal part at each moment of its career.[5]

To say what properties a persisting object has at a classical moment of time both endurantism and perdurantism must relativize possession of properties to times. The endurantist can do it in a number of ways that bring with them somewhat distinct metaphysics of temporal modification, each coupled with a corresponding semantic of temporal predication.[6] We can abstract from these details and put the guiding idea as follows:

Definition 5. Enduring object o has Φ at t (i.e., at $\boldsymbol{o}_{\perp t}$) in classical spacetime $=_{df}$ o bears Φ-at to t.

The perdurantist, in her turn, must endorse the following analysis, or some analog:

Definition 6. Perduring object o has Φ at t (i.e., at $\boldsymbol{o}_{\perp t}$) in classical spacetime $=_{df}$ o's t-part has Φ.

To illustrate, consider Pif, a dog that, as we normally say, is angry at noon and calm at midnight. The endurantist underwrites this talk by making Pif bear two tenseless relations *angry-at* and *calm-at* to, respectively, noon and midnight. For the perdurantist, Pif is a 4D entity extended both in space and time. It persists by

---

[5] As noted above, these formulations are opinionated and gloss over some controversial issues. First, there are exotic counterexamples, e.g., objects enduring according to (3), but having temporal parts according to (4). Similarly, an object might be a *temporally extended simple* that has no temporal parts. Some authors take exotica of this sort seriously enough to motivate a more fine-grained classification of different ontologies of persistence distinguishing *locational* endurance and perdurance (where the disagreement boils down to the issue of whether or not objects are temporally extended) from their *mereological* counterparts (where the disagreement is about possession of temporal parts). See, in particular, Gilmore (2006, 2008), where these distinctions are developed in detail and amply illustrated. We will abstract from the exotic cases below and focus on natural combinations of locational and mereological views.

[6] For details, see Lewis (1988), Haslanger (2003), and Balashov (2010, 18–22, 74–77).

having distinct temporal parts at every moment of its existence. When we say that Pif is angry at noon and calm at midnight what we really mean is that Pif's noon part is simply angry and his midnight part simply calm (see Fig. 2.2).

Obviously, endurance and perdurance represent two very different metaphysical and semantic views. The question of whether ordinary material objects endure or perdure continues to dominate the debate about persistence. Special relativity adds new features to it.

2.3 Enduring and Perduring Objects in Special Relativistic (Minkowski) Spacetime

The spacetime of special relativity (Minkowski spacetime) does not support the notion of absolute simultaneity and the associated partition of spacetime events into equivalence simultaneity classes. Instead it embodies an absolute metrical relation between events known as the Interval,[7] which imposes partial order on them.[8] Global chronological precedence thus gives way to local relations of timelike and lightlike separation. Simultaneity becomes a frame-relative notion, and moments of time (i.e. hyperplanes of simultaneity) in different reference frames crisscross (see Fig. 2.3 below).[9]

As we have seen, in classical spacetime, locations of persisting objects, their parts, and temporary properties were indexed to moments of absolute time (more precisely, to t-slices of the objects' paths). A natural adaptation of this strategy to Minkowski spacetime suggests further relativization to inertial frames of reference[10] resulting in the replacement of the classical 't' with a two-parameter index 't^{F}' referring to moments of time in a given inertial reference frame F. As before, one could begin with a three-place relation 'p is a part of o at a temporally unextended region $R_{\perp}$.' Temporally unextended regions of interest are now 't^{F}-slices' – spacelike intersections of time hyperplanes with the objects' paths in Minkowski spacetime. Where p, o and a t^{F}-slice $\boldsymbol{o}_{\perp t^{\mathrm{F}}}$ of o's path $\boldsymbol{o}$ stand in such a relation, we shall say that p is a *spatial part* (s^{F}-part) of o at $\boldsymbol{o}_{\perp t^{\mathrm{F}}}$:

Definition 7. $p_{\perp}$ is a *spatial part* (*s^{F}-part*) of o at t^{F} $=_{df}$ $p_{\perp}$ is a part of o at $\boldsymbol{o}_{\perp t^{\mathrm{F}}}$.

---

[7]Expressed in a given inertial reference frame as $I = c^2\Delta t^2 - \Delta \mathbf{r}^2$

[8]The sense in which Minkowski spacetime is partially ordered is the sense in which its points can be ordered by the relation $R^{+}(q, p) \equiv c^2[t(q) - t(p)]^2 - [\mathbf{r}(q) - \mathbf{r}(p)]^2 \geq 0 \wedge t(q) - t(p) \geq 0$, which is reflexive, antisymmetric and transitive.

[9]For useful non-technical introductions to the geometrical structure of Minkowski spacetime see Geroch (1978) and Balashov (2010, ch. 3).

[10]A move made by Sider (2001, 59, 84–86); Rea (1998); Sattig (2006, §§ 1.6 and 5.4); and defended by Balashov (2010, §5.2,) but strongly resisted by Gibson and Pooley (2006, 160–165) and, to some extent, by Gilmore (2008).

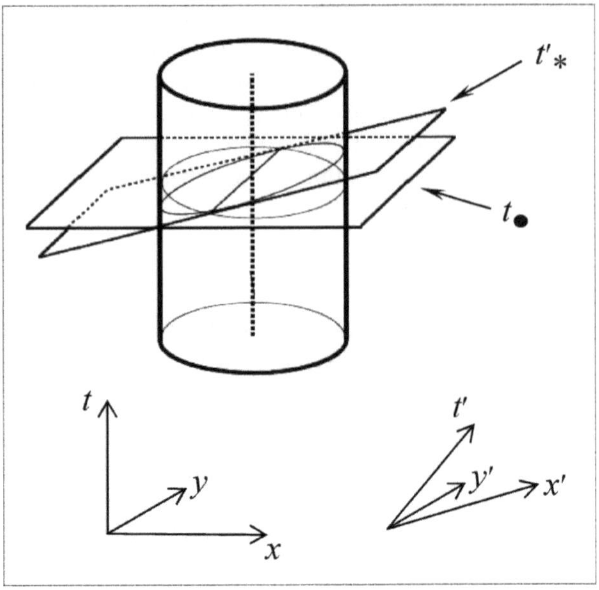

Fig. 2.3 A persisting spherical object in Minkowski spacetime

And we explicate the notion of temporal parthood as follows:

Definition 8. $p_{\|}$ is a *temporal part* (t^{F}-part) of o at t^{F} $=_{df}$ (i) $p_{\|}$ is located at $\boldsymbol{o}_{\perp t^{\mathrm{F}}}$ but only at $\boldsymbol{o}_{\perp t^{\mathrm{F}}}$, (ii) $p_{\|}$ is a part of o at $\boldsymbol{o}_{\perp t^{\mathrm{F}}}$, and (iii) $p_{\|}$ overlaps at $\boldsymbol{o}_{\perp t^{\mathrm{F}}}$ everything that is a part of o at $\boldsymbol{o}_{\perp t^{\mathrm{F}}}$.

These notions can then be employed to give a tentative analysis[11] of relativistic endurance and perdurance:

Definition 9. *o endures* in Minkowski spacetime $=_{df}$ (i) o's path is temporally extended,[12] (ii) o is located at every t^{F}-slice of its path, (iii) o is located only at t^{F}-slices of its path.

Definition 10. *o perdures* in Minkowski spacetime $=_{df}$ (i) o's path is temporally extended, (ii) o is located only at its path, (iii) the object located at any t^{F}-slice of o's path is a proper t^{F}-part of o at that slice.

As before, these definitions must be supplemented with an account of the relativization of temporary properties of persisting objects to their locations (in the case of endurance), or the locations of their t^{F}-parts (in the case of perdurance). Such locations are, of course, t^{F}-slices of the objects' paths, which can be usefully labeled with the same two-parameter index that figures in the above definitions:

[11] Important refinements will be made in Sect. 2.4.

[12] That is, includes at least two timelike separated points.

Definition 11. Enduring object o has Φ at t^{F} (i.e., at $\boldsymbol{o}_{\perp t^{\mathrm{F}}}$) in Minkowski spacetime $=_{df}$ o bears Φ-at to t^{F}.

Definition 12. Perduring object o has Φ at t^{F} (i.e., at $\boldsymbol{o}_{\perp t^{\mathrm{F}}}$) in Minkowski spacetime $=_{df}$ o's t^{F}-part has Φ.

Thus, while in the classical framework objects have properties at absolute moments of time (more precisely, at absolute time slices of the objects' paths), in the Minkowskian framework possession of temporary properties is relativized, in effect, to times-in-frames (more precisely, to frame-relative time slices of the objects' paths). This brings new features. Consider, for example, an object whose path is a 'cylindrical' region in Fig. 2.3 (with one dimension of space suppressed). Even if the object does not change its *proper shape* (i.e. the shape it has in its rest frame), it exemplifies different shapes at time slices of its path drawn in different reference frames, such as (x, y, t) and (x', y', t'). The endurantist will say that the object is located at both slices and bears the *spherical-at* relation to $t_{\bullet}$, a moment of time (i.e. a time plane) in the frame (x, y, t) hosting one of the slices and the *oblong-at* relation to t'_{*}, in the frame (x', y', t'), hosting the other slice. The perdurantist will say that the object is located at its path and has two distinct t-parts, the $t_{\bullet}$-part and the t'_{*}-part, with different corresponding shapes. This is, of course, none other than the familiar relativistic effect of Lorentz contraction dressed in modern metaphysical clothes. Geometrically speaking, the effect is grounded in different (non-parallel) orientations of time hyperplanes, containing time-slices of the object's paths, in different reference frames – a distinctly relativistic phenomenon absent from the geometry of classical spacetime.

The implications of this phenomenon are more dramatic than it may appear. Lewis (1988) has famously said that nothing can be bent and straight in the same respect. This seems to imply, a fortiori, that nothing can be both *bending* and *keeping straight*. But there is a sense in which this is not true in relativistic spacetime. Consider a granite block moving with velocity v (which is a considerable fraction of the speed of light) and suspended from vertical threads moving along with it (Fig. 2.4).[13] At a certain moment all threads are cut and the block starts to fall, continuing at the same time its inertial horizontal motion. Figure 2.5 represents a series of snapshots showing the block at some stages in this process.[14] Figure 2.6 represents a similar series of snapshots taken in the original rest frame of the block.

The block remains straight in the first series but becomes progressively bent in the second. How could it be? There may, initially, be two worries about it. First,

[13]The essential details of the scenario come from Sartori (1996, 185–190), where it is used to illustrate one of the lesser-known 'paradoxes' of special relativity, first introduced by Rindler in (1961). My exposition of the case comes from Balashov (2010, 198–200). Thanks to Oxford University Press for permission to use this material.

[14]Figures 2.4–2.6 are *not* spacetime diagrams but series of merely spatial 'snapshots' taken at different moments of time in two reference frames.

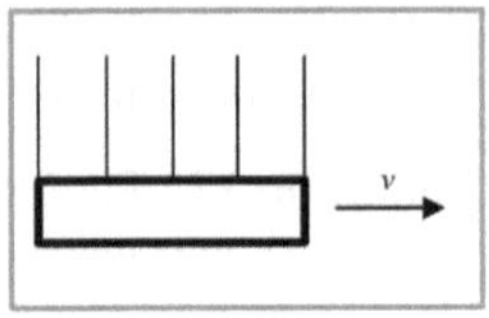

Fig. 2.4 Granite block in horizontal motion

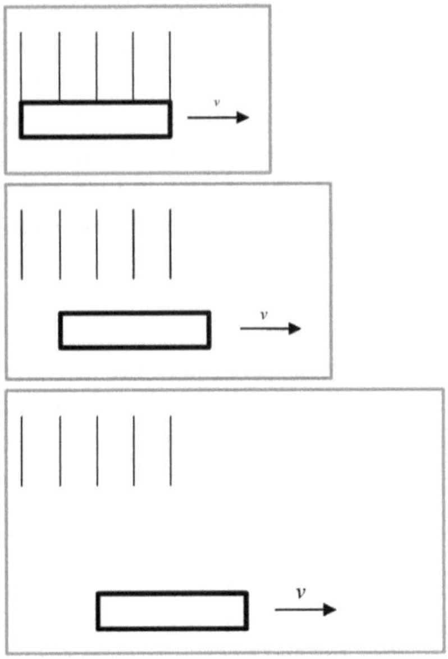

Fig. 2.5 Granite block in free fall, continuing to move horizontally

the block is made of granite and thus simply cannot bend. (If you think granite is insufficiently rigid, pretend that the block is made of *super*granite.) Second, the block cannot *both* remain straight *and* undergo bending (here is where we may come up against Lewis's dictum).

These worries are, of course, misplaced. The block does both things, i.e., is both bending and keeping straight over the same stretch of its career (loosely speaking). And it bends no matter how rigid its material is. Moreover, it always bends in the same way. How so? The key lies in the relativity of simultaneity. The threads suspending the block are cut *simultaneously* in the 'laboratory frame' resulting in free fall of all segments of the block (Fig. 2.5). In the original rest frame of the block, however, the cutting events occur *successively* (Fig. 2.6). When the rightmost thread is cut the part of the block previously held by it begins to fall immediately. But the rest of the block remains horizontal. By the time the next thread is cut the segment of the block just underneath it still 'does not know' that the rightmost part is already in free fall and, hence, does not have a chance to exert a sheer force that could stop the bending of the right end of the block. Why? Because the cutting events are simultaneous in the laboratory frame, hence, spacelike separated from each other. Therefore, no physical influence can propagate from one such event to

Fig. 2.6 Granite block in 'free fall' with snapshots taken in its rest frame

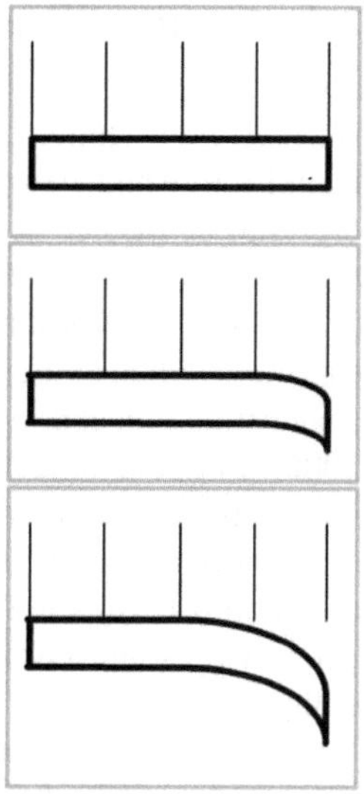

the next. Nothing can stop a given segment of the block from free fall, once the thread holding it is cut. Accordingly, nothing can stop the block from bending. The strength of the material is beside the point.

Along with some other 'paradoxes,' this scenario is sometimes taken to show that there are no rigid bodies in special relativity, that is to say, no bodies that can keep their shape invariant, even in the idealized limit.[15] (Thus supergranite is of no help.) Shape and other arrangements in 3D space are, in this theory, merely *perspectival* phenomena. But there must be something permanent standing behind all the different perspectives, such as those shown in Figs. 2.5 and 2.6. What stands behind them is, of course, a 4D invariant shape of the path of the persisting object.[16] If this object perdures then it is temporally as long as its path and fits exactly in it. This fact could then be used to explain the unity behind many perspectivally restricted shapes of the object's temporal parts (see Balashov 2010, ch. 8). If the object endures such an explanation is unavailable (or so I argue in *ibid.*), but one can still derive comfort from the notion that a single enduring 3D object can fill its 4D path by exhibiting different 3D shapes – as drastically different as *bent* and *straight* – at its rampantly crisscrossing locations slicing its path at various angles in spacetime. Indeed, according to our understanding of relativistic endurance so far, the object is located at *every* t^F-slice of its path.

But it has been argued that this leads to problems, just around the corner. I discuss these arguments in the next section, where I also draw some morals for the broader understanding of relativistic persistence.

[15]See, e.g., Sartori (1996, 184–185).

[16]I make no attempt to depict it.

2.4 Corner Slices and Shrinking Chairs

As they now stand, our accounts of relativistic endurance and perdurance (see Definitions 9 and 10) embrace a very liberal view of location allowing each enduring object to be located at *every* t^F-slice of its path and each perduring object to have a t^F-part at *every* t^F-slice of its path, as per clauses (ii) and (iii) of the corresponding definitions. In classical spacetime, liberalism of this sort appears unproblematic, especially when combined with a very natural understanding of the notion of path of a persisting object as a *union*[17] of regions at which the object is located (see Gilmore 2006). Suppose objects endure. If we start by saying that an enduring object o is wholly present at all absolute moments of time from a certain range Δt and arrive at the notion of its path by taking the union of the instantaneous spacetime regions at which o is thus multilocated then it is anything but surprising that o is located at every (absolute) time-slice of its path. This is reassuring, even if not particularly enlightening.

Things are importantly different in relativistic spacetime. Suppose o endures and is located at each of a continuous family of instantaneous regions forming its path, but at *no* other region (Fig. 2.7). Then each member of this family supplies, quite trivially, a legitimate location of o. But this is not true of any 'slanted' instantaneous slice of o's path, such as $\boldsymbol{o}_{\perp}*$. The same holds, mutatis mutandis, of perdurance. Suppose o perdures, and each of the continuous family of instantaneous slices of its path hosts o's temporal part. This does not automatically grant the same privilege to the 'slanted' slice $\boldsymbol{o}_{\perp}*$. For all we know, $\boldsymbol{o}_{\perp}*$ may fail to contain a temporal part of o. Imagine Unicolor, a persisting object one of whose essential properties is to be *uniformly colored* (cf. Smart 1987, 63–64). Suppose further that Unicolor uniformly changes its color with time in a certain inertial reference frame F. Consider a t^{F*}-slice of Unicolor's path that is at an angle to hyperplanes of simultaneity in F. Whatever (if anything) is located at such a slice is not uniformly colored and, hence, must be distinct from Unicolor, even though it is filled with the (differently colored) material components of Unicolor.[18]

Admittedly, cases such as the Unicolor are metaphysically recherché (what in reality grounds Unicolor's mysterious essential property *being uniformly colored*?) and could probably be set aside. However, according to Gilmore (2006, 212–213) and Sattig (2012, forthcoming), the feature of Minkowski spacetime that underlies such cases leads to a more tangible problem. Gilmore argues that this problem eventually undermines the viability of relativistic endurance. Sattig argues that the problem affects relativistic perdurance as well as endurance, albeit for different reasons, and gives additional support to his double-layered ontology of ordinary objects. The common set-up of both arguments is as follows (see Gilmore 2006,

---

[17]Or perhaps a *sum*. This depends on whether regions are taken to be set-theoretical or mereological notions. We adopt the first strategy, primarily for convenience, not as a matter of principle.

[18]For another illustration of the same point, see Gilmore (2006, 210–211)

Fig. 2.7 Crisscrossing locations of persisting objects, or their t^F-parts, in Minkowski spacetime

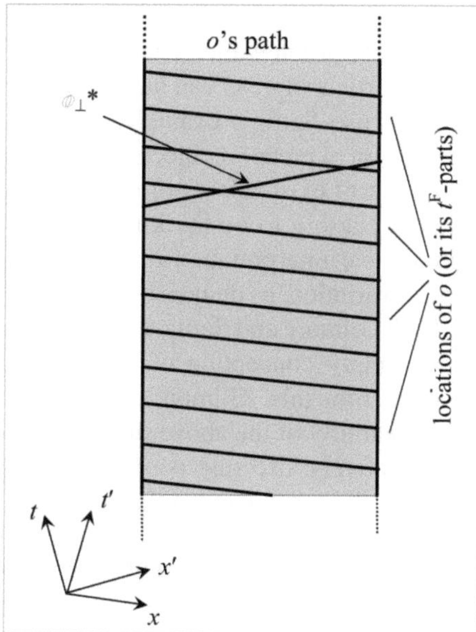

Fig. 2.8 'Corner slice' (see Gilmore 2006, 212–213)

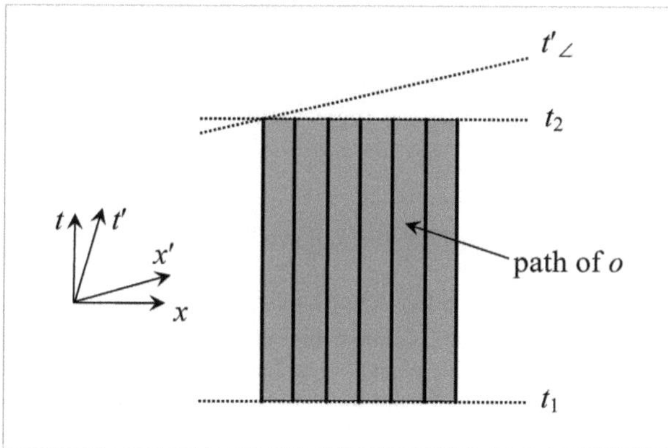

212–213). A persisting object o composed of many particles pops into existence at time t_1 and pops out of existence at t_2, in a frame (x, t). Its path $\boldsymbol{o}$ is a shaded region in Fig. 2.8.[19] Both t_1- and t_2-slices of $\boldsymbol{o}$ are good candidates for hosting o (if o endures) or o's temporal parts (if o perdures), and so are all the t-slices between t_1 and t_2 in the frame (x, t). But consider a 'corner slice' $\boldsymbol{o}_{\perp t'_\angle}$ drawn through a corner of $\boldsymbol{o}$ at the time $t'_\angle$ in the frame (x', t'). Being a temporally unextended slice of $\boldsymbol{o}$ it must be a location of o, or its temporal part, according to clauses (ii) and

[19]Strictly speaking, o's path is not a continuous hyper-rectangle but a densely packed 'multifilament region'. We ignore this complication here.

(iii) of our accounts (9) and (10) of relativistic endurance and perdurance so far. But this is problematic. The $t'_{\angle}$-slice of $\boldsymbol{o}$ is a single point[20] hosting, at most, a single particle of o, so can hardly qualify as a suitable location of o, or its temporal part. To use Sattig's example, suppose o is a chair. According to our ordinary conception of material objects, a chair, in particular, cannot shrink to a point without going out of existence. Ordinary objects cannot undergo radical variation in shape without ceasing to be the kind of objects they are. According to a very intuitive geometrical interpretation of special relativity,[21] however, they do undergo such radical variation, as demonstrated by the corner-slice scenario.

Both Gilmore and Sattig agree that scenarios of this sort create a tension between our ordinary conception of persistence and relativity. But they derive different lessons from this. Gilmore argues that corner-slice scenarios cast doubt on the very tenability of the above statement of endurance in Minkowski spacetime, while not negatively affecting perdurance.[22] Sattig, on the other hand, uses the 'point-shaped chair' problem to reinforce his case for a 'double-layered' ontology of ordinary material objects, with a view of resolving the tension described above.[23] Their disagreement about the proper lessons of the scenarios, however, interests me less than their common attitude toward such scenarios. I believe, they both overreact to them. I will show it by looking more critically at the details of two somewhat different versions of the corner slice/shrinking chair case: 'abrupt' and 'gradual.' I will argue below that abrupt scenarios involve violation of conservation laws of physics, whereas the relativistic considerations underlying the arguments in question presuppose their validity. This undermines the consistency of abrupt

---

[20]Or so we assume; alternatively, it could be a one-dimensional line or a two-dimensional surface, with the same effect.

[21]Amply illustrated in Fig. 2.8 and other figures in this paper.

[22]See Gilmore (2006). Gilmore himself takes the case to demonstrate, first and foremost, the need to allow enduring objects to be located, not just at flat time-slices, but at *arbitrary maximal* spacelike slices of their paths in relativistic spacetime, including curved such slices, a move raising further objections developed by Gibson and Pooley; see Gibson and Pooley (2006, 186). I argue against admitting curved slices as legitimate locations of persisting objects in Minkowski spacetime on independent grounds in Balashov (2008, Section 5; 2010, Section 5.2.).

[23]Sattig's neo-Aristotelian ontology, systematically developed in (forthcoming) and a number of earlier papers, regards ordinary objects as 'double-layered compounds of matter and form.' The centerpiece of his theory is the thesis that the material and the formal 'layers' of ordinary objects ground two different perspectives on them, which generate divergent truth conditions of various claims about objects. Both perspectives – the material (or sortal-abstract) and the formal (or sortal-sensitive) – are equally important, and both are found in ordinary discourse. Some of our thinking about ordinary objects tracks their underlying matter (e.g., when we reflect that two distinct objects cannot occupy the same region of space, or spacetime), while other intuitions track sortal-sensitive 'careers' of objects, whose various stages may include materially distinct subjects (e.g., when we re-identify a certain cat composed of a particular mass of matter today with a certain cat composed of a numerically different mass of matter tomorrow). Sattig argues – systematically, rigorously, and persuasively – that the availability of these two perspectives holds key to resolving various problems, including the problem of corner slices/point-shaped chairs (if the latter is a problem). For details, see Sattig (2012; forthcoming, Chapter 8).

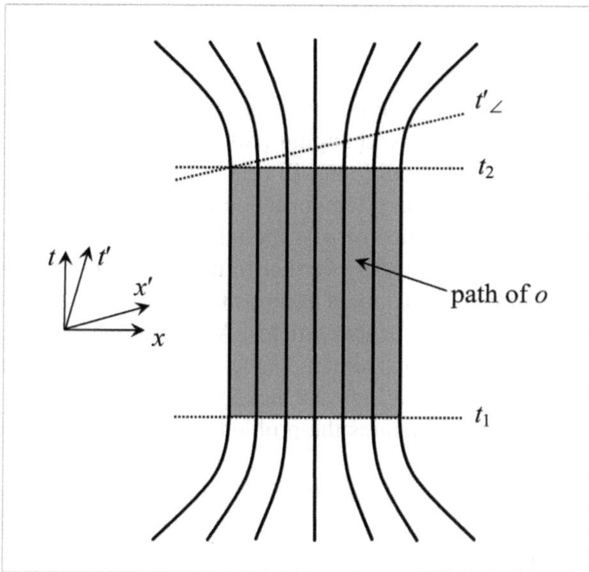

Fig. 2.9 Gradual 'corner-slice' scenario

scenarios. Gradual scenarios are more complicated. They conform with the physical laws but crucially involve *vagueness* of material composition. I believe that a proper account of the vagueness factor takes the sting from the problem of corner slices/shrinking chairs.

The 'abrupt' version is essentially as above. One can resist the arguments based on it by simply denying the possibility of abrupt corner slices/shrinking chairs scenarios. More carefully, the careers of the objects represented in them violate the conservation laws of physics (because the careers represent objects as popping into and out of existence), while the whole line of reasoning based thereon and motivating pessimism about the viability of relativistic endurance (in Gilmore's case) or about the prospects of familiar single-layered ontologies (in Sattig's case), assumes the physics of relativity which requires strict validity of conservation laws. The incoherence of this sort makes physically impossible states of affairs, such as that depicted in Fig. 2.8, irrelevant to the discussion in hand, even if they are not impossible *tout court*.

This motivates a transition[24] from the abrupt to a gradual version of the scenario. Suppose that, instead of popping in and out of existence, initially scattered particles come to compose object o at t_1 and stop doing so at t_2, when they 'break up' and begin to separate (Fig. 2.9). What do we now say of the $t'_\angle$-slice of o's path? It still

[24]Suggested by Gilmore in personal correspondence and developed in some detail in Sattig (2012; forthcoming, Chapter 8).

appears to contain a single point, so the problem recurs, but conservation laws are now respected.

Let us consider the situation more carefully. The 'break up' of o's particles ending its career cannot be instantaneous. It must grounded, perhaps in a complicated way, in the rapidly changing pattern of their causal interaction. In all likelihood, the grounding conditions will be vague, resulting in an extended interval of 'fading away,' with no sharp temporal boundaries, such as t_2. Hence it is not so clear, after all, that the $t'_\angle$-slice of $\boldsymbol{o}$ is ineligible to be one of o's locations (or a location of its temporal part). Any verdict to this effect will depend on the fine details of the relevant theory of spatial composition, the nature of the object in question, and the exact trajectories of its particles. And even when all that is taken into account, the answer will perhaps remain vague. Thus drawing the path of o in the form of a clear cut rectangle (as in Figs. 2.8 and 2.9) is misleading. But it is precisely such clear cut drawing that generates the problem of corner slices/point-shaped chairs in the first place.

What is the real upshot of these considerations? One should recognize that on any view of vagueness, some t^{F}-slice of $\boldsymbol{o}$ or other will *not* be eligible (perhaps, on some precisification) to serve as o's location (or a location of its temporal part), or at least not determinately so eligible. The notion of *eligibility* must thus be written into an official account of relativistic persistence. But considerations of eligibility, stemming from widespread worries about the vagueness of material composition, cannot be neglected even in the classical setting. They arise, for example, whenever we ask whether a progressively scattering composite object *still exists* at a certain moment of absolute time. If we think that this question does not have a determinate answer then considerations of vagueness must be taken into account in the explication of the notion of the object's *path* even in classical spacetime. Relativity does not add anything new to this step. What appears to be new emerges at the next step: *after* the path of a persisting object in relativistic spacetime has been assembled from its eligible momentary locations indexed to a particular reference frame (which already presumes coming to terms with vagueness), one apparently gains unrestricted freedom to slice the path thus produced at various angles, including those generating 'corner slices.' The freedom comes from rampant crisscrossing of time hyperplanes in Minkowski spacetime. The question is whether one can exploit it at will, in the way suggested.

I submit that one cannot. 'Unbridled crisscrossing' must be rejected in favor of 'disciplined crisscrossing,' and considerations ruling over the process at this stage are essentially the same as those at play at its first step, that of assembling the path of a persisting object from its eligible momentary locations in a particular reference frame. The same sort of vagueness may inflict both of them, but if so, it must be dealt with in the same way. And the need to deal with it is as urgent in classical spacetime as it is in Minkowski spacetime. To see this, return to Fig. 2.9 and consider the evolution of o in (x', t'). From the physical point of view, (x', t') is a legitimate frame of reference, which represents o as moving as a whole while progressively

Fig. 2.10 Progressive shedding of particles by a moving object

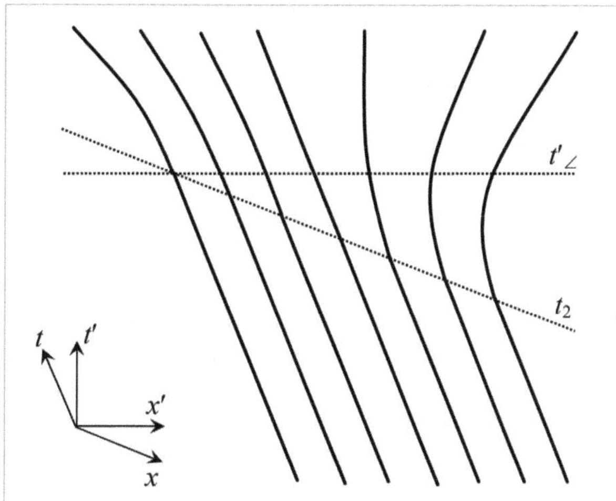

shedding particles until the process reaches the corner slice $\boldsymbol{o}_{\perp t'_\angle}$ (Fig. 2.10).[25] How many particles could o shed without ceasing to exist? Maybe just a few, or maybe the majority of them. Exactly at what point in (x', t') did o go out of existence? More likely than not, before $t'_\angle$; but there is hardly more to be said. Perhaps there is no general answer to such questions, and the answer depends, in each case, on the nature of the object under consideration. But when the evolution of o is viewed from this perspective it becomes clear that (i) questions of this sort must be settled *before* one attempts to draw the boundaries of o's path, and (ii) *exactly the same* questions would arise if spacetime were classical and time planes in (x', t') represented *absolute* time planes.

The real lesson of the corner-slice/shrinking chair scenarios is, therefore, that questions of *locational eligibility* are metaphysically prior to questions about the exact boundaries of o's path in relativistic spacetime.[26] This motivates the following modifications to our earlier accounts of relativistic endurance and perdurance:

Definition 9′. *o endures* in Minkowski spacetime $=_{df}$ (i) o's path is temporally extended, (ii) o is located at every o-eligible t^{F}-slice of its path, (iii) o is located only at t^{F}-slices of its path.

Definition 10′. *o perdures* in Minkowski spacetime $=_{df}$ (i) o's path is temporally extended, (ii) o is located only at its path, (iii) the object located at any o-eligible t^{F}-slice of o's path is a proper t^{F}-part of o at that slice.

---

[25]For simplicity, Fig. 2.10 does not represent the first episode of the original scenario, when the initially scattered particles come to compose o in the first place. But similar considerations apply, mutatis mutandis, to such 'coming into existence' episodes as well.

[26]Cf. Gibson and Pooley (2006, 186–187), who develop a very similar suggestion.

An intuitive picture underlying these accounts is as follows:

Certain particles come together to compose an object o at time t_1 in a particular reference frame (x, t) and stop composing it at t_2. By anyone's lights, a complete description of the process requires a well-developed theory of composition addressing, among other things, the issue of vagueness. The very same resources are needed to give an account of a similar process in the classical framework.

All the momentary locations of o (or the locations of its temporal parts) in frame (x, t) comprise o's *partial path* $\boldsymbol{o}(x, t)$. The very same particles that compose o (or the t-parts of o) at all moments $t \in [t_1, t_2]$ in (x, t) *may* or *may not also* compose o (or o's t'-part) at a particular 'slanted' slice of $\boldsymbol{o}(x, t)$ corresponding to a moment of time t' in another frame. Whether or not they do is a question whose answer requires the very same metaphysical resources as the answer to the first question.

Finally, in the spirit of relativity, there is nothing special about the initial choice of the frame (x, t). One could start with assembling a partial path of o in (x', t'), $\boldsymbol{o}(x', t')$, and *then* raise a question about whether any particular t-slice of $\boldsymbol{o}(x', t')$ is eligible to host o as well.

The *full* path of o is then simply the union of all its partial paths in all inertial frames of reference. In some idealized cases it will be clear-cut. In more realistic cases it will have a well-delineated *core* along with a possibly ragged 'penumbra'. How the core is stitched together with the penumbra is a question that cannot be addressed here. But in light of the above considerations it should be clear that this question too has nothing distinctly relativistic about it.

Acknowledgements I am very grateful to Cody Gilmore and Thomas Sattig for their comments on the draft of this paper.

References

Balashov, Y. (2008). Persistence and multilocation in spacetime. In D. Dieks (Ed.), *The ontology of spacetime* (Vol. 2, pp. 59–81). Amsterdam: Elsevier.

Balashov, Y. (2010). *Persistence and spacetime*. Oxford: Oxford University Press.

Bittner, T., & Donnelly, M. (2004). The mereology of stages and persistent entities. In R. Lopez de Mantaras & L. Saitta (Eds.), *Proceedings of the European conference of artificial intelligence* (pp. 283–287). Amsterdam: IOS.

Geroch, R. (1978). *General relativity from A to B*. Chicago/London: The University of Chicago Press.

Gibson, I., & Pooley, O. (2006). Relativistic persistence. In J. Hawthorne (Ed.), *Philosophical perspectives: Metaphysics* (Vol. 20, pp. 157–198). Oxford: Blackwell.

Gilmore, C. (2006). Where in the relativistic world are we? In J. Hawthorne (Ed.), *Philosophical perspectives: Metaphysics* (Vol. 20, pp. 199–236). Oxford: Blackwell.

Gilmore, C. (2008). Persistence and location in relativistic spacetime. *Philosophy Compass, 3/6*, 1224–1254.

Haslanger, S. (2003). Persistence through time. In M. J. Loux & D. W. Zimmerman (Eds.), *The Oxford handbook of metaphysics* (pp. 315–354). Oxford: Oxford University Press.

Hudson, H. (2001). *A materialist metaphysics of the human person*. Ithaca: Cornell University Press.

Lewis, D. (1988). Rearrangement of particles: Reply to Lowe. *Analysis, 48*, 65–72.
Parsons, J. (2007). Theories of location. In D. W. Zimmerman (Ed.), *Oxford studies in metaphysics* (Vol. 3, pp. 201–232). Oxford: Oxford University Press.
Rea, M. (1998). Temporal parts unmotivated. *Philosophical Review, 107*, 225–260.
Rindler, W. (1961). Lenght contraction paradox. *American Journal of Physics, 29*, 365–366.
Sartori, L. (1996). *Understanding relativity*. Berkeley: University of California Press.
Sattig, T. (2012). Ordinary objects in the relativistic world. *Chronos: Proceedings of the Philosophy of Time Society XI* (2010–2011), 28–61.
Sattig, T. (forthcoming). *The double lives of objects*.
Sattig, T. (2006). *The language and reality of time*. Oxford: Clarendon.
Sider, T. (2001). *Four-dimensionalism: An ontology of persistence and time*. Oxford: Clarendon.
Smart, J. J. C. (1987). Space-time and individuals. In J. J. C. Smart (Ed.), *Essays metaphysical and moral: philosophical papers* (pp. 61–77). Oxford: Blackwell.

Chapter 3
Parthood and Composition in Quantum Mechanics

Claudio Calosi and Gino Tarozzi

3.1 Introduction

The present paper is an attempt to explore the notions of parthood and composition, that play a crucial role in both physics and metaphysics, within the context of non relativistic quantum mechanics. The question of what mereological theory quantum systems are models of is virtually unexplored in literature.[1] We will argue that they are a model of the so called closed extensional mereology (CEM) and we will consider whether they could be models of some stronger closure mereological theory. Let us immediately spell out the importance of dealing with such a topic. First of all there is a growing interest in the philosophy of physics for questions about part, wholes, boundaries and so on, especially when it comes to quantum mechanics. This interest is witnessed for example in Healey (2013) and Field (2014). Also Morganti (2013) dedicates an entire chapter on parts and wholes. Second, we have reason to believe that some quantum features will be retained in our next fundamental physical theory. If so, the analysis put forward in the paper shows how parthood and composition behave (or are likely to behave) at (some) fundamental level. This is something that should not be overlooked by metaphysicians. Finally, as it will be clear in due course, some of the arguments presented have consequences for hotly debated metaphysical issues such as the fate of the *Unrestricted Composition* principle or questions about metaphysical fundamentality. The plan of the paper is as follows. In Sect. 3.2 we review some

[1]But see Calosi et al. (2011) and (Healey 1991, 2013) for some important insights.

C. Calosi (✉) • G. Tarozzi
Department of Basic Sciences and Foundations, Urbino University, Urbino, Italy
e-mail: claudio.calosi@uniurb.it; gino.tarozzi@uniurb.it

C. Calosi and P. Graziani (eds.), *Mereology and the Sciences*, Synthese Library 371,
DOI 10.1007/978-3-319-05356-1_3,

basic facts about quantum composite systems and some mereological principles, and then in Sect. 3.3 we give a detailed account of the mereology of quantum systems. We will offer a brief conclusion in Sect. 3.4.

3.2 Primers in Quantum Mechanics and Formal Theories of Parthood

In this section we review some important facts about quantum composite systems (Sect. 3.2.1) and about some mereological principles (Sect. 3.2.2). This review is not intended to be complete,[2] but rather to introduce some notions that we will use extensively throughout the paper.

3.2.1 Quantum Mechanics of (Composite) Systems

Our purpose here is to collect some important facts about quantum composite systems that some reader may not be familiar with. Firstly we briefly review some technical material. To every quantum system S is associated a separable (not necessarily finite dimensional)[3] Hilbert space H over the complex field. Different linear operators can be defined over H. We are interested in three particular operators, i.e. *Hermitian operators*, *projective* (or *projection*) and *density* operators. Their importance will be clear later on. Let us start from the first.[4]

Definition 1. An operator A is *Hermitian* if for every vector $|v\rangle$, $\langle u|Av\rangle = \langle Au|v\rangle$

Recall that a function (vector) $|u\rangle$ is called an eigenfunction (eigenvector) of an operator A corresponding to eigenvalue a if $|u\rangle \neq |0\rangle$ and $A|u\rangle = a|u\rangle$. Three important facts can be proven: (i) eigenvalues of Hermitian operators are real; (ii) representing any hermitian operator as a square matrix, it turns out that the sum of the diagonal elements of such matrix is the sum of the operator's eigenvalues; (iii) eigenvectors corresponding to distinct eigenvalues are mutually orthogonal.

Definition 2. An operator A is *idempotent* if $A = AA = A^2$

With this in hand it is possible to give a simple definition of projective operators.

---

[2]We refer to Beltrametti and Cassinelli (1981, pp. 61–77) and Jauch (1968, pp. 175–182) for quantum composite systems and to Casati and Varzi (1999, pp. 29–49) and Varzi for mereological principles. See also Simons (1987) and the Appendix of this volume.

[3]We will always deal with finite dimensional cases. We will omit the reference thereon.

[4]We will use the so called Bra-ket notation for the inner product.

Definition 3. A *projective* operator, which we shall denote by P_i, is a linear operator that is both Hermitian and idempotent.

Projective operators are operators that project vectors into subspaces of H that are spanned by the operators eigenvectors, so that the set of projective operators is in one to one correspondence to subspaces of the Hilbert space. Then we just need to define density operators. To do that we introduce the notions of a *trace class* and *trace* of an operator first.

Proposition 1. *An operator A belongs to the trace class if (i) for every $|u\rangle$, $\langle u|Au\rangle \geq 0$, and (ii) $\sum_i \langle u_i|Au_i\rangle$ is finite*

The *trace* of an operator A, denoted by $Tr(A)$ is the number defined via:

$$Tr(A) = \sum_i \langle u_i|Au_i\rangle \tag{3.1}$$

Now the last piece:

Definition 4. A *density* operator,[5] which we shall denote by D_i, is a trace class operator of trace 1.

There are various relations that hold between the different operators we have defined. Some of them are of crucial importance.

Proposition 2. *For any projection P_i that projects onto a n-dimensional subspace $Tr(P_i) = n$*

From Definition 4 and Proposition 2 it follows that:

Proposition 3. *Every projection operators that projects onto a 1-dimensional subspace is a density operator.*

The other crucial relation, this time between Hermitian and projection operators, is ensured by the so called *Spectral Theorem* (Jauch 1968, pp. 53–54; Beltrametti and Cassinelli 1981, pp. 293–296).[6] Informally it says that any Hermitian operator can be decomposed into a weighted sum of projective operators:

Theorem 1. *Let A be an Herminitan operator defined over a finite dimensional Hilbert space. Then there are real numbers $a_1, \ldots, a_n$ and projective operators $P_1, \ldots, P_n$ such that $A = \sum_{i=1}^{n} a_i P_i$, where numbers $a_1, \ldots, a_n$ are eigenvalues of A.*

Why have we spent so much time and effort to provide these definitions? Simply because these operators play a crucial role in the quantum mechanical description of physical systems. Let us see how.

---

[5]It is sometimes also called statistical operator or density matrix.

[6]See also Hughes (1992, p. 50).

In general, given a quantum system S, with associated Hilbert space H, every state of the system S is represented by a density operator defined over H, and every observable O of S is represented by a Hermitian operator. In the absence of superselection rules the converses also hold. Given Proposition 3 and the *Spectral Theorem* it also follows that projection operators P_i such that $Tr(P_i) = 1$, i.e. projection operators that project onto 1-dimensional subspaces, represent states and that every observable is associated with a weighted sum of projection operators.

Quantum mechanics provides an algorithm to establish the probability that an observable represented by the Hermitian operator A lies in a particular subset E of the real line[7] when the state of S is D. What is this algorithm? Recall that, according to the *Spectral Theorem* every operator A determines for every subset E a projection operator that we will write as $P_A(E)$. The aforementioned probability is then the number $Tr(DP_A(E))$ We have said that states of quantum systems are represented by density operators. They form a convex set. Let us be clearer. Let $D_1, \ldots, D_n$ be density operators and $w_1, \ldots, w_n$ real positive numbers. Then:

Definition 5. D is the *convex sum* of $D_1, \ldots, D_n$ if (i) $D = \sum_1^n w_i D_i$ with $i = 1, \ldots, n$ and (ii) $\sum_1^n w_i = 1$

In such cases D is still a density operator and thus represent a possible state of the system. In general we have then that:

$$Tr(DP_A(E)) = \sum_i w_i Tr(D_i P_A(E)) \tag{3.2}$$

Equation (3.2) is the general quantum mechanical algorithm we were looking for. It gives the probability of observable represented by A having a value that lies in E given the general convex sum state D. We will see that this take a more perspicuous form in particular cases. To see this we need an important distinction, namely that of *pure* and *mixed* states. Go back to Definition 5.

Proposition 4. *A state D that cannot be written as a convex sum of other states is called a pure state. A general convex sum is instead a nonpure or mixed state.*

We have already seen that projectors onto one dimensional subspaces are states. Furthermore the following important consequence can be proven:

Proposition 5. *A state D is a pure state iff it is a projection operator that projects onto a 1-dimensional subspace.*

It follows that the set of projective operators that projects onto 1-dimensional subspaces is in a one-to-one correspondence with the possible pure states of the system. Now, 1-dimensional subspaces are spanned by a single vector. Hence pure states can be represented by vectors. Suppose $D = P^{|u\rangle}$ where the right hand side should be read as the projective operator that projects onto the 1-dimensional

[7] Technically a Borel subset of R. We will omit this specification.

subspace spanned by the vector $|u\rangle$. Then every vector that is a scalar multiple of $|u\rangle$ represents the same pure state.[8]

Suppose now that $|u\rangle$, $|v\rangle$ represent two distinct pure states. The linearity of the Hilbert space allows us to generate another pure state that is called the linear superposition of states $|u\rangle$, $|v\rangle$, namely $c_1|u\rangle + c_2|v\rangle$ provided that $|c_1|^2 + |c_2|^2 = 1$. In general we have the following.

Proposition 6. *Let* $\{|u_1\rangle, \ldots, |u_n\rangle\}$ *represents a set of pure states. Then the linear superposition of such state defined as* $\sum_1^n c_i|u_i\rangle$, *where* $\sum_1^n |c_i|^2 = 1$ *is again a pure state of the system.*

Needless to say this should not be confused with the convex sum in Definition 5.

So far we have talked about pure states. What is the relation between pure and mixed states? It turns out that every state D can be written as a weighted (not necessarily finite) sum of pure states, that is:

$$D = \sum_i w_i P_i \tag{3.3}$$

where P_i are projectors onto 1-dimensional subspaces, that we know represents pure states,[9] and $\sum_i w_i = 1$

There is an important fact about Eq. (3.3) that will play a critical role in some of our arguments. This equation ensures a decomposition of mixed states into pure ones. However this decomposition is not unique. This entails that "an ignorance interpretation" of mixed states cannot be consistently maintained. Here is a brief argument.

Consider a system S in a mixed state $D = c_1P_1 + c_2P_2$. According to the ignorance interpretation S is really in one of the pure states P_1, P_2, we just do not know which one. However, given the non uniqueness of decomposability we can write $D = c_3P_3 + c_4P_4$ where the four pure states that appears in the two decompositions are distinct. Then, repeating the previous argument it follows that according to the ignorance interpretation S is really in one of the pure states P_3, P_4 which contradicts the conclusion that it was either in P_1 or P_2. Thus an ignorance interpretation of mixed states is untenable.

We have pointed out that in particular cases Eq. (3.2) takes a more perspicuous form. Suppose a system S is in a pure state. Then, by Proposition 5 it is represented by a projection operator that projects onto a 1-dimensional subspace. Let $|\psi\rangle$ be a vector belonging to that subspace. It is possible to choose a basis for the Hilbert

---

[8]Sometimes pure states are said to be represented by normalized vectors. This is indeed that vector $|v\rangle = a|u\rangle$ such that $\langle v|v\rangle = 1$. This choice of representing states has some conventional latitude.

[9]This justifies the claim that the set of states of a system forms a convex set of which the extremal points are the pure states Hughes (1992, p. 143).

space that contains $|\psi\rangle$. Then, the probability that the value of an observable A lies in E when $D = P^{|\psi\rangle}$ is given by:

$$\langle\psi|P_A(E)\psi\rangle \tag{3.4}$$

This is probably what could be found in every (introductory) text in quantum mechanics. But what if we want to compute the probability that observable A has a definite value λ_i rather than lying in an arbitrary Borel set E? We will answer this question in the simple case in which the observable in question has a discrete point spectrum, i.e. the support of the probability measure defined by P_A is simply $\lambda_1, \ldots, \lambda_n$. These are the eigenvalues of the observable A. We will furthermore make the simplifying assumption that there is no degeneracy, i.e. all eigenvalues are distinct. Then to every eigenvalue λ_i corresponds an eigenvector[10] $|\varphi_i\rangle$. Substituting in Eq. (3.4) we get:

$$\langle\psi|P_A(\lambda_i)\psi\rangle = \left\langle\psi|P^{|\varphi_i\rangle}\psi\right\rangle \tag{3.5}$$

Now, $\left|P^{|\varphi_i\rangle}\psi\right\rangle = \langle\varphi_i|\psi\rangle\,|\varphi_i\rangle$. Substitute it in Eq. (3.5) to get:

$$\langle\psi|\,\langle\varphi_i|\psi\rangle\,|\varphi_i\rangle = \langle\varphi_i|\psi\rangle\,\langle\psi|\varphi_i\rangle = |\,\langle\varphi_i|\psi\rangle\,|^2 \tag{3.6}$$

By linearity and symmetry of the inner product. Equation (3.6) gives us a very simple expression for the probability of finding value λ_i corresponding to eigenvector $|\varphi_i\rangle$ when the state of the system is $|\psi\rangle$. Actually we can come up with even a simpler expression. This is because we can write $|\psi\rangle$ as $|\psi\rangle = \sum_i c_i\,|\varphi_i\rangle$ where $c_i = \langle\varphi_i|\psi\rangle$ so that the probability in question is simply given by $|c_i|^2$ as can be seen from direct substitution in the right hand-side of Eq. (3.6). Let us stop and consider a simple example.

Suppose a system S is in pure state $|\psi\rangle = c_1\,|\varphi_1\rangle + c_2\,|\varphi_2\rangle$ where $|\varphi_1\rangle, |\varphi_2\rangle$ are the two eigenvectors of an operator A belonging to the eigenvalues λ_1 and λ_2 respectively.[11] Then the probability of finding value λ_1 when measuring the observable represented by A is simply $|c_1|^2$. The same goes for λ_2. This leads us to a point that will be crucial in some of the arguments we will present in the next section. Suppose the state S is in an eigenfunction of a particular observable O represented by operator A, i.e. $|\psi\rangle = c\,|\varphi\rangle$ where $|\varphi\rangle$ is the eigenfunction of O belonging to eigenvalue λ. Then the probability of S having value λ for the observable O is, according to the previous argument:

$$|c|^2 = 1 \tag{3.7}$$

---

[10]This is ensured by our simplifying assumption.

[11]Note that since $|\psi\rangle$ is a pure state by assumption, it follows that $|c_1|^2 + |c_2|^2 = 1$.

since $|\psi\rangle$ is by assumption a pure state. Hence it will be certain that S has the property $O = \lambda$. This is sometimes referred to as the *Eigenfunction-Eigenvalue* link in literature, namely

Proposition 7. *A system S has the property of having $O = \lambda$ iff S is in an eigenfunction of the observable O belonging to the eigenvalue λ*

So far we have talked about simple systems. Let us then move to composite ones.

Suppose systems S_1 and S_2 form a composite system S. We have seen that to every quantum system corresponds an Hilbert space. What is the Hilbert space corresponding to S, or better, how can we construct the Hilbert space H of S out the Hilbert spaces H_1 and H_2 of S_1, S_2? The Hilbert space in question is called the tensor product of the Hilbert spaces of the component systems and is indicated with $H = H_1 \otimes H_2$. More precisely consider the Cartesian or topological product $H_1 \times H_2$, i.e. the set of ordered pairs $(|\varphi\rangle, |\psi\rangle)$, $|\varphi\rangle \in H_1, |\psi\rangle \in H_2$. Then we have that:

Definition 6. H is the tensor product space of H_1, H_2 iff there exists a map $\otimes : H_1 \times H_2 \rightarrow H$ that satisfies:

(i) $|\varphi\rangle \otimes |\psi\rangle, |\varphi\rangle \in H_1, |\psi\rangle \in H_2$ spans H,[12] and
(ii) $\langle\varphi_1 \otimes \psi_1|\varphi_2 \otimes \psi_2\rangle = \langle\varphi_1|\varphi_2\rangle\langle\psi_1|\psi_2\rangle$ for all vectors in H_1, H_2

Clause (ii) simply says that the inner product on the tensor product space is defined in terms of the inner products of the Hilbert spaces of the component systems, since $\langle\varphi_1|\varphi_2\rangle$ is defined over H_1, whereas $\langle\psi_1|\psi_2\rangle$ is defined over H_2. Since by clause (i) the set $|\varphi\rangle \otimes |\psi\rangle$ spans H, the inner product defined in clause (ii) can be extended by linearity to the whole of H. This construction can be generalized to n subsystems (Jauch 1968, p. 186). In this case we have:

Definition 7. H is the tensor product space of $H_1, H_2, \ldots, H_n$ iff there exists a map $\otimes : H_1 \times H_2 \times \ldots \times H_n \rightarrow H$ that satisfies:

(i) $|\varphi\rangle \otimes |\psi\rangle \otimes \ldots \otimes |\vartheta\rangle$, with $|\varphi\rangle \in H_1, |\psi\rangle \in H_2, \ldots, |\vartheta\rangle \in H_n$, spans H
(ii) $\langle\varphi_1 \otimes \psi_1 \otimes \ldots \otimes \vartheta_1|\varphi_2 \otimes \psi_2 \otimes \ldots \otimes \vartheta_2\rangle = \langle\varphi_1|\varphi_2\rangle\langle\psi_1|\psi_2\rangle \ldots \langle\vartheta_1|\vartheta_2\rangle$

Clauses (i) of Definitions 6 and 7 tell us that the relevant sets span the Hilbert space. This means, restricting our attention to the first case, that every vector $|\phi\rangle \in H$ can be written as a linear combination of vectors in the relevant sets, i.e.:

Theorem 2. $|\phi\rangle = \sum_1^{\dim H_1} \sum_1^{\dim H_2} c_{ij}|\varphi_i\rangle \otimes |\psi_j\rangle$

This is a version of the so called *Schmidt Decomposition* theorem.

The important thing about this theorem is that it entails that not all the vectors, though they can be written as a *linear combination* of vectors of the form $|\varphi\rangle \otimes |\psi\rangle$, can be written simply as $|\varphi\rangle \otimes |\psi\rangle$ directly. We will see that vectors that *cannot* be written this way represent *entangled states*, and we will see particular examples

[12]Note that all the following notations are equivalent: $\varphi \otimes \psi$, $|\varphi\rangle \otimes |\psi\rangle$, $|\varphi\rangle|\psi\rangle$, $|\varphi\psi\rangle$.

of them. Here is another, very brief, equivalent way to put the same point. The topological product is a proper subset of the tensor product.

We have constructed the Hilbert space for the composite system. How can we construct the operators? Given the linearity of Hilbert space we can define linear operators by the way they transform the vectors in the basis (i) of Definition 6[13] and then extend it to the whole Hilbert space. Let then A_1, A_2 be two linear operators defined on H_1, H_2 respectively. Then we define a new linear operator over H via:

$$(A_1 \otimes A_2)(\varphi_i \otimes \psi_j) = A_1|\varphi_i\rangle \otimes A_2|\psi_j\rangle \tag{3.8}$$

Among all possible operators defined via Eq. (3.8) we are interested in particular ones, namely $A_1 \otimes I_2, I_1 \otimes A_2$ where I_1, I_2 are the identity operators in H_1, H_2. Operator $A_1 \otimes I_2$ defined over H represents the same physical quantity of A_1 defined over H_1. In the first case we are simply considering the subsystem as a part of the composite one.

This leads us to the last topic of the section, namely that of relating the states of the composite and the component systems. First we impose some sort of consistency requirement, that is we require that if we perform a measurement of any observable on the subsystem $S_1(S_2)$, whether we measure it as an individual system, or as a part of the composite system S the result turns out to be the same. Recall that (i) Eq. (3.2) gives us the probability of finding a value for an observable as $Tr(DP_i)$ for every projection operator that the spectral theorem determines for an observable A_i and that (ii) for every operator, $A_1 \otimes I_2$ represents the same observable as A_1.[14] Then we can simply write the consistency requirement as:

$$\begin{aligned} Tr(D_1 P_1) &= Tr(D(P_1 \otimes I_2)) \\ Tr(D_2 P_2) &= Tr(D(I_1 \otimes P_2)) \end{aligned} \tag{3.9}$$

Possible solution(s) to Eqs. (3.9) tell us important facts about composite systems. The first important thing that can be proven is the following:

Proposition 8. *The state of the total system D uniquely determines the states D_1, D_2 of the component systems. This is true regardless of D being a pure or a mixed state.*

The converse however does not hold. The states D_1, D_2 can determine D only if they are *pure* states. In particular we have:

Proposition 9. *Let D_1, D_2 be two pure states. Then D is uniquely determined by D_1, D_2 and it is in the pure state given by $D = D_1 \otimes D_2$, i.e. $D_1 \otimes D_2$ is the only solution to Eq.* (3.9).

---

[13]The same goes for the most general cases in Eq. (3.7).

[14]The same goes for $I_1 \otimes A_2$.

Let us see what happens when D is a pure state. Then we have two different cases. The first one is the simplest:

Proposition 10. *Let D be in the pure state $D = |\varphi\rangle \otimes |\psi\rangle$, where $|\varphi\rangle \in H_1$ and $|\psi\rangle \in H_2$. Then $D_1 = |\varphi\rangle$, $D_2 = |\psi\rangle$.*

Note that this, together with Proposition 9 entails that the component states are in pure states iff the state of the composite system can be written as $|\varphi\rangle \otimes |\psi\rangle$. But we have already pointed out discussing the *Schmidt Decomposition* theorem that not every vector in H can be written in that form. Then we have the following important result:

Proposition 11. *Let D be a pure state of the general form $D = \sum_{i,j} c_{ij}\varphi_i \otimes \psi_i$ that cannot be written as $|\varphi\rangle \otimes |\psi\rangle$. Then D_1, D_2 are still uniquely determined by D but are mixed states.*

We call the states in Proposition 11 *entangled states* for their component systems do not behave independently but rather exhibit important correlations.[15] It is also said that the total state is not *factorizable*.

Let us conclude this section by giving an example of such entangled states. Let S_1, S_2 be two identical spin subsystems, where spin eigenstates are given by the up and down states $|\uparrow\rangle, |\downarrow\rangle$. Then, in general we will have that

$$\begin{aligned} D_1 &= |\varphi\rangle = (c_1|\uparrow\rangle + c_2|\downarrow\rangle) \\ D_2 &= |\psi\rangle = (c_3|\uparrow\rangle + c_4|\downarrow\rangle) \end{aligned} \tag{3.10}$$

Whence

$$\begin{gathered} D_1 \otimes D_2 = |\varphi\rangle \otimes |\psi\rangle = \\ c_1c_3|\uparrow\rangle|\uparrow\rangle + c_1c_4|\uparrow\rangle|\downarrow\rangle + c_2c_3|\downarrow\rangle|\uparrow\rangle + c_2c_4|\downarrow\rangle|\downarrow\rangle \end{gathered} \tag{3.11}$$

Now, suppose that the system is in the pure state

$$D = \frac{1}{\sqrt{2}}(|\uparrow\rangle|\uparrow\rangle + |\downarrow\rangle|\downarrow\rangle) \tag{3.12}$$

Now, the question is. Can the state represented by Eq. (3.12) be written in the form of Eq. (3.11)? The answer is clearly no. To see this note that this would entail

[15]It is possible to define a correlation coefficient $\rho(A_1, A_2, D)$ between any two observables A_1, A_2 given state D. Then it is possible to define entangled systems those systems for which $\rho \neq 0$. On the other hand if the composite system is in the state of (3.28) $\rho = 0$, there are no correlations and the components systems behave independently.

$$c_1c_3 = \frac{1}{\sqrt{2}} = c_2c_4$$
$$c_1c_4 = 0 = c_2c_3 \tag{3.13}$$

That cannot be both satisfied (Calosi et al. 2011, p. 1746). Equation (3.11) does indeed represent a classic example of an entangled state. With this we conclude our crash course in the quantum mechanics of composite systems. We now move to the next topic.

3.2.2 Mereologies

Mereology is the formal theory of parthood and of parthood relations. Different mereological theories of different strength can be obtained by regimenting the primitive notion of parthood[16] with different axioms. We are interested here in only a few of them. Let us write

$$(Parthood) x \prec y \tag{3.14}$$

For x is part of y. Then using first order logic with identity we can define the important mereological notions of *Proper Parthood*, *Overlap* and *Underlap* via:

Definition 8. (*Proper Parthood*) $x \prec\prec y =_{df} x \prec y \wedge x \neq y$

Definition 9. (*Overlap*) $O(x, y) =_{df} (\exists z)(z \prec x \wedge z \prec y)$

Definition 10. (*Underlap*) $U(x, y) =_{df} (\exists z)(x \prec z \wedge y \prec z)$

Informally a proper part of something is a part of that something that is distinct from the whole it is part of, two things overlap if they share a part and underlap if there is something of which they are both parts.

In what follows we briefly develop different formal mereological theories of different strengths that we will discuss throughout the paper. Formulas are intended to be universally closed unless otherwise specified. Sometimes the different axioms regimenting mereological notions are divided in *lexical axioms*, those axioms that allegedly capture the meaning itself of the notion of part, *decomposition principles*, that take from the whole to the parts that make up the whole, and *composition principles*, that take from the parts to the whole they are part of.[17] Among the lexical axioms it is customary to include *Reflexivity*, *Anti-symmetry* and *Transitivity*:

---

[16]We follow Casati and Varzi (1999) and Varzi (2014) in such a choice. For a more comprehensive account of theories of parthood see the Appendix to this volume.

[17]It is a matter of dispute where the lines should be drawn. See Varzi (2014).

Axiom 1. *(Reflexivity R)* $x \prec x$

Axiom 2. *(Anti-Symmetry AS)* $x \prec y \wedge y \prec x \rightarrow x = y$

Axiom 3. *(Transitivity T)* $x \prec y \wedge y \prec z \rightarrow x \prec z$

These axioms are familiar enough not to deserve an informal rendering. They make parthood a partial order. Let us call M, for *Ground Mereology* the mereological theory comprising only Axioms 1–3.[18]

We are interested in two decomposition principles in particular, namely the so called *Weak* and *Strong Supplementation Principle*(s). Here is a formal rendering of the former:

Axiom 4. *(Weak Supplementation Principle WSP)*
$x \prec\prec y \rightarrow (\exists z)(z \prec y \wedge \sim O(z, x))$

Informally *WSP*[19] says that if something has a proper part then it has a part that is discrete from it. Call this latter part the mereological remainder. *WSP*, together with R and T, entails AS. We call a mereological theory that comprises R, T and *WSP Minimal Mereology*, or MM for short. We can strengthen *WSP* simply by changing the antecedent. This is what's done in the *Strong Supplementation Principle*:

Axiom 5. *(Strong Supplementation Principle SSP)*
$\sim y \prec x \rightarrow (\exists z)(z \prec y \wedge \sim O(z, x))$

Informally it says that if something fails to include something else among its parts than there is a mereological remainder between the two. It is called *Strong* for it can be proven that *SSP* entails *WSP*, whereas the converse does not hold (Simons 1987, p. 29). We call the theory that comprises R, T and *SSP Extensional*[20] *Mereology* or EM.

Finally we have the composition principles. Consider x and y. Then, we might want to require that if they do underlap they have a minimal underlapper. This minimal underlapper is called the mereological sum of x and y and is defined as that thing that overlaps exactly those things that overlap either x or y. It corresponds to the following *Binary Sum Principle*:

Axiom 6. *(BinarySum Principle BSP)*
$U(x, y) \rightarrow (\exists z)(\forall w)(O(z, w) \leftrightarrow O(z, x) \vee O(z, y))$

We call z the mereological sum of x and y and we write $z = Sum(x, y)$, or $Sum(\varphi(x))$ for the mereological sum of all those entities that satisfy the open formula φ. In the presence of *SSP* the entity whose existence is asserted in the consequent of Axiom 6 is unique. If you add *BSP* to MM you obtain the so called

---

[18]We are following, and will continue to follow, the terminology employed in Varzi (2009a).

[19]This ensures that nothing has a unique proper part, or, in other words, that every composite object has at least two proper parts.

[20]The reason of this very name will be clear later on.

Closure Minimal Mereology, or CMM. If you add it to EM you obtain *Closure Extensional Mereology* CEM instead.

One of the most interesting composition principles is the infinitary extension of *BSP*. Informally it says that any two non empty sets of objects whatsoever have a mereological sum. It is called *Unrestricted Composition Principle* and can be rendered as follows:

Axiom 7. *(Unrestricted Composition UC)*
$(\exists w)(\varphi(w)) \rightarrow ((\exists z)(\forall w)(O(z, w) \leftrightarrow (\exists v)(\varphi(v) \wedge O(w, v))))$

Informally for every non empty set of φ-ers there exists a mereological sum of those φ-ers, defined as that entity z that overlaps all and only those things that overlap some φ-er.[21] Important insights on the relations between *UC* and *SSP* are given in Varzi (2008). It is uncontroversial that *UC* entails *BSP*. The mereological theory obtained by adding *UC* to EM is called *General Extensional Mereology*, GEM, and is sometimes referred to as *Standard Mereology*, or even *Classical Mereology*. It is a powerful theory, almost as powerful as standard set theory, a result proven in Tarski (1956).

So we have developed different mereological theories. The logical relations holding among some of them can be easily summed up[22]:

$$M \subset MM \subset EM \subset CEM \subset GEM \tag{3.15}$$

Although it is a matter of philosophical dispute, we believe that it is possible, and indeed rather probable, that different ontological domains may be models of different mereological theories.[23] Thus a natural question arises. Of what mereological theory are quantum systems a model of? It is to this question and its important consequences that we now turn on.

3.3 The Mereology of Quantum Systems

We now explore the neglected question of what mereological theory quantum systems are a model of. As we already pointed out in the introduction this is the first systematic attempt to deal with such an important question. We provide new arguments in favor of the claim that they are a model of *Extensional Mereology* EM and (possibly) even stronger closure mereologies, i.e. mereological theories

[21]Hovda (2009) and Varzi (2009) argue that *Unrestricted Composition* and *Weak Supplementation* together entail *Strong Supplementation*, insofar as the notion of parthood is not radically distorted. Rea (2010) replies to such arguments.

[22]For a more detailed diagram see Casati and Varzi (1999, p. 48), though, as Varzi (2014) points out, that work contains a mistake, for the result of adding *UC* to MM is not equivalent to GEM.

[23]See Sider (2007) for an argument to the contrary.

with some composition principles (Sect. 3.3.1). We then go on to discuss possible objections to such arguments (Sect. 3.3.2). The discussion of some objections in later sections will bring to the foreground consequences of the arguments presented that are central both to issues in philosophy of physics and hotly debated topics in analytic metaphysics.

3.3.1 Quantum Models of Mereological Theories

We want to suggest new arguments for the following claim: quantum systems are models of an extensional mereology. Whether they are models of closure mereologies too is more controversial. We want to tackle this question rather systematically so we will proceed step by step.

Let us start from *Reflexivity*. Are quantum systems part of themselves? We do not have really an argument in favor of a particular answer. We just don't see why they cannot be taken to be as part of themselves, that is, we believe that there is some conventional latitude in answering such a question. Famously Rescher (1955) points out that there are non reflexive uses of the notion of parthood in biology. It could be the case that the uses Rescher has in mind carry over into the quantum domain. According to such uses an electron, (or a proton, or a pion) would not count as part of itself. We believe this can hardly be considered an argument. It seems to us to rest on some sort of linguistic intuition. Apart from these considerations of ours, this is hardly a problem. The following is in fact a straightforward consequence of R, AS, T:

$$x \prec y \leftrightarrow x \prec\!\prec y \vee x = y \tag{3.16}$$

This shows that we could have taken *Proper Parthood* as primitive. Then we could have defined *Parthood* in its terms.[24] This is actually what's done in Simons (1987). But *Proper Parthood* is not reflexive. Then we could simply argue that when we utter sentences like "an electron is not part of an electron" we are simply speaking loosely and we should properly say that an electron is not a *proper part* of an electron. Thus the whole issue seems to boil down to a choice of a preferred primitive (Varzi 2009a).[25] We conclude that *Reflexivity* is safe.

Transitivity is next. Suppose we have a quantum system S_1 that is part of a quantum system $S_3 = Sum(S_1, S_2)$. This is in turn part of the quantum system $S = Sum(S_3, S_4)$. Now, the question is: is S_1 part of S? To see this note that it follows from Sect. 3.2.1 that the Hilbert space associated with S is:

$$H = H_3 \otimes H_4 = H_1 \otimes H_2 \otimes H_4 \tag{3.17}$$

[24] Via: $x \prec x =_{df} x \prec\!\prec y \wedge x \neq y$

[25] We will address a more specific objection about *Reflexivity* in the following section.

Where H_n is the Hilbert space associated with the nth system. Even the simple Eq. (3.17) shows that the Hilbert space for S is constructed via the Hilbert space of S_1. This should be evidence enough for the latter to be a part of the former. But there's more. Given the *Schmidt Decomposition* theorem it follows that every state D in H can be written as:

$$D = \sum_{1}^{\dim H_3} \sum_{1}^{\dim H_4} c_{ij} |\xi_i\rangle \otimes |\vartheta_j\rangle \tag{3.18}$$

Where $\{|\xi_i\rangle\}, \{|\vartheta_j\rangle\}$ are basis for H_3, H_4 respectively. The same holds for every $|\xi_i\rangle \in H_3$, i.e. we can write:

$$|\xi_i\rangle = \sum_{1}^{\dim H_1} \sum_{1}^{\dim H_2} g_{hk} |\varphi_h\rangle \otimes |\psi_k\rangle \tag{3.19}$$

We can then substitute this expression into Eq. (3.18). This shows that every state of the composite system S has some "contribution", so to speak, from the state of S_1. Since probabilities for observable values of S depend on the state of the system via Eq. (3.2) it follows that the system S_1 somehow contributes to values of observables of S. And this, we contend, does show that it can be considered part of S. This argument is very general, and works for both entangled and non entangled states. Let us give a simpler version of the argument. Assume that all the states in question are factorizable. Hence:

$$D = |\xi\rangle \otimes |\vartheta\rangle = (|\varphi\rangle \otimes |\psi\rangle) \otimes |\vartheta\rangle \tag{3.20}$$

The vector $|\varphi\rangle \in H_1$ represents the state of S_1, so that we can safely conclude that it is part of S. That is all for *Transitivity*. *Transitivity* has been traditionally regarded as the most controversial lexical, partial ordering axiom.[26] This argument shows that, as long as the entities in question are regarded as quantum mechanical systems, transitivity holds. And it holds in virtue of the way composite systems are described within quantum mechanics. Since, arguably at the fundamental level, everything can be considered a quantum system, the argument shows that, at the fundamental level, parthood is transitive.

What about the *Strong Supplementation Principle*? It is better to spend a few words on the *Weak Supplementation Principle WSP* first. According to *WSP* there are no composite entities with a single proper part. Do quantum systems constitute a counterexample to such principle? Suppose they do. Then there will be a quantum system S with a single proper part S_1. Now, S and S_1 are different so that their physical description should be given in terms of density operators in different Hilbert spaces. Let H_1 be the Hilbert space associated with S_1. What is the Hilbert

[26] See Rescher (1955).

space H associated with S? It could not be H_1 for the reasons we just pointed out. It should actually be a tensor product space. But since there is no other proper part that makes up for the difference between S and S_1 there is no other Hilbert space H_n such that $H = H_1 \otimes H_n$. This simply means that we cannot give a quantum mechanical description of S. It is indeed a general assumption of this work that it is always possible to give a quantum mechanical description of quantum systems. Hence we conclude that quantum systems do not constitute a counterexample to *WSP*.

From this analysis and the results in Sect. 3.2 it follows that quantum systems are a model of both M and MM. Are they a model of EM too? To answer this question we should turn to the *Strong Supplementation Principle SSP*. We believe this is a crucial point.[27] It shows, once again, that at the fundamental level parthood obeys an extensionality principle, contrary to widespread agreement. We will discuss some of the implications later on in the paper.

Recall that informally the principle states that if something fails to include something else among its parts then there is a mereological remainder, i.e. a part that does not overlap the former. Suppose now systems S_1, S_2 compose system S. Then S_1 fails to include S among its parts. The same goes for S_2. According to *SSP* it must be the case that there is a proper or improper part that is disjoint from $S_1(S_2)$. There is indeed a natural candidate, namely $S_2(S_1)$. But are S_1 and S_2 disjoint? The following argument tries to establish that they are. Suppose S is in a general state:

$$D = \sum_{i,j} \lambda_{ij} \varphi_i \otimes \psi_j, \varphi_i \in H_{1,} \psi_j \in H_2 \tag{3.21}$$

Which is an equivalent way to write Theorem 2. We know from Proposition 8 that D determines uniquely the states D_1 and D_2 of S_1, S_2 respectively. The state of S_1 is given[28] by:

$$D_1|\varphi\rangle = \sum_{i,j} c_{ij} \langle \varphi_i|\varphi\rangle \varphi_j, c_{ij} = \sum_k \lambda *_{ij} \lambda_{jk}, \varphi \in H_1 \tag{3.22}$$

Equation (3.22) does not give the state of S_1 as a convex combination but this could be achieved by the *Spectral Theorem* (Beltrametti and Cassinelli 1981, p. 67). The state $D_1|\varphi\rangle$ in Eq. (3.22) is sometimes called the *reduced state* and D_1 the *reduced density matrix*. The important thing to note is that in the expression of the reduced state there do not appear terms that belong to H_2. This is indication that S_1 and S_2 are indeed disjoint. For if they were to share a part we would expect some "contributions" from Hilbert space H_2, or from a subset of the Hilbert space H_2 to such reduced state. This argument is very general and it is valid for both factorizable

---

[27]Note that Healey (2013) arrives at this very same conclusion. However he provides an entirely different argument.

[28]We denote with * the complex conjugate of a complex number.

and non factorizable states if valid at all. In the simplest case, where all states are factorizable, the argument takes an even simpler form. Suppose $D = |\varphi\rangle \otimes |\psi\rangle$. Then the reduced state is simply $D_1 = |\varphi\rangle \in H_1$. The argument does generalize to cases with more than two subsystems but, as it stands, it could be charged of begging the question. This is because the argument seems to implicitly presuppose that the Hilbert spaces H_1, H_2 are such that $H_1 \cap H_2 = \emptyset$. But isn't this the case in virtue of the fact that S_1 and S_2 do not overlap, which is what we are trying to prove?

Suppose then explicitly that they do share a part. Call that part S_3. Since we are working under the hypothesis that $S_1(S_2)$ fails to include $S_2(S_1)$ among its parts it follows that S_3 is a proper part of S_1. Then we can take S_1 to be composed by S_3 and S_4. We now know by *Weak Supplementation* that S_3 and S_4 are disjoint so that $H_2 \cap H_4 = \emptyset$. Then we could calculate the reduced state of S_4 via Eq. (3.22) again and be sure that there are no "contributions" from H_2, so that we can safely conclude that S_4 and S_2 are disjoint and *SSP* is therefore safe.

So we have just argued that quantum systems are a model of EM. It is now time to address some the most delicate mereological principles, namely the two composition principles. Let us start from the former, that is way less controversial. The *Binary Sum Principle BSP* states that if there exists a sum of two entities there is a minimal sum.

Recall that the minimal, or mereological sum of two entities x, y is defined, via Axiom 6, as that thing that overlaps all and only those things that overlap either x or y. Let S_1, S_2 be two quantum systems, to which the two Hilbert spaces H_1, H_2 are associated and suppose that they underlap, i.e. there is something of which they are both parts. Is there a minimal sum? Recall how we "built" composite systems in Sect. 3.2.1. It was a system S such that each state of S could be written as:

$$D = \sum_{i,j} \lambda_{ij}\varphi_i \otimes \psi_j, \varphi_i \in H_{1,}\psi_j \in H_2 \tag{3.23}$$

According to Eq. (3.23) every state of S, and thus every probability of every observable having some values, is "built" up from all the resources of H_1, H_2 and nothing else, that is to say by all and only the states of S_1 and S_2. Thus S should count as a minimal underlapper. It seems that quantum systems are models of CEM too. *BSP* is however a conditional principle, that is, it is *assumed* in the antecedent that the two systems in question have a sum. But, it seems natural to ask, are there any necessary and sufficient conditions for a set of entities to have a mereological sum? And if so, what are they?

These last questions are rough variants of the infamous *Special Composition Question* raised in Inwagen (1990).[29] The principle of *Unrestricted Composition* can be seen as a radical response to that question. There are no such conditions, *every set* of entities whatsoever has a mereological sum. There is another rather

[29] See Markosian (2008) for a good introduction.

radical answer, that is sometimes called *Mereological Nihilism*.[30] It is something like the following. There are no such conditions, *no set* of entities whatsoever ever has a mereological sum. It corresponds to adding a mereological axiom to the point that there are only mereological atoms, i.e. entities with no proper parts:

Axiom 8. *(Mereological Nihilism MN)* $\sim (\exists y)(y \prec\!\prec x)$

Between these two extremes there are the so called *moderate* answers.[31] They maintain in general that there ia a set of necessary and sufficient conditions Ψ that has to be met in order for a set of entities to have a sum. Needless to say, there are many different candidates for Ψ. These moderate answers have been discussed mostly within analytic metaphysics. No serious candidate from physics have been advanced to fill in the role of Ψ. Here we will discuss whether such a physical candidate is worth considering. The moderate answers can be thought of as admitting the following among the mereological axioms:

Axiom 9. *(Restricted Composition RC)*
$((\exists w)(\varphi(w)) \wedge (\forall w)(\varphi(w) \rightarrow \psi(w)))$
$\rightarrow ((\exists z)(\forall w)(O(z, w) \leftrightarrow (\exists v)(\varphi(v) \wedge O(v, w)))$

Apart from its difficult formal rendition, Axiom 9 just says that there exists a mereological sum of some φ-ers iff some conditions Ψ are met. What is the answer to the *Special Composition Question* in the quantum domain? From what we have seen it seems possible to rule out *MN*. But what about *UC* and *RC*?

Probably the most influential[32] arguments in favor of *UC* are the ones in Lewis (1986, pp. 211–213) and in (Sider 2001, pp. 134–139). They both crucially depend on the *semantic theory of vagueness*, i.e. roughly the thesis that there is no ontological vagueness and vagueness is simply semantic indecision, to borrow the terminology from Lewis (1986). This very point has been criticized as begging the question in Koslicki (2003), Simons (2006), and Elder (2008). Moreover, arguments using the semantic theory of vagueness are particularly slippery in the quantum domain, for it has been defended that quantum objects are indeed vague objects, for example in French and Krause (1996). Though vagueness of identity is not by itself an obvious explanation of vagueness of composition Simons (2006) notes that this would weaken the aforementioned arguments for we couldn't require the non vagueness of composition without begging the question, once the non vagueness of identity is given up.[33] We don't want to enter this discussion here. Rather we want to tackle the question differently. We want to investigate whether physics has something to tell about this question, that is we want to address whether quantum

[30]For a discussion and a defense see Rosen and Dorr (2002).

[31]Markosian (2008) advances the *Brutal Composition* thesis. It is roughly the thesis that composition is a brute fact, i.e. there is no simple and general answer to the *Special Composition Question*

[32]But see also Rea (1998) for a different argument.

[33]The argument is however controversial. For a critique of the argument see Darby (2010).

mechanics tells us about composition without invoking vagueness.[34] We are afraid that, despite the fact that quantum mechanics offers us important and deep insights that have been too often overlooked by analytic metaphysicians, we haven't been able to locate a strong quantum mechanical argument that could favor one answer over the other as far as composition goes.[35]

Consider the following promising argument. It can be proven that:

Proposition 12. *Every pure state* $|\varphi\rangle$ *is an eigenvector of some operator* A *representing an observable* O*, i.e. for every* $|\varphi\rangle$ *there is an* A *such that* $A|\varphi\rangle = a|\varphi\rangle$

As we have pointed out in Sect. 3.2.1 the *Eigenfunction-Eigenvalue* link states that if a particular system is in an eigenfunction of some observable O belonging to a specific eigenvalue a then such system has the property of having $O = a$. It follows that for any pure state we have a system S that has the property $O = a$ for some O and some a.

Now, we bring into play a fairly non radical metaphysical principle that we label *Instantiation Principle*[36] (*IP*):

Proposition 13. *(Instantiation Principle IP) For every existing property P there is something that instantiates P*

Let us then propose a tentative classification of properties, into *Mereologically Reducible Non Relational Properties*, *Mereologically Reducible Relational Properties* and *Mereologically Irreducible Inherent*[37] *Properties*. This classification is loosely inspired by Teller (1989, pp. 214–215), and in particular by Morganti (2009a, p. 227, 2009b, pp. 1029–1031).

Proposition 14. *(Mereologically Reducible Non Relational Property MRNR): A property P of a composite object x is a Mereologically Reducible Non Relational Property, or a MRNR-Property for short, iff there are properties* $P_1, \ldots, P_n$ *of the component parts such that P is (somehow) reducible to* $P_1, \ldots, P_n$.

[34]Consider another suggestion. Composition occurs in all of those cases in which QM dictated the use of a tensor product space. This seems at first sight question begging. Doesn't QM dictate the use of such space when dealing with composite systems? This could be however one of this cases in which there is some sort of circularity, but it is virtuous rather than vicious. For it could be the case that we can come up with results about the relations between tensor product spaces and "simple" spaces that do have profound consequences. We cannot pursue this approach here. But we think it is something worth exploring. This suggestion was offered us by Matteo Morganti.

[35]See Healey (2013) for a pragmatic take on how physics treats composition. This is, in our opinion, one of the most insightful papers dealing with composition in physics. Needles to say the fact that we have not been able to locate a quantum mechanical argument in favor of either *UC* o *RC* (or neither) does not mean that there isn't one.

[36]In Calosi et al. (2011, p. 1753) it is called the *Aristotele-Armstrong Principle*

[37]We follow Morganti (2009a) in using such terminology. He envisages it as an extension of the proposal found in Teller (1986).

The classic example of a *MRNR-Property* would be a classical additive property such as having a particular mass. The property of x having a mass $= m$ is reducible to the fact that the component parts of x have masses whose values add up to m.

Then we have:

Proposition 15. *(Mereologically Reducible Relational Property MRR) A property P of a composite object x is a Mereologically Reducibile Relational Property, or a MRR-property for short, if there are relations* $R_1, \ldots, R_m$ *holding between some of the component parts of x such that P is (somehow) reducible to* $R_1, \ldots, R_m$.

An example of a *MRR-property* could be that of a "shape of a composite object". It could be argued that this property could be somehow reduced to the spatial relations (and shape properties) holding between the component parts of the object.

Proposition 16. *(A Mereologically Irreducible Inherent Property MII) A Property P of a composite object x is a Mereologically Irreducible Inherent property, or a MII-Property for short, if there are no properties* $P_1, \ldots, P_n$*, of the component parts nor relations* $R_1, \ldots, R_m$ *holding among those component parts such that P is (somehow) reducible to* $P_1, \ldots, P_n$ *or* $R_1, \ldots, R_m$.

This classification helps us in establishing necessary ontological commitments required by the *Instantiation Principle IP*. The crucial thing to note is that in the case of both *MRNR* and *MRR*-properties the ontological commitment to the composite objects *can* be paraphrased away, so to speak, once we have a prior commitment to the component parts. This is not the case with the *MII*-properties. In this last case the commitment to the composite object *cannot* be paraphrased away.

Now, prepare a system in such a way that the quantum pure state $D = |\varphi\rangle$ assigned to the composite system S is an entangled state. It follows from Proposition 12 that this state is an eignefunction of some observable O belonging to eigenvalue a. By *IP* the property $O = a$ has to be instantiated by something. Suppose moreover that D is a non factorizable state. The question is: is $O = a$ a *MRNR*, a *MRR* or a *MII*-property?

It is fairly easy to argue that it isn't a *MRNR*-property. This is because by Proposition 11 the states of the component systems are mixed states, since the system is, by assumption, in an entangled state. So, by the *Eigenfunction-Eigenvalue* link they simply do not have any determinate state dependent properties. They could if they were "really" in pure states. But this is essentially to maintain what we called an ignorance interpretation of quantum *mixed* states. And we have already argued that this is untenable. What about $O = a$ being a *MRR*-Property? Maybe an example will be of some help. Consider two particles in the singlet state:

$$D = |\varphi\rangle = \frac{1}{\sqrt{2}}(|\uparrow\rangle_1|\downarrow\rangle_2 - |\downarrow\rangle_1|\uparrow\rangle_2) \tag{3.24}$$

This state exemplifies the property "having total spin = 0". We have just argued that it is not a *MRNR*-property. The reduced states for particles 1, 2 can be in fact calculated via Eq. (3.22) and they are mixed states. Hence particles 1, 2 do not have any definite spin properties. The question is whether "having total spin = 0" is a *MRR*-property. In the particular example at hand this very question can be phrased more perspicuously along the following lines: can the property "having total spin = 0" be reduced to the relation holding between particles 1, 2 of "having opposite spin"?

There are at least two arguments suggesting it cannot. Schaffer (2010, p. 54), building on Healey (1991, p. 420), argues that reducing such properties to relations between component parts would amount to a loss of the unity of properties. This is because "having total spin = 0" is a property that could be instantiated by a various number of systems with many different component parts. If this property is instead reduced to relations among those parts then it is really not the same property when attributed to systems with different parts.[38] In Morganti (2009a) Morganti proposes a different and very interesting argument. He claims that the non reducible character of such properties *explains* the peculiarity of quantum statistics.[39] If any of these arguments is on the right track there is some ground to maintain that $O = a$ is a *Mereologically Irreducible Inherent Property*. But we have seen that in such a case the ontological commitment to the composite system dictated by the *Instantiation Principle* cannot be paraphrased away. The upshot of the argument seems then to be the following: *it is a sufficient condition for quantum systems* $S_1, \ldots, S_n$ *to have a mereological sum* S *to be in an entangled non factorizable state.*

We believe this is an important insight that quantum mechanics offers into the question of composition. Does it favor *UC*, *RC* or neither? A discussion of the argument is in order. It should be clear that, in its present form, it favors neither. Whether it could be extended is a substantive question that deserves a careful, independent investigation that goes beyond the scope of this paper. We want however to point out some important facts.

It seems to us that the argument could not be reformulated as to lend support to *RC*. It could seem at first sight, that "being in an entangled, non factorizable state" is the relevant condition Ψ appearing in the Axiom 9 that should be met in order to ensure the existence of a mereological sum. But the argument, if valid, is only able to show that such a condition is *sufficient* for the existence of a mereological sum, and not also *necessary*. And clearly it would seem radical to maintain that entities have a sum iff they happen to be in an entangled state. As we have seen, composite systems in factorizable states are possible. This does not mean that the quantum domain is not a model of a mereological theory containing Axiom 9, but only the weaker conclusion that entanglement is not the relevant condition Ψ. However that fact that "being in an entangled, non factorizable state" is sufficient for composition has important and profound metaphysical consequences. For example it rules out some moderate answers, such as *Organicism*, to the *Special Composition Question*. *Organicism*, a view most notably held by metaphysicians like Merricks

---

[38]For a critique of this argument see Morganti (2009c, p. 277).

[39]In a word the fact that the number of possible distributions W of N particles to which M possible states are available is $W = (N + M - 1)!N!(M - 1)! < NM$ where NM is the classical statistic.

and Van Inwagen, roughly maintains that the x-s compose a y iff (i) either there is only one x or (ii) the x-s constitute a life. But surely there are quantum systems that happen to be in an entangled state, yet do not constitute any life. So, *Organicism* is untenable.

As far as *UC* goes, there could be more room for arguing. Go back to our classification of properties. We have claimed that in the cases of *MRNR* and *MRR*-properties the ontological commitment to the composite object *can* be paraphrased away. We did not employ modal notions accidentally. This is because, even if the ontological commitment to the composite object can be paraphrased away it does not mean that it *should*. It could even actually be argued that quantum properties are best understood in terms of inherent properties even in those cases in which the system is not in an entangled state. This is the position defended in Morganti (2009a). This could then in turn be used to build an argument similar[40] to the one we proposed in favor of *UC*. If such an argument could be developed it would be, not just different from the ones usually found in literature, but probably even stronger, for it will have its roots in one of our best scientific theories. We will return to this issue later on.

Another way to build an argument in favor of *UC* could be via a restriction of the notion of parthood. We could define a notion of φ-parthood where the adverbial modifier is to be intended as "being parts of a system in an entangled state", that is we could define a notion of"entangled part" and then ask what mereological axioms this new notion of φ-parthood obeys. Healey (forthcoming) calls this notion *Thread*. A substantive problem seems to lurk. Suppose the quantum system S is composed by subsystems S_1 and S_2, and suppose furthermore that S is in an entangled state. Then we say that $S_1(S_2)$ are entangled parts of S. An interesting feature of entanglement is that it is *monogamous*, that is to say that if two quantum systems are entangled then neither of them can be entangled with any other system. Then, it could be argued that φ-parthood violates *Unrestricted Composition UC*, for given any other system S_4, it could never be entangled with $S_1(S_2)$, thus composing with it some other system, as *UC* requires. This argument is however flawed. The monogamous character of the entanglement guarantees that S_1, S_2 are the *only* entangled φ-parts of S. And they do have a sum. These considerations do not add up to an argument, but they do gesture towards some. We are afraid fully developed arguments will have to wait for another time.

So, let us briefly sum up what we have argued for. We have argued that (i) quantum systems are a model of an extensional mereology, (ii) in particular quantum systems are a model of *Closure Extensional Mereology* CEM, (iii) entanglement is a sufficient condition for the existence of a mereological sum and (iv) there could be

---

[40]This claim is maybe too bold. Morganti (2009a) addresses explicitly only the statistics of those systems that are *assumed* to have a sum. Moreover his proposal does away with the *Eigenfunction-Eigenvalue* link. Thus, a possible argument in favor of *UC*, that builds on those premises would probably look different from the one we just sketched.

grounds to develop the arguments from entanglement to unrestricted composition. In the next section we deal with possible objections to the arguments we put forward.

3.3.2 *Objections*

In this section we discuss some possible objections to our arguments. We start from very general objections and then move on to some more specific ones. We are sure there will be many others. These are the ones that we felt pressing. We should add that our formulation of the objections is not intended to be either complete or exhaustive.

3.3.2.1 No Individuals Objection

The arguments rest upon the assumptions that quantum systems are individuals. Mereology was indeed called in the seminal work Leonard and Goodman (1940), a calculus of individuals. The existence of individuals is quite controversial in the quantum domain. It could be argued that Quantum Mechanics supports the view that there are *no individuals at all*. Ladyman (2007) flirts with such an approach, if it does not endorse it explicitly. If so the application of a mereological framework is simply wrongheaded.

Individuality, identity, discernibility and so on are among the hottest topics in the debate in the foundations of quantum mechanics. It is impossible to give even a little flavor of the discussion here.[41] The individuality objection however misses the point. It is true that, historically, formal mereology was developed with a specific nominalistic flavor whose ontological underpinning were somewhat classical individuals. However all that the arguments require is that there are composite objects with (proper) parts. If these parts and composite objects are individuals, in what sense they are individuals and what the criteria of such individuality are, are different questions entirely. This is not something that could be settled by any mereological theory. Rather it is an empirical question for natural science to decide. This answer to the *No Individuals Objection* prepares the way, so to speak, for another general objection.

3.3.2.2 Entangled Systems Are Not Composite Systems

In our answers to the previous objection we have admitted that the arguments depend crucially upon the assumptions that there are quantum composite systems,

[41] See French and Krause (2006), Saunders (2006), Ladyman (2007), Muller and Saunders (2008), and Hawley (2009) among the others. An interesting proposal is the one put forward in Morganti (2009a) that we already mentioned.

i.e. systems that have parts. In the end, even if mereology can be disentangled from a theory of individuality and of individuals, it still remains a theory of the parthood relations. But, the objection goes on, it is rather contentious that quantum systems are composite systems after all. In particular consider a quantum system in an entangled state, such as the state described by Eq. (3.24). Since it is impossible to attribute any state dependent property to the quantum systems S_1, S_2 we should actually refrain from admitting S_1 and S_2 in our ontology. But if this is the case the quantum system S is not a composite system after all. And it is well known that if there are no proper parts mereology just boils down to a theory of identity, and loses much of its specific interest.

There are several things to say in reply. The first one is that the objection, as it stands, simply consider entangled systems. Of all our arguments only the last one about composition crucially rests on an interpretation of entanglement. So all the other arguments are safe. But we believe that the objection is not only limited in scope, but also unsound. Let us see why. It is true that systems in entangled states are such that state dependent properties cannot be attributed to component systems independently. But state dependent properties are only some of the properties that we can attribute to quantum systems. Among the others we can attribute, there are properties which are state independent, such as mass or charge (Albert 1992, p. 49). And we can attribute them to S_1 and S_2 separately. This should be good enough a reason to admit them in our ontology, or so we contend. Moreover, and with this we conclude, the objection entails a rather radical consequence. Consider again system S in state D of Eq. (3.24), and grant for the sake of argument that the objection we are considering is on the right track and so S should not be considered a composite system. Suppose then that a measurement of the observable associated with the operator $S_{x1} \otimes I_2$, where $S_x = \frac{1}{2}\begin{pmatrix} 0 & 1 \\ 1 & 0 \end{pmatrix}$ is the Pauli matrix for the x-component of spin, is made. Such a measurement makes the state of S collapse, let us say onto the state:

$$D = |\uparrow\rangle_1 |\downarrow\rangle_2 \tag{3.25}$$

But the state represented in Eq. (3.25) is separable. Thus we can attribute individual spin properties to S_1 andS_2. In this case we could then admit S_1 and S_2 in our ontology. It follows that the so called measurement problem is even worse than originally thought. A measurement does not only change the properties of systems rather than simply revealing them, it also *brings into existence new entities.*[42] If we had component systems in our ontology right from the start we did not have to face such a radical consequence instead. As in the previous case our answer to the

[42]This is rather strong. It could be argued that this is not an example of "creation" of new entities, rather some process of "localization" of a quantum field in two particles. It is impossible to render justice to such a claim here. It is however a significant claim. It was suggested to us again by Matteo Morganti.

Entangled Systems are not Composite Systems objection made use of an implicit assumption that can be questioned. Let us turn to this last general objection.

3.3.2.3 Quantum States Do Not Represent Properties

The arguments we have proposed do have some bite, if any at all, only under the assumption that quantum states represent properties of quantum systems. This is not however the only possible interpretation. It is also possible to maintain that they do not describe properties of quantum systems at all, but rather encode our knowledge of these properties. This is sometimes called the statistical view of quantum states. It has been recently beautifully discussed and defended in Harrigan and Spekkens (2010). The statistical view would undermine most of the arguments we have presented simply because it would not be possible to infer anything about the nature of quantum systems via their quantum states. It would also easily deal with the measurement problem.

Let us first note that some realistic interpretation of quantum states seems indispensable for asking the kinds of question we are asking in the present work. That said, our reply to such an objection is twofold. The statistical interpretation is not at all common in the foundations of quantum theory. To endorse it just to respond to the arguments above seems too high a price to pay. Naturally, those who are inclined to endorse such a view on independent grounds would not be moved by such a remark. But there is a stronger one to make. Recently, Pusey et al. (2012) have argued that the statistical view is not able to reproduce quantum mechanical measurement outcomes for particular measurements. Thus, they conclude, quantum states represent properties of quantum systems. This result takes care of the objection.

We now turn to more specific objections concerning *Strong Supplementation* and *Composition* respectively.

3.3.2.4 Against Strong Supplementation

Suppose x_1, x_2 compose y_1, y_2 such that $y_1 \neq y_2$. This mereological model violates the *Strong Supplementation Principle SSP*. This is because $y_1(y_2)$ fails to include $y_2(y_1)$ among its parts and yet there is no mereological remainder between the two. *SSP* entails the uniqueness of composition.[43] It can in fact be proven that the following is a theorem:

Theorem 3. $(\exists z)(z \prec\prec x \vee z \prec\prec y) \rightarrow (x = y \leftrightarrow (\forall z)(z \prec\prec x \leftrightarrow z \prec\prec y))$

Disregard the antecedent. Then Theorem 3 says that sameness of composition is both a *necessary* and *sufficient* condition for identity. This is why it is known as the *Extensionality theorem*. This is one of the most crucial issues, together with issues

[43]This is the reason why a mereological theory comprising *SSP* is called an Extensional Mereology.

about composition. Let us see then how a possible objection might go. Consider two quantum systems S_1 and S_2 with associated Hilbert spaces $H_1 = H_2$, where $\{|\varphi_1\rangle, |\varphi_2\rangle\}, \{|\psi_1\rangle, |\psi_2\rangle\}$ are basis for H_1, H_2 respectively. Let the states of S_1 and S_2 be the following ones:

$$\begin{aligned} D_1 &= \frac{1}{2}(P^{|\varphi_1\rangle} + P^{|\varphi_2\rangle}) \\ D_2 &= \frac{1}{2}(P^{|\psi_1\rangle} + P^{|\psi_2\rangle}) \end{aligned} \tag{3.26}$$

And let us write:

$$\begin{aligned} |\vartheta_1\rangle &= |\varphi_1\rangle \otimes |\psi_2\rangle \\ |\vartheta_2\rangle &= |\varphi_2\rangle \otimes |\psi_1\rangle \\ |\vartheta_3\rangle &= \frac{1}{\sqrt{2}}(|\varphi_1\rangle \otimes |\psi_2\rangle + |\varphi_2\rangle \otimes |\varphi_1\rangle) \\ |\vartheta_4\rangle &= \frac{1}{\sqrt{2}}(|\varphi_1\rangle \otimes |\psi_2\rangle - |\varphi_1\rangle \otimes |\psi_2\rangle) \end{aligned} \tag{3.27}$$

Then the state of the composite system S *could* be:

$$\begin{aligned} D &= \frac{1}{2}(P^{|\vartheta_1\rangle} + P^{|\vartheta_2\rangle}) \\ D* &= P^{|\vartheta_3\rangle} \\ D** &= P^{|\vartheta_4\rangle} \end{aligned} \tag{3.28}$$

But, the objection goes on, $D, D*$ and $D**$ represent states with different properties.[44] So quantum systems in $D, D*$, and $D**$ are different, by Leibniz's law, and hence composition is not unique against *SSP*. It is surely true that states (3.26) fail to determine the state of the composite system uniquely. This is because they are mixed states. And it could be checked that D, $D*$ and $D**$ are indeed solutions to Eq. (3.9). That said, what the objection overlooks is that the fact that the composite system *could* be in those states, does not entail that *there are different composite systems in different states*. The modal component of the claim cannot be downplayed. It is well known that in such cases appeal to Leibniz's law is problematic at best. If there were different co-located composite systems in different states in spite of being constituted by exactly the same parts, two different measurements of the same observable carried out, to use a terminology that is somehow question begging, on the same content of the same spatio(temporal) region could give two different results. These two measurements could in fact record

[44] Recall our discussion of the *Quantum States Do Not Represent Properties* objection.

different categorical properties of numerically distinct physical systems constituted by exactly the same parts. It is difficult to see how to do physics in such context. It is probably in this spirit that in Beltrametti and Cassinelli (1981, p. 259) is written that from the physical point of view the nature of the physical subsystems should determine the nature of the compound system they form.

We believe furthermore that the quantum domain is not particularly friendly to some other classic anti-extensionalist arguments. The *locus classicus* is the literature on the so called problem of material constitution.[45] Let *St* be a statue composed by a quantity *Cl* of clay. Then *St* and *Cl* have the same parts but are not the same thing since they have different modal properties such as "could possibly survive the loss of a single small part". Note that these are not categorical properties so that our previous response does not apply. Actually it is usually granted that, as far as categorical properties go, there is none which *St* has and *Cl* hasn't. These arguments, that seem to have some sort of intuitive force when applied to commonsense ontology, lose much of their force once removed from the framework they have been formulated within. Consider a hydrogen atom and suppose for the sake of the argument that the electron and the proton compose the atom.[46] How could we even reformulate the anti-extensionalist argument in such case? It seems that hydrogen atom functions here as the *St*-"counterpart", and the proton and electron as the *Cl*-"counterpart".[47] Can the hydrogen atom survive the loss[48] of one of its proper parts? It doesn't seem the case. The *Strong Supplementation Principle* is not so easily violated.

Before passing to the last objection let us address some of the wide reaching consequences of both the extensionality argument and the replies to its possible objections. First of all, as we already pointed out various times, they show (if valid) that at the fundamental level parthood is extensional. Foes of extensionalism (and there are many) should be worried. This result has its roots in the description

[45]We refer to Rea (1997) for classic papers and detailed references.

[46]Healey (2013) contends such a claim. Moreover he takes this to support the fact that quantum systems violate unrestricted composition. His argument is roughly the following. The solution of the Schrödinger equation for the hydrogen atom depends on a decomposition of the system into subsystems corresponding to relative and center of mass subsystems of the electron and proton. Yet the subsystems taken together do not have a mereological sum. It is hard to evaluate briefly this argument. As far as we can see it depends crucially on admitting that the center of mass is a somewhat mereological part *simpliciter* of the hydrogen atom. But this assumption is questionable. It could be regarded as an example of a φ-part instead where the relevant adverbial modifier φ should be cashed out along the following lines: "decomposed subsystem necessary for the solution to the dynamical equation representing the evolution of the system". But we should not expect that the axioms regimenting the notion of mereological parthood *simpliciter* regiment that of φ-parthood too.

[47]It should be clear that we are not employing the technical notion of counterpart used in quantified modal logic and metaphysics of modality here.

[48]It would be a substantive and fascinating question whether this could be taken as an argument in favor of *mereological essentialism*, roughly the view that parts are essential to their wholes. For a defense of mereological essentialism see Chisholm (1973).

of composite systems offered by one of our best confirmed physical theory. This is not something that can be discard so easily. There are furthermore two other important consequences. Quantum mechanics is often credited to have refuted the following metaphysical reductionist claim about parts and wholes: the whole is nothing more than the sum of its parts. When phrased this way this claim is so vague that is difficult to see how and whether it could be either supported or refuted. The argument from extensionality shows that if by that claim we mean that it is not possible to give an identity criterion for wholes in terms of their component parts, then quantum mechanics has not refuted the reductionist stance. There is however another way the reductive claim could be intended. It could be taken to mean that the properties of composite systems supervene on the properties of the component parts taken separately. In this case, as we argued in Sect. 3.3.1, when a quantum composite system is in an entangled state there is at least a property, namely that represented by the density operator of the whole system, that does not supervene on those of its component parts taken separately. But this is not a threat to mereological extensionalism. Or so we contend. The mereological sum of some φ-ers is that entity that overlap all and only those things that overlap a φ-er (as the second consequent of axiom 3.7 states), and nothing else. The notion of mereological sum is silent about whether the sum of the φ-ers is itself a φ-er. It is actually silent about all the properties of the alleged sum. Hence, surely, it does not say that they all have to supervene on the properties of the component parts. This failure of supervenience has been recently used, most notably by Schaffer (2010), in favor of yet another metaphysical claim, namely that the whole is more fundamental than its parts. This issues deserves an independent and careful analysis. However we want to suggest that the extensionality argument could have something to say about this other metaphysical problem too. Schaffer argues roughly that the whole is more fundamental because, as we showed, the state of the component system always determines that of its parts whereas the converse does not hold. The extensionality argument on the other hand shows that it is possible to give an identity criterion for wholes in terms of the parts. This could be taken to be evidence enough for the existence of some sort of dependence of composite systems on their component ones. If both Schaffer's argument and the argument from extensionality have some bite, we would have found grounds to argue that, contrary to much of the current literature, the relation of metaphysical dependence, is not anti-symmetric after all. We do not intend this to be a fully-fledged argument. As we said already this is a very interesting topic that needs further and careful investigation. We simply wanted to point out that the arguments we put forward could have far reaching consequences for hotly debated questions in analytic metaphysics.

We have arrived with this at the last objections, the ones about composition.

3.3.2.5 Against Composition

The argument about composition we have proposed is vulnerable to at least two serious objections as far as we can see. Let us start from the first. The argument moves from the recognition that every pure state is an eigenfunction of some operator. This, via the *Eigenfunction-Eigenvalue* link, ensures that every system that we can prepare in a pure state has a certain property with probability = 1. The problem is that not every Hermitian operator represents a physically relevant property. So the system in question could be in an eigenfunction of some property that has physically no explanatory power. This is a serious objection. We offer some considerations to counter it and leave it to the reader to weigh them. It is true that not all the operators represent physically interesting properties but some of them do. This is the case of a system in state described by Eq. (3.24), i.e. the property of "having total spin = 0" is physically relevant. Some other examples include the total momentum operator or the relative position operator. Also, anyone who subscribe to the so called abundant theory of properties[49] (see Lewis 1986, pp. 59–69) should not be moved by the objection. Probably, however, the best answer would be to develop the argument we have proposed in order to generalize them also to factorizable states. If such an extended argument could be given it would undermine the objection for every state will then fall under its scope.

Another serious objection can be put forward along the following lines. The argument has hidden the complications that arise when passing from a bipartite case of entanglement, i.e. when the compound systems is composed of two subsystems only, to cases of multi-partite entanglement. In these cases a notion of full separability can be spelled out simply by generalizing the notion of bipartite separability, i.e.:

Definition 11. Let subsystems $S_1, \ldots, S_n$ compose system S. The state $|\psi_{1,\ldots,n}\rangle$ of S is fully separable iff $|\psi_{1,\ldots,n}\rangle = |\psi_1\rangle \otimes \ldots \otimes |\psi_n\rangle$

However a violation of this condition does not, by itself, guarantee cases of "true n-partite entanglement" (Horodecki et al. 2007: 31). We could define this last notion via:

Definition 12. A state $|\psi_{1,\ldots,n}\rangle$ is an example of a true n-partite entanglement iff all bipartite partitions produce mixed reduced density matrices

This would entail that there are no bipartite cuts such that the resulting state is a product, i.e. factorizable, state. But this falls short of guaranteeing that in all the cases of non full separability we could not trace out a subsystem and leave the rest in a product state. For example we could have, for a non fully separable state $|\psi_{123}\rangle$:

$$|\psi_{123}\rangle = |\omega\rangle \otimes |\zeta_2\rangle, |\omega\rangle = \frac{1}{\sqrt{2}}(|\varphi_1\rangle|\vartheta_1\rangle + |\varphi_2\rangle|\vartheta_2\rangle) \tag{3.29}$$

where $\{\varphi_1, \varphi_2\}, \{\vartheta_1, \vartheta_2\}, \{\zeta_1, \zeta_2\}$ are basis for the Hilbert spaces of S_1, S_2 and S_3 respectively, with $|\omega\rangle \in H_{12}$, and where $H_{nm\ldots z}$ is the Hilbert space associated with

---

[49]We are afraid this is not the most comfortable position for a naturalistically inclined metaphysician.

the quantum system composed by S_n, S_m, ..., S_z. Then, it could be noted that our argument, if valid, would not guarantee that there is a mereological sum of systems S_{12} and S_3 since we could trace out S_3 in such a way as to result in S_{12} and S_3 being in a product state. We cannot enter the details of full and partial separability here. We can however offer some remarks to weaken the objection. The first one amounts to simply pointing out that there are no conclusive criteria nor experimental tests for partial separability yet, and thus it would be hasty to draw any conclusion. The second remark is that our argument, if valid, guarantees that there is a mereological sum of S_1, S_2 and S_3 on the one hand, namely S_{123} for the state represented in Eq. (3.29) is not fully separable by assumption, and a mereological sum of S_1 and S_2 on the other, namely S_{12}. But isn't $S_{123}$the sum of S_{12} and S_3?

Finally, we have envisaged the possibility to develop the composition argument in such a way as to cover also factorizable states. If this could be done both the objections will be undermined. Then we would have an argument for *Unrestricted Composition* that comes from fundamental physics. Some would probably argue that this by itself, ensures that it would be stronger than the ones found in the metaphysical literature. We really cannot make justice to such claims here. We want to say at least one further thing though. This argument would undermine a recent objection in Elder (2008). There it is argued that arbitrary sums would not count as "Aristotelian objects" for they would not have any property. If we are correct this objection will miss the point, at least in the quantum mechanical case. Since every pure state is an eigenfunction of some observable the sum would have at least the property that is represented by that very eigenfunction. We grant however that the composition argument is among the more controversial ones we proposed.

3.4 Conclusion

This paper is an exploration of how parthood and composition behave in non relativistic quantum mechanics. This is already an explicit acknowledgments of its limitations. For if we were to consider relativistic and quantum field theories we would probably find that these notions behave differently. It is a hot debated topic in philosophical literature whether mereology is part of the so call formal ontology. If this were the case parthood would be regimented by the same mereological theory independently of any ontological domain. We cannot even begin to argue against such a view here. We simply find plausible, and indeed we find it probable, that different ontological domains are models of different mereological theories. In this paper we have argued that the quantum domain is a model of an extensional mereology, namely *Closure Extensional Mereology* CEM. Whether it can be also a model of the *General Extensional Mereology* GEM is more controversial and an thorough investigation of such an issue would have to wait for another occasion. There is another metaphysical issue that deserves careful, independent scrutiny that is intimately related to some of the issues we have touched upon here.

Recently it has been authoritatively argued, especially in Schaffer (2010), that quantum mechanics favors a particular metaphysical thesis known as *Priority Monism*. Roughly speaking *Priority Monism* is the claim that the universe is more fundamental than any of its parts. Actually it is the only fundamental object, the only object that is truly metaphysically independent. Leaving questions about the consistency of the notion itself of metaphysical dependence,[50] the argument crucially depends on the ubiquity of entangled states and on the claim that when a composite system is in an entangled state it is more fundamental than its component parts. We believe that questions about composition and the extensionality of composition we put forward could have a direct and important bearing on these issues. This is because the extensionality of parthood and composition seems to suggest there is some sort of dependence of any quantum composite system on its component parts. Moreover the arguments in favor of *Monism*, as they stand, seem to us not to consider carefully the complications that arise when passing from a fairly uncomplicated case of bipartite entanglement to the complicated n-multipartite entanglement cases.[51] Naturally these considerations do not add up to an argument, they do not even gesture towards an argument. They simply advance some reservations and doubts. But we are afraid that, in this case too, careful analysis and detailed arguments will have to wait for another occasion.

Acknowledgements We are very grateful to Matteo Morganti, Jonathan Schaffer and Vincenzo Fano for careful readings and criticisms of various drafts of the paper. Without their help this work would have been substantially worse.

References

Albert, D. (1992). *Quantum mechanics and experience*. Cambridge: Harvard University Press.

Beltrametti, E., & Cassinelli, G. (1981). *The logic of quantum mechanics*. Reading: Addison-Wesley.

Calosi, C., Fano, V., & Tarozzi, G. (2011). Quantum ontology and extensional mereology. *Foundations of Physics, 41*(11), 1740–1755.

Casati, R., & Varzi, A. (1999). *Parts and places: The structure of spatial representation*. Cambridge: MIT.

Chisholm, R. (1973). Parts as essentials to their wholes. *Review of Metaphysics, 26*(4), 581–603.

Darby, G. (2010). Quantum mechanics and metaphysical indeterminacy. *Australasian Journal of Philosophy, 88*(2), 227–245.

Elder, C. (2008). Against universal mereological composition. *Dialectica, 62*(4), 433–454.

Field, C. (2014). Consistent quantum mechanics admits no mereotopology. *Axiomathes, 24*, 9–18.

French, S., & Krause, D. (1996). Quantum objects are vague objects. *Sorotites, 6*, 21–33.

French, S., & Krause, D. (2006). *Identity in physics: A historical, philosophical and formal analysis*. Oxford: Oxford University Press.

[50] See Schaffer (2003).

[51] See also Morganti (2009c).

Harrigan, N., & Spekkens, R. (2010). Einstein, incompleteness and the epistemic view of quantum states. *Foundations of Physics, 40*, 125–157.
Hawley, K. (2009). Identity and indescernibility. *Mind, 118*, 101–119.
Healey, R. (1991). Holism and nonseparability. *Journal of Philosophy, 88*(8), 393–421.
Healey, R. (2013). Physical composition. *Studies in History and Philosophy of Modern Physics, 44*, 48–62.
Horodecki, R., Horodecki, P. I., Horodecki, M. I., & Horodecki, K. (2007). Quantum entanglement. http://arxiv.org/pdf/quant-ph/0702225.pdf.
Hovda, P. W. (2009). What is classical mereology? *Journal of Philosophical Logic, 38*(1), 55–82.
Hughes, R. I. G. (1992). *The structure and the interpretation of quantum mechanics*. Cambridge: Harvard University Press.
Jauch, J. (1968). *Foundations of quantum mechanics*. Reading: Addison-Wesley.
Koslicki, K. (2003). The crooked path from vagueness to four-dimensionalism. *Philosophical Studies, 114*, 107–134.
Ladyman, J. (2007). On the identity and diversity of objects in a structure. *Proceedings of the Aristotelian Society Supplementary Volume, 81*, 21–43.
Leonard, H. S., & Goodman, N. (1940). The calculus of individuals and its uses. *Journal of Symbolic Logic, 5*, 45–55.
Lewis, D. (1986). *On the plurality of worlds*. Oxford: Blackwell.
Markosian, N. (2008). Brutal composition. In J. Hawthorne, T. Sider, & D. W. Zimmerman (Eds.), *Contemporary debates in metaphysics* (pp. 341–364). Oxford: Blackwell.
Morganti, M. (2009a). Inherent properties and statistics with individual particles in quantum mechanics. *Studies in History and Philosophy of Modern Physics, 40*, 223–231.
Morganti, M. (2009b). A new look at relational holism in quantum mechanics. *Philosophy of Science, 78*(5), 1165–1176.
Morganti, M. (2009c). Ontological priority, fundamentality and monism. *Dialectica, 63(3)*, 271–288.
Morganti, M. (2013). *Combining science and Metaphysics, contemporary physics, conceptual revision and common sense*. London: Palgrave Mcmillan.
Muller, F., & Saunders, S. (2008). Discerning fermions. *British Journal for the Philosophy of Science, 59*, 499–548.
Pusey, M., Banett, J., & Rudolph, T. (2012). On the reality of quantum state. *Nature Physics, 8*, 476–479.
Rea, M. (1997). *Material constitution*. Boston: Rowman & Littlefield.
Rea, M. (1998). Temporal parts unmotivated. *Philosophical Review, 107*, 225–260.
Rea, M. (2010). Universalism and extensionalism: A reply to Varzi. *Analysis, 70*(3), 490–496.
Rescher, N. (1955). Axioms for the part relation. *Philosophical Studies, 6*, 8–11.
Rosen, G., & Dorr, C. (2002). Composition as fiction. In R. Gale (Ed.), *The Blackwell guide to metaphysics* (pp. 151–174). Oxford: Blackwell.
Saunders, S. (2006). Are quantum particle objects? *Analysis, 66*, 52–63.
Schaffer, J. (2003). Is there a fundamental level? *Nous, 37*, 498–517.
Schaffer, J. (2010). Monism: The priority of the whole. *Philosophical Review, 119*(1), 31–76.
Sider, T. (2001). *Four-dimensionalism: An ontology of persistence and time*. Oxford: Clarendon.
Sider, T. (2007). Parthood. *Philosophical Review, 116*, 51–91.
Simons, P. (1987). *Parts: A study in ontology*. Oxford: Clarendon.
Simons, P. (2006). Real wholes, real parts: Mereology without algebra. *Journal of Philosophy, 103(12)*, 597–613.
Tarski, A. (1956). On the foundations of boolean algebras. In *Logic, semantics metamathematics* (pp. 321–341). Oxford: Clarendon.
Teller, P. (1986). Relational holism and quantum mechanics. *British Journal for the Philosophy of Science, 37*, 71–81.
Teller, P. (1989). Relativity, relational holism and the bell's inequalities. In J. Cushing & E. McMullin (Eds.), *Philosophical consequences of quantum theory: Reflections on Bell's theorem* (pp. 298–223). Notre Dame: Notre Dame University Press.

Van Inwagen, P. (1990). *Material beings*. Ithaca: Cornell University Press.
Varzi, A. (2008). The extensionality of parthood and composition. *The Philosophical Quarterly, 58*, 108–133.
Varzi, A. (2009). Universalism entails extensionalism. *Analysis, 69(4)*, 599–604.
Varzi, A. (2014). Mereology. *Stanford encyclopedia of philosophy*.

Chapter 4
Continuity of Motion in Whitehead's Geometrical Space

Vincenzo Fano and Pierluigi Graziani

4.1 Introduction

In his book on the continuum White (1992, 185) notes that, in contrast with the Aristotelian tradition, according to which it is not possible for parts of the continuum to be non continuous, the modern topological approach considers the continuum as a supervenient property of non continuous entities. Even if the modern approach is perfectly coherent (Grünbaum 1952) its lack of intuitiveness justifies, at least partially, the ongoing attempts to reconsider the continuum in different terms (see for example Roeper 2006). In the first half of the twentieth century the hope of constructing scientific concepts out of an application of mathematical logic to sense data (Russell 1914) was still alive. This construction would have preserved epistemological order in the logical order. The basic idea was to start from actual experiences and then go on to build scientific abstract notions via exact definitions (Carnap 1928). The prospected fourth volume of *Principia Mathematica* on the foundations of geometry by Russell and Whitehead should have adopted this general framework. The book was never actually written but part of its contents are known through chapters 2 and 3 of the IV part of Whitehead's *Process and Reality* (1929).

In those chapters, in line with the aforementioned principle according to which the logical order should respect the epistemological one, there is an attempt to construct the notion of point as an abstraction from the notion of continuum, rather than defining the latter as an aggregate of points. These pages are Whitehead's second attempt, after those made in the third part of *An Enquiry Concerning the Principles of Human Knowledge* (1919) and in chapter IV of *The Concept of Nature* (1920). While Whitehead's approach was greeted with interest by scholars such as Miller (1946), Ushenko (1949), Lawrence (1950) and Mays (1951), it was criticized

V. Fano (✉) • P. Graziani (✉)
Department of Basic Sciences and Foundations, Urbino University, Urbino, Italy
e-mail: vincenzo.fano@uniurb.it; pierluigi.graziani@uniurb.it

C. Calosi and P. Graziani (eds.), *Mereology and the Sciences*, Synthese Library 371,
DOI 10.1007/978-3-319-05356-1_4,

by philosophers such as Grünbaum (1953). According to the latter, a positivistic point of view that attempts to reduce scientific concepts to an experiential basis would be simply untenable.[1] It is indeed difficult to provide an accurate evaluation of a method that has not been clearly formulated and developed. It is for this reason that the formal renditions of Whitehead's ideas given by Clarke (1981) and Gerla (1995) are of crucial importance. It is however noteworthy that, independently of Whitehead and following Leśniewski, the positivistic attitude was developed by Tarski (1929) and Grzegorczyk (1960), although these approaches are far less intuitive than those of Whitehead and Gerla.[2] Another independent approach is the one developed in Sambin (2003), which does not however take into account the alignment between the order of concept construction and their epistemic access.

There have been several other attempts to reformulate the concept of continuum in more intuitive terms, such as the intuitionistic one in Brouwer (1930) and Troelstra (1983), and the one in Roeper (2006). The first is somewhat imprecise. The second, though precise, lacks much of its alleged intuitiveness. Finally the third, even if it does not assume the notion of point, presupposes that of intervals with exact abutments. This, as Arsenijević and Kapetanović (2008) have shown, renders Roeper's attempt substantially isomorphic to the standard approach. In this paper we will take up Grünbaum's challenge. We will address the question whether Whitehead's approach, as formulated by Gerla, could be a viable alternative for contemporary science, in particular mathematical physics. We are not interested in the foundations of mathematics in itself, but rather in its role in the foundations of physics.

Since the Copernican revolution modern science has taught us not to take our perceptual evidence for granted. This has led many philosophers and scientists to endorse either an instrumentalist or a Platonist attitude towards the counter-intuitive notions used in mathematical physics. Both attitudes, if blindly endorsed, are dangerous for a correct collocation of natural science within the broader framework of contemporary culture. The instrumentalist attitude tends to belittle the epistemological value of natural science thus opening the way to irrationalism. The Platonist standpoint, besides having been refuted many times in history (since most of the theoretical entities that have been introduced have been shown to be either non existent or different from what expected) runs the risk to encourage a conservative attitude towards scientific research thereby placing scientific knowledge in an uncontrollable hyperuranic world. For these reasons, the possibility of providing a logical formulation of the notions used by mathematical physics which respects

[1] See however the responses to Grünbaum given by Shamsi (1989) and Ringel (2001). Grünbaum (1953, 220) himself acknowledges that in *Process and Reality* Whitehead does not require all the regions used in the *extensive abstraction* method to be perceivable. Grünbaum does not seems to fully understand this new approach, which has become clear in the formulation of Gerla and Tortora (1992, 1996). This new rendition is thus immune from Grünbaum's criticism.

[2] Biacino and Gerla (1996) claim that the two approaches are substantially equivalent. Their arguments do not strike us as conclusive. For a clear analysis of the contributions by Leśniewski, Tarski and Grzegorczyk see Gruszczyński and Pietruszczak (2009).

the order of epistemic access to them, thus rendering them more intuitive, should be considered a step forward in the attempt at reaching a critical attitude towards the theoretical entities of science.

Building on the above assumptions we will consider in the following section to consider some cases in which the standard Cantor-Grünbaum's approach reveals its strength. In Sect. 4.3 we will examine the formal notion of connection[3] as proposed by Whitehead-Gerla, and finally in Sect. 4.4 we will argue that this concept, though more intuitive, fails in some of the cases we have envisaged in Sect. 4.2.

4.2 Motion as a Supertask

In *Phys* 223a21-30, 239b 11-14 and 263a4-b9 Aristotle formulates and discusses Zeno's *Dichotomy Paradox*. This paradox can be cached in modern terms along the following lines:

1. Suppose an object O moves from a to b, two distinct spatial regions, with constant velocity. Suppose further, for the sake of simplicity, that the distance between a and b is 1 m and that it takes O 1 s to go from a to b. Then O's velocity = 1 m/s. In general standard kinematics tells us that O will take 1/M of a second to cover a spatial distance in *ab* of 1/M meters.
2. Suppose that the space between a and b is composed of an uncountable set of points.
3. Then it follows that O, to get from a to b, has to pass an infinite series of adjacent spatial intervals. The first interval will have length $= 1/2\,\text{m}$, the second $= 1/4$, the third $1/8$ and so on. We can indicate such a series as (4.1):

$$\frac{1}{2}, \frac{1}{4}, \frac{1}{8}, \ldots, \frac{1}{2^n} \tag{4.1}$$

4. Since O moves with finite velocity it will take it a finite amount of time to cover each and every spatial interval of the series (4.1).
5. An infinite sum of finite entities cannot but be infinite. Thus it will take O an infinite amount of time to get from a to b. O will never get to its destination.

To a mathematically sophisticated mind the fallacy in Zeno's argument is pretty obvious. It is the point 5 above. Nowadays we have the mathematical resources to establish that the infinite sum of the members of series (4.1) = 1 rather than infinite.

[3]It should be noted that Gerla and Miranda (2008) provide a formal rendition also of the notion of *extensive abstraction*, one based upon the notion of inclusion. They further address the formal rendering of the relation of connection that Whitehead takes from De Laguna (1922). We will focus only on the latter, which is both more intuitive and effective.

However in *Phys* 263a4-b9, Aristotle notes another aspect of Zeno's argument that is quite independent from the fact that it would take an infinite time to cover an infinite sum of spatial intervals. He observes that in general it is not possible to carry out an infinite set of acts for the simple reason that infinity does not have any final term. In other words, it is not possible for O to go from a to b because it would have to perform an infinite series of acts, and infinity does not have a final term, so O could never get to any destination. This holds independently of the length of spatial intervals.

Aristotle is probably addressing here the problem that modern philosophers of supertasks raise against the standard solution of Zeno's Dichotomy, i.e. the one that is centered around the fact that $S_n = 1 - 1/2^n$ for n that goes to infinity approaches 1.[4]

In modern terms the physical supertask problem[5] is twofold: On the one hand it is hard to understand how an infinite number of movements can be made in a finite amount of time, independently of whether their sum has a finite length. On the other hand, the fact that S_n tends to 1 when n tends to infinity is a fact that concerns the members of the series and not its final term; 1 does not belong to the series.

Thus, proving that S_n tends to 1 does not amount to proving that O gets to its destination. Aristotle tries to solve this problem with the following argument. The interval ab that O has to cover can be understood in two different ways: Either as a continuum according to the definition given in *Phys* 227a 10ff, that is as something which lacks internal limits, or as divided into infinite intervals.

In the first case the infinite intervals are only potential, whereas in the second case they are actual. There are no problems in passing infinite intervals potentially, whereas it is obvious that this cannot be done if the intervals are actual. If the intervals are potential, O passes them only accidentally.

One of the major features of modern science is that it has dropped the distinction between potentiality and actuality. This is not because we cannot make room in our ontology for possible entities, as many philosophers do, but rather because a possible entity is always only vaguely individuated. Consider for example the difference between these two simple definitions:

(i) The set of clothes that is in my closet.
(ii) The set of clothes that could be in my closet.

The first set is perfectly determined whereas the second is not defined exactly. Thus, even if the Aristotelian response is reasonable, it is not adequate when compared to more rigorous standards. It is true that many philosophers of physics

---

[4]In order to appreciate the importance of the succession S_n for Zeno's Dichotomy consider again series (4.1): $1/2, 1/4, 1/8, \ldots$. Its generic term is $1/2^n$. As n approaches infinity the series measures how much time is left for O to get to b. We know that $ab = 1\,\text{m}$, so $S_n = 1 - 1/2^n$ represents the space covered by O. It is easy to prove that, if n goes to infinity, S_n approaches 1 since $1/2^n$ tends to 0. Thus an infinite sum of intervals is not necessarily infinite and O can get to its final destination.

[5]We do not address the logical problem of supertask.

maintain that scientific theories do not avoid modal notions altogether. Just to give an example think of the possibility, if not the existence, of dispositional properties. Contemporary physics seems however to provide, at least at first sight, a rigorous way to handle such properties, for example the use of differential equations, whereas it is rather unclear how to handle the Aristotelian concept of potentiality in a rigorous manner.[6]

The distinction between actual and potential infinite division has been precisely rendered in a certain sense by Grünbaum via the distinction between *staccato-run* and *legato-run*.

To understand the first notion, let us go back to our object O going from a to b, which are 1 m apart, in 1 s. Suppose we describe O's motion as follows: in the first quarter of a second O covers half a meter, then it stops for $1/4$ of a second; in the subsequent $1/8$ of a second it covers a distance of $1/4$ of a meter, then it stops for $1/8$ of a second, and so on. It is clear that O would cover an infinite amount of intervals detached from one another, i.e. actually, as Aristotle would say.

The legato-run on the other hand is the uninterrupted movement. In this case the problem of accomplishing an infinite number of tasks does not arise because strictly speaking it is a unique motion. This statement needs a more detailed justification. The notion of task can be understood in two different ways. A more general and a more specific one. The former refers to any change in the world (Laraudogoitia 2009); the latter to the fact that a physical variable has extremal values in different times. The latter is due to Black (1951, 1954), and, since we are here concerned with the physical problem of supertask only it seems the more suitable one. So we will endorse it.

Black defines *act* (the task) as something that is separate from its environment because it (act) has a defined beginning and an end. In his 1954 response to the debate that followed his 1951 paper, Black polishes this definition observing that, if an object O is characterized by a variable m, we say that O has accomplished an act (task) during the interval $t_1 - t_2$, iff the variable m has extreme values at t_1, t_2.

It follows from this definition that a ball that is bouncing between two parallel walls accomplishes a task every time it travels from one wall to the other. It also follows that O accomplishes a unique task and not an infinite series of tasks, since none of the variables that characterize it behave in such a discontinuous way.

This is exactly Grünbaum's distinction between legato and staccato-run. The supertask problem arises only for the latter and O's motion is not a staccato-run. Therefore a legato-run is not at all a supertask. One could nonetheless ask: is a staccato-run physically possible? To this question the answer couldn't be definitive. Nonetheless, if motion is not continuous, it is not clear how the body could arrive at its final destination.

---

[6]It may be possible to characterize rigorously the Aristotelian notion of "infinite divisibility", where the modality involved in the second word is crucial, in mereological terms, through a statement of atomlessness.

We could still raise the objection that we do not actually divide the interval *ab*, and that all statements about the series S_n do not necessarily hold for b, that is not in that series. As Laraudogoitia (2009, sec. 3) rightly points out, to solve this problem in a conclusive way it would be necessary to resort to some sort of *continuity principle*: if space is continuous there is nothing in between the infinite series of intervals in *ab* and the final point b.

Let us discuss this idea more carefully. In topology[7] we say that:

Definition 1. A sequence of points x_1, x_2, x_3 in the topological space X converges to x in X iff, for every neighborhood U of x there exists a positive natural number N such that x_n belongs to U for every $n > N$.

As it is clear from this definition, even if we have proven that the series of points converge to O's final destination, it still does not follow that O gets there because the convergence point might not belong to the sequence. Now, let us define:

Definition 2. A separation of a topological space X is a couple of open and disjoint subsets A, B of X, such that $A, B \neq \oslash$ and $A \cup B = X$.

Definition 3. X is connected if and only if it does not admit any separations.

With these definitions in hand we can argue that if the interval covered by O is not connected, O would have to accomplish more than one task. We know that the representation of space as an interval of real numbers renders the interval connected, so that the problem of 'jumping' to the final destination does not arise.

We should note however that we had to use the assumption that space is somehow composed of an uncountable set of points, an assumption that we mentioned explicitly in our formulation of Zeno's Dichotomy.

If we maintained, on the other hand, that space is an infinite and dense set of points, it would not follow that it is also a connected set. Actually it can be proven that rational numbers, an infinite and dense set, are totally disconnected, i.e. the only connected subsets of Q are just its singletons (Munkres 2000, 149).

This means that there could be an infinite number of separations in O's way, and its movement would thus turn out to be a supertask, i.e. a staccato-run. The hypothesis underpinning mathematical physics is however that space is not adequately represented by a set of points that is solely dense, but rather by an effectively connected set, such as that of real numbers. If so we could say that O neither makes any jumps nor is its arrival separated from its course.[8]

---

[7] See for example Munkres (2000, 98).

[8] As far as continuity of space is concerned, it should be taken into account that recent attempts to unify gravity and Quantum Theories have yielded to the hypothesis that space, at the Planck scale, is not continuous after all, but rather discrete. These hypotheses are still under scrutiny. If they turned out to be correct, the problem of motion as a supertask would vanish.

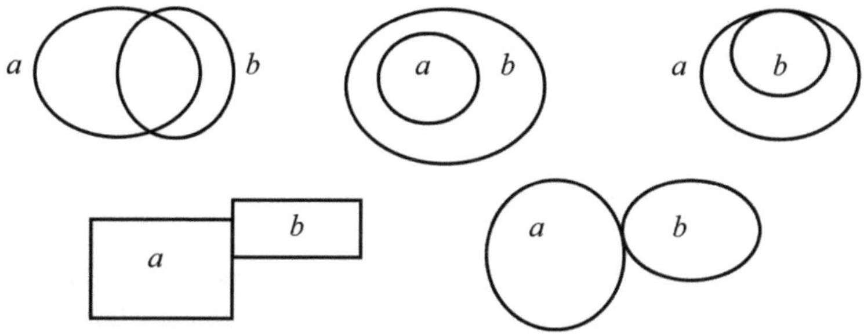

Fig. 4.1 Regions that are connected in the plane

Fig. 4.2 Regions that are not connected in the plane

4.3 Whitehead-Gerla's Perspective

Let us now add[9] a predicate C to First Order Logic with identity, which can be informally understood as *being connected* (see Fig. 4.1 and Fig. 4.2). Let us refer to this predicate as *connection*. It is a relational predicate holding between entities that we can label *regions*. We define C on a set R of regions.

We require that C obeys the following axioms[10]:

Axiom 1. (C1, Symmetry) $C(x, y) \rightarrow C(y, x)$

Axiom 2. (C2, Locality) $\exists y \neg C(x, y)$

Axiom 3. (C3, Mediate Connection) $\exists z(C(x, z) \wedge C(y, z))$

Axiom 4. (C4, Reflexivity) $C(x, x)$

C1 informally says that C is symmetric, C4 that it is reflexive. C2 informally states that there isn't a region that is connected with all other regions. From a general point of view this amounts to saying that C applies to a domain of entities that is not immersed in a geometrical space, contrary to what typically happens in mathematical physics.

[9]We follow the clear exposition given in Pecoraro (2006). We refer to this for many details about the formal rendition of the numerous assumptions proposed by Whitehead.

[10]All the formulas are intended to be universally closed for any free variable unless otherwise specified.

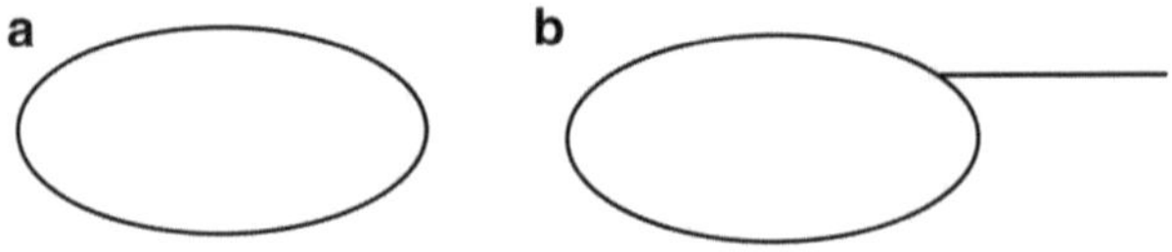

Fig. 4.3 Regular open set (**a**) and non-regular open set (**b**)

C3 states that all regions, even those that are not connected, are *mediately connected*, i.e. for any two regions, even disconnected regions, there is a region that is connected to both.

Note moreover that C is not transitive, for, if it were, it would follow from C3 that two regions whatsoever are always connected, which contradicts C2. We do not introduce an axiom stating that all regions are self-connected so as to keep our approach as general as possible. However, we will only refer to self-connected regions when dealing with possible models. The models of such theory are not necessarily topological spaces since neither infinite union nor infinite intersection are guaranteed. We can however understand regions as open regions, rather than closed ones, i.e. regions without boundaries. For the sake of clarity let us introduce the following definition:

Definition 4. $In(x, y) =_{df} C(z, x) \rightarrow C(z, y)$.

Relation *In* is intuitively the *inclusion* relation among regions, that is, writing $C(x)$ for the set of all regions connected to x $In(x, y) =_{df} C(x) \subseteq C(y)$. Let us now add two more axioms:

Axiom 5. (C5, Antisymmetry) $(In(x, y) \wedge In(y, x)) \rightarrow x = y$

Axiom 6. (C6, Infinite Divisibility) $\exists x \exists y (In(x, z) \wedge In(y, z) \wedge \neg C(x, y))$

C5 states that if two regions are mutually included, so to speak, they are one and the same region. C6 affirms that every region includes two regions that are not connected. Since *In* is clearly reflexive and transitive, it is a partial order for regions. Let us then define:

Definition 5. $Pin(x, y) =_{df} In(x, y) \wedge \neg(x = y)$,

which we call *proper inclusion*.

It follows from C4 and C6 that every region properly includes another. Thus there does not exist a minimal region, i.e. a region such that no other region is properly included in it. It is worth noting that a maximal region does not exist either: it is in fact easy to prove that a region includes all other regions only if it is connected to all of them, contra C2.

It can be observed that the set of regular open connected non empty intervals of a Euclidean space in which we interpret $C(x, y)$ as "the intersection of the complement of x with the complement of y is not the empty set" is a model for axioms C1–C6. Thus they are coherent. Let us recall that regular (Fig. 4.3a above) here means that such regions do not contain pieces of different dimensionality (Fig. 4.3b above).

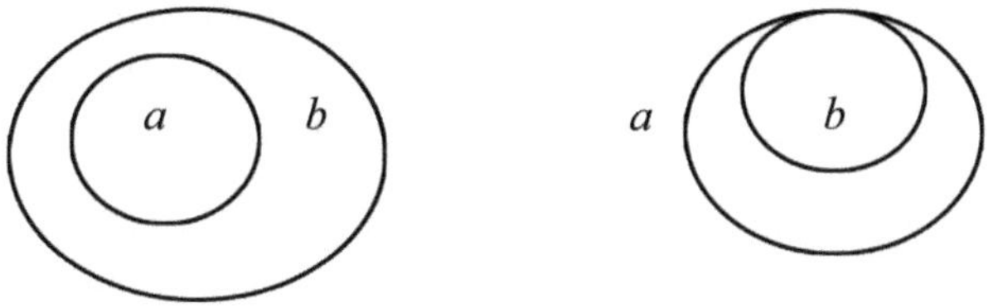

Fig. 4.4 Tangential inclusion and not tangential inclusion

Let us introduce a further definition:

Definition 6. $S(x, y) =_{df} \exists z(In(z, x) \wedge In(z, y))$,

which defines the notion of *overlap*.[11]

We can now proceed to the definition of *point* by introducing the notion of *not tangential inclusion*:

Definition 7. $NTIn(x, y) =_{df} C(z, x) \rightarrow S(z, y)$.

Practically, to connect to x you need to *overlap* y (Fig. 4.4).

Let us introduce the notion of *Abstractive Set*:

Definition 8. A set of regions X is an abstractive set $AG(X)$ iff X is totally ordered by the relation of not tangential inclusion and a minimal element does not exist with respect to that order.

This means not only that does $(X, AG(X))$ has no minimal element, but also that no region exists, including those that do not belong to X, that is included in all the regions of X. We could say that this set of nested regions already defines an abstractive element. If we started up with three-dimensional regions, that element could be a point, a line or a surface. It is however possible to prove that there exists an infinite number of different abstractive sets that single out the same abstractive element.

A further definition is therefore needed:

Definition 9. An abstractive set X *covers* an abstractive set Y iff for every $x \in X$ there exists an element $y \in Y$ such that $Pin(x, y)$.

X covers Y then if, in the *nestification* of the latter there exists a region such that every region of Y that is included in it is also included in a region of X. Let us write $Cov(X, Y)$ for X covers Y. Such a relation is reflexive and transitive, and through it we can define the *relation of equivalence between abstractive sets*:

Definition 10. $X \equiv Y =_{df} Cov(X, Y) \wedge Cov(Y, X)$.

[11]Here we do not enter into the question of *dissections*. For this we defer to Gerla and Tortora (1996) and Pecoraro (2006, 40ff). We also defer the reader to Pecoraro (2006) for many other details about the formal renditions of Whitehead's assumptions that are not relevant for the present work.

We have therefore eliminated the problem of different abstractive sets by singling out the same geometrical place. A geometrical element is defined via an equivalence class of the relation *Cov* between abstractive sets of regions. We will refer to it as $[X]$.

As we were saying, the geometrical loci of a structure of three-dimensional regions could be surfaces, lines and points. To define the latter notion we need to introduce the *incidence* relation between geometrical elements:

Definition 11. $Inc([X], [Y]) =_{df} Cov(Y, X)$

$[X]$ is incident in $[Y]$ iff Y covers X. The relation *Inc* is anti-symmetric and thus it is an order for geometrical loci. We can finally give the following definition:

Definition 12. A point is the minimal element of the set of geometrical loci ordered by the incidence relation. We will write $[P]$ for point P, $[Q]$ for point Q and so on.

We now need to establish what it means for a point to be *situated* in a region. In order to do that we may say that:

Definition 13. An abstractive set X is a member of a geometrical place $[X]$, which we will write as $X \in [X]$, iff X belongs to the equivalence class of $[X]$.

A member of a geometrical element is a way of defining such an element. We now have:

Definition 14. A point $[P]$ is situated in a region x, and we will write $Sit([P], x)$, iff there exists an abstractive set $X \in [P]$ such that $x \in X$.

In practice in order for a point to be situated in a region there has to be a way of defining that point that refers to the region. Our axiomatic system guaranteed that its models, let us call them WG-spaces, contain infinite regions. It is however indeterminate how many points there are in one region. To ensure that they are infinite we add the following axiom:

Axiom 7. (C7) $x \neq y \rightarrow \exists [P] \exists [Q] ([P] \neq [Q] \wedge Sit([P], x) \wedge Sit([Q], y))$

Informally C7 says that for every two distinct regions x and y there have to be two distinct points situated in x and y respectively. C7 ensures that in every region an infinite number of points is situated.

It is now possible to consider an interesting theorem proven by Gerla and Miranda.[12] Let us refer to the open, regular, non empty, connected set of a Euclidean three-dimensional space as R. Interpret $C(x, y)$ as "the intersection between the complement of x and the complement of y is distinct from the empty set". Then:

Theorem 1 (Gerla and Miranda Theorem). *The set of points defined via the Whitehead-Gerla procedure coincides with the points of* R^3.

[12]For details see Gerla and Miranda (2008) and (Pecoraro 2006, 47).

This theorem seems to suggest that the relation between regions and points defined via the Whitehead-Gerla method is a good inversion of the standard procedure that defines regions from points. Hence it could make us think that it is an adequate instrument for modern science. If this were indeed the case, then we should be able to solve the supertask problem in such a space when dealing with motion. Let us therefore examine *WG-spaces* from this perspective.

4.4 Whitehead-Gerla Approach and the Supertask Problem

As we have pointed out, in the standard approach, that we might label the *Cantor-Grünbaum's approach*, an object O that moves from a to b passes through an uncountable set of points. Nonetheless, its motion is not a composition of an uncountable set of actions, a performance which would probably be physically impossible to carry out in a finite amount of time. Given the connection relations of the intervals of real numbers that describe ab, this motion is rather a single task, namely a *legato-run*, to use Grünbaum's terminology. It follows that Zeno's Dichotomy argument, which we find in Aristotle (*Phys* 263a4-b9), does not constitute a threat for O's motion, provided that this notion is represented as suggested by Cantor-Grünbaum's approach.

That said, we realize that this approach is far from being intuitive. Indeed, the logical order in which concepts are constructed does not follow the order of the epistemological access to those concepts. Moreover in the Cantor-Grünbaum's approach continuity is a property emerging in a somewhat mysterious way from entities which are not continuous. For this reason, many scholars attempted to invert the order, and to assume continuous entities as primitives. The main purpose of developing a geometry which built upon the notion of an open, connected region, along the lines of Whiteheads's foundational attempt, was that of restoring the above mentioned epistemological order. Let us try to consider matters from the Whitehead-Gerla perspective and see what happens to motion. We will consider the standard description (Cantor-Grünbaum's approach) of motion and see how it behaves within the WG-space. We will try to show how the connection of these two ideas creates tensions. All this will lead us to reflect on the idea of movement as a central concept for the construction of the above epistemological order.

In general the structure of regions endowed with the relation of connection does not produce a topological space. This means that the standard topological concept of connectedness plays no role in this context; hence we must look for continuity somewhere else. Moreover, recall that points are not regions; it is easy to define the notion of self-connectedness for regions by appealing from the primitive already introduced. We could say that a finite set of regions $x_1 \dots x_n$ is a *path* iff for every x_i, x_{i+1} $S(x_i, x_{i+1})$ holds. We say that a region is self-connected iff for every two regions that are included in it there is a path through them (Gerla 1995, 1023–1024). But these definition does not hold for points, since they are not regions.

In order to see whether the Whitehead-Gerla approach is an adequate instrument for mathematical physics, let us introduce the usual notion of material point, that is a body whose dimensions are negligible.[13] The notion of material point is essential for the physico-mathematical representation of the world, insofar as it plays an important role not only in mechanics. However, we will examine only the case of the motion of a material point in classical mechanics.[14]

It must be pointed out that, given the WG-definition of point in (Definition 14), points are not regions. So they cannot be the locations of objects. Moreover, the notion of self-connection just defined is not applicable to them. Let us furthermore note that the relation of being situated in a region is not the set-theoretic relation of membership. Now, in mathematical physics we cannot help resorting to material points, that is objects that are located at a point. This does not mean that there are material points in the world, but simply that using those points to describe material objects we capture important features of those objects. Now, since WG-points are not regions we have to construct some sort of injective function $L(M_i, [P_i])$ that locates material points at WG points.

We consider the set of points P, i.e. the loci that identify the equivalence classes of abstractive sets which are lower bounds of a chain of nested regions. We know that the regions are infinite (C6), and so the points are infinite (C7). We take the set of all subsets of P including the empty set $\oslash$ and call it *PP*. It is a topological space, in that the union, which is also infinite, and the intersection are closed. *PP*, is a totally disconnected set, because only its singletons are both closed and open. We may also provide *PP* with additional structures in order to avoid this conclusion, but the way forward is not obvious.

Following Whitehead-Gerla's approach, we say that a geometric element $[Y]$ is infimum with regard to a set of points X, if and only if there is a geometric element $[Z]$ in which all the points belonging to X are located and for which $Inc([Z], [Y])$. An infimum geometric element concerning a set that contain two points $X = ([P], [Q])$ is, by definition, a segment. By considering C7 we know that in a segment are located infinite points.

By considering these ideas, it seems reasonable to assume, following the standard Cantor-Grünbaum's approach and contemporary physical-mathematical point of view, that, in the system WG, the motion of a material point from $[P]$ to $[Q]$ passes through the geometric element that we have called "segment". Recall that we are speaking about the motion of a material point. We can then reasonably assume that the point will occupy all points situated in the segment during his motion. We know that the set of these points is totally disconnected. It follows that we can not appeal to any Continuity Principle in order to claim that the motion of a material point from $[P]$ to $[Q]$ is not a staccato-run. So this motion becomes a physical supertask. We can see that every point belonging to the segment is defined by an independent

[13]See, for example, Landau and Lifshitz (1976, V.I,1).

[14]On the contemporary debate in philosophy of science concerning this topic see: Arntzenius (2004, 2011) and Field (2014).

abstractive set, so it is easy to find a variable that characterizes it and that in turn takes an extreme value. For example, each point has at least one region different from all those that define the other points of the segment.

From a kinematic point of view the physical supertasks are not impossible (Laraudogoitia 2009). However, it remains very whether that they are possible from a dynamic point of view. The motion from $[P]$ to $[Q]$ seems to be a infinite series of departures and arrivals from one point to another. Even if it were possible to achieve it in a finite time, it would require an infinite force.

If this were indeed the case, motion would become an infinite set of jumps, that is a supertask. Now, supertasks are not impossible, but they are surely controversial processes. This means that if we try to preserve some standard (Cantor-Grünbaum's approach) mathematical physics notions in WG-space and to use the notion of WG-point to do standard mathematical physics, we find ourselves in a great conceptual tension. All this leads us to believe that we need a new and different characterization of movement in WG-space.

From a physical point of view it is surely possible to accomplish an infinite number of acts, if and only if some sort of continuity holds between them. But this does not hold, as far as we know, for WG-points. For this reason the embedding of the classical standard representation of motion between points in a WG-space runs the risks of rendering this representation impossible.

It could be objected that the object countenanced by the Whitehead-Gerla approach are located just at regions, and that their motions have to follow paths, in the technical sense of the word. If this were the case then the problem we have outlined would vanish. But mathematical physics uses the notion of material points systematically. So it remains to be assessed whether and how it is possible to do physics with this different way of thinking about objects and motion. And the burden of the proof is on Whitehead-Gerla.

We could also say that when we resort to the notion of material point in mathematical physics, we are not speaking strictly, that is, the material point can be located in an extended region, but this fact is physically irrelevant. Thus, strictly speaking, there would exist no material points, and again our problem would vanish. It should be noted however that in the mathematical representation of the physical world, the material points are necessary. And if the points of this representation are taken to be WG-points, motion turns out to be a hypertask. For the above reasons, and without entering the ontological domain, we consider the WG approach as unfit to our standard and contemporary physical-mathematical representation of the world.

We therefore conclude that, despite of its being more intuitive than the standard Cantor-Grünbaum's approach, WG does not seem an adequate instrument for contemporary mathematical physics.

Acknowledgements We are very grateful to Adriano Angelucci, Caludio Calosi, Massimo Sangoi and anonymous referees for their comments on the draft of this paper.

References

Arntzenius, F. (2004). Is quantum mechanics pointless? *Philosophy of Sciences, 70*, 1447–1457.

Arntzenius, F. (2011). Gunk, topology and measure. In D. De Vidi, M. Hallett, & P. Clark (Eds.), *Logic, mathematics, philosophy. Vintage enthusiasms. Essays in honor of John L. Bell* (The Western Ontario Series in Philosophy of Science, Vol. 75, pp. 327–343). Dordrecht/New York: Springer.

Arsenijević, M., & Kapetanović, M. (2008). The great struggle between Cantorians and Neo-Aristotelians: Much ado about nothing. *Grazer Philosophische Studien, 76*, 79–90.

Biacino, L., & Gerla, G. (1996). Connection structures: Grzegorczyc's and Whitehead's definition of point. *Notre Dame Journal of Formal Logic, 37*, 431–439.

Black, M. (1951). Achilles and the tortoise. *Analysis, 11*, 91–101.

Black, M. (1954). Is Achilles still running? In M. Black (Ed.), *Problems of Analysis: Philosophical Essays*. Ithaca: Cornell University Press.

Brouwer, L. E. J. (1930). *Die Struktur des Kontinuums*. Komitee zur Veranstaltung von Gastvortrgen auslndischer Gelehrter der exakten Wissenschaften, Wien

Carnap, R. (1928). *Der logische aufbau der welt*. Leipzig: Felix Meiner Verlag.

Clarke, B. L. (1981). A calculus of individuals based on connection. *Notre Dame Journal of Formal Logic, 22*, 204–218.

De Laguna, T. (1922). Point, line, and surface as sets of solids. *The Journal of Philosophy, 19*, 449–461.

Field, C. (2014). Consistent quantum mechanics admits no mereotopology. *Axiomathes 24*, 9–18.

Gerla, G. (1995). Pointless geometries. In F. Buekenhout (Ed.), *Handbook of incidence geometry* (pp. 1017–1031). Amsterdam: Elsevier.

Gerla, G., & Miranda, A. (2008). Mathematical features of Whitehead's point-free geometry. In M. Weber, & W. Desmond (Eds.), *Handbook of Whiteheadian process thought* (Vol. II, pp. 119–130). Frankfurt: Ontos Verlag.

Gerla, G., & Tortora, R. (1992). La relazione di connessione in A. N. Whitehead. Aspetti matematici. *Epistemologia, 15*, 351–364.

Gerla, G., & Tortora, R. (1996). Dissezioni, intersezioni di regioni in A.N. Whitehead. *Epistemologia, 19*, 289–308.

Grünbaum, A. (1952). A consistent conception of the extended linear continuum as an aggregate of unextended elements. *Philosophy of Science, 19*, 280–306.

Grünbaum, A. (1953). Whiteheads method of extensive abstraction. *British Journal for Philosophy of Science, 4*, 215–226.

Gruszczyński, R., & Pietruszczak, A. (2009). Space, points and mereology. On foundations of point-free euclidean geometry. *Logic and Logical Philosophy, 18*, 145–188.

Grzegorczyk, A. (1960). Axiomatizability of geometry without points. *Synthese, 12*, 228–235.

Landau, L. D., & Lifshitz, E. M. (1976). *Mechanics*, (Vol. I, 3rd ed.). Oxford: Butterworth-Heinemann.

Laraudogoitia, J. P., (2009). Supertasks. In E. N. Zalta (Ed.), *The Stanford encyclopedia of philosophy* (Fall 2013 ed.). http://plato.stanford.edu/archives/fall2013/entries/spacetime-supertasks/.

Lawrence, N. (1950). Whitehead's method of extensive abstraction. *Philosophy of Science, 17*, 142–163.

Mays, W. (1951–52). Whitehead's theory of abstraction. *Proceedings of the Aristotelian Society, 52*, 905–918.

Miller, D. L. (1946). Extensive continuum. *Philosophy of Science, 13*, 144–149.

Munkres, J. R. (2000). *Topology*. New Jersey: Prentice Hall.

Pecoraro, A. (2006). *Formalizzazione della geometria senza punti di A. N. Whitehead*. PhD thesis, University of Salerno.

Ringel, C. M. (2001). Whitehead's theory of extension. *http://www.math.uni-bielefeld.de/ ringel/opus/extension.pdf*.

Roeper, P. (2006). The aristotelian continuum. A formal characterization. *Notre Dame Journal of Formal Logic, 47*, 211–232.

Russell, B. (1914). The relation of sense-data to physics. In B. Russell (Ed.), *Mysticism and logic and other essays* (pp. 145–179). London: George Allen and Unwin.

Sambin, G. (2003). Some points in formal topology. *Theoretical Computer Science, 305*, 347–408.

Shamsi, F. A. (1989). Whitehead's method of extensive abstraction. *Indian Philosophical Quarterly, 16*, 125–161.

Tarski, A. (1929). *Les fondements de la géométrie des corps* (Annales de la Société Polonaise de Mathématiques, pp. 29–34). (Reprinted in, Foundations of the geometry of solids. In *Logic, Semantics, Metamathematics. Papers from 1923 to 1938*, by Tarski, A., 1956, Oxford: Clarendon Press).

Troelstra, A. S. (1983). Analysing choice sequences. *Journal of Philosophical Logic, 12*, 197–260.

Ushenko, A. P. (1949). Einstein's influence on philosophy. In A. Schilpp (Ed.), *Albert Einstein:Philosopher-scientist* (Vol. VII in the library of living philosophers, pp. 609–645). New York: MJF Books.

White, M. J. (1992). *The continuous and the discrete*. Oxford: Clarendon.

Whitehead, A. N. (1919). *An enquiry concerning the principles of human knowledge*. Cambridge: Cambridge University Press.

Whitehead, A. N. (1920). *The concept of nature*. Cambridge: Cambridge University Press.

Whitehead, A. N. (1929). *Process and reality*. New York: The Free Press.

Part II
Mathematics

Introduction to Part II: Mereology and Mathematics

Mathematics is arguably the discipline that, at least historically, bears the closest relationship to mereology. Lesniewski's original formulation was supposed to constitute a nominalistic alternative to set-theory. The same spirit is found in Leonard and Goodman's seminal paper on the calculus of individuals. In 1935 Tarski was already able to prove that classical mereology was a complete Boolean algebra modulo the zero element. It follows that the relation of parthood axiomatized by classical mereology has the same properties of set-theoretic inclusion (modulo the existence of the null-set). Lewis (1991) is a thorough exploration of the possibility of reconstructing the entire set-theoretic universe in mereological terms. Even from these cursory remarks it is possible to see that the relation between mereology and set-theory has always occupied a somewhat privileged status.

Another example of a privileged relation between mereology and foundational mathematical theories is the relation with topology. Since the works of De Laguna (1922) and Whitehead (1929) there have been attempts to either define the topological relation of connection in terms of parthood, as in Whitehead (1929), or to define the parthood relation in terms of connection, as in Clarke (1981). It is still an open question whether and at what price these attempts are satisfactory (though the consensus, we might add, seems to point to the contrary). It is not by chance that the papers in this section continue this long and distinguished tradition.

In the first one (*Multi-Valued Logic for a Point-Free Foundation of Geometry*) Coppola and Gerla investigate new prospects for foundational approaches to Whiteadean point-free spaces that were also the focus of the last work of the previous section. They consider, following Whitehead's own lead, inclusion spaces and connection spaces. It turns out that (i) there is no minimal element in inclusion spaces, i.e. the definition of point is (elegant but) empty, and (ii) whereas in connection spaces it is possible to define the relation of inclusion, the converse does not hold, for the relation of connection is not invariant under the automorphisms of inclusion structures. Despite these difficulties for the inclusion based approach they are able to show in the rest of the paper that both desiderata, i.e. non-emptiness of definition of point, (ii) definability of topological connection in terms of inclusion, are met if only the inclusion relation is taken to be a graded notion. Thus, they claim, shifting to a multi-valued logic provides an adequate foundation for point-free geometry.

The second work (*The Relation of Supremum and Merelogical Sum in Partially Ordered Sets*) by Gruszczyński and Pietruszczak focuses on the problem of establishing under which conditions the mereological notion of sum and the set-theoretic notion of supremum coincide. They show that (i) for every *quasi-partially ordered set*, i.e. a set that satisfies reflexivity and transitivity but not anti-symmetry, that also satisfies weak supplementation supremum and sum do coincide, (ii) that coincidence does not obtain without further assumptions for *partially ordered sets*, which also obey anti-symmetry. Coincidence can be restored by adding different axioms governing the supremum or the mereological sum. In particular they show that some axioms concerning the existence of the mereological sum are sufficient to ensure such a coincidence. These existence axioms could have profound philosophical consequences in that they are (both) weaker than the classical principle of *unrestricted composition*, and thus could be used to ground moderate, restricted answer to the infamous *Special Composition Question*, i.e. under which condition a non-empty set of entities has a mereological sum.

Finally the paper by Hovda (*Natural Mereology and Classical Mereology*) is an investigation of the (alleged) disagreement between two philosophical orientations regarding mereological issues. On the one hand *naturalism* holds that something is part of another iff the composite object is an organic, natural union to which the part in question partakes. On the other hand *formalism* maintains that it exists a single parthood relation that is formally and mathematically well-behaved in that it obeys the axioms of classical mereology. There is at first sight tension between the two because those axioms include the *unrestricted composition* principle which sanctions all kinds of mereological sums, not just natural organic ones. Hovda's insightful suggestion is that if naturalists admit the existence of sets of objects they are already committed to, and if formalists agree to a distinction between natural and non-natural objects and sets of objects, then the ontological disagreement between the two philosophical orientations vanishes. The development of the formal apparatus that is able to flesh out this suggestion is not a trivial task for there seem to be two different requirements pulling in opposite directions: whereas on the one hand one has to extend the relation of natural parthood defined over a set of natural objects in such a way that this relation obeys the laws of classical mereology, on the other hand the domain of such objects need to be extended too in order to admit arbitrary sums. The bulk of the paper is the development of such an apparatus.

There are countless introductions to the mathematics used in this section. A first, must-read introduction to set-theory is Halmos (1974). A more detailed and extensive survey is Enderton (1977). A more recent treatment can be found in Devlin (1993). For a good philosophical take see Potter (2004). As a first introduction to fuzzy logic (see Nguyen and Walker (2000)). For both fuzzy logic and fuzzy set theory there is Klir and Yuan (1995). As far as topology goes it is difficult not to mention the classic (Munkres (1975)).

References

Clarke, B. L. (1981). A calculus of individuals based on connection. *Notre Dame Journal of Formal Logic, 22*, 204–218.

De Laguna, T. (1922). Point, line, and surface as sets of solids. *The Journal of Philosophy, 19*, 449–461.

Devlin, K. (1993). *The joy of sets: Fundamentals of contemporary set-theory* (2nd ed.). New York: Springer.

Enderton, H. B. (1977). *Elements of set-theory*. London: Academic.

Halmos, P. (1974). *Naive set-theory*. New York: Springer.

Klir, J.G., & Yuan, B. (1995). *Fuzzy sets and fuzzy logic: Theory and applications*. Upper Saddle River: Prentice Hall.

Lewis, D. (1991). *Parts of classes*. Cambridge: Blackwell.

Munkres, J. R. (1975). *Topology*. Upper Saddle River: Prentice Hall.

Nguyen, H. T., & Walker, E. A. (2000). *A first course in fuzzy logic* (2nd ed.). Boca Raton: Chapman and Hall/CRC.

Potter, M. (2004). *Set theory and its philosophy: A critical introduction*. Oxford: Oxford University Press.

Whitehead, A. N. (1929). *Process and reality*. New York: The Free Press.

Chapter 5
Multi-valued Logic for a Point-Free Foundation of Geometry

Cristina Coppola and Giangiacomo Gerla

5.1 Introduction

Łukasiewicz's many-valued logic (see Hájek 1998), Chang and Keisler's continuous logic (1966) and Pavelka's fuzzy logic (1979) define very interesting chapters of formal logic. Recently, under the name 'continuous logic', these researches were reconsidered to extend the powerful tools of model theory to classes of structures which are not definable in classical first order logic. This since these structures assume as primitive a real-valued function. Examples are the metric spaces, the normed spaces, the probabilities (see for example Yaacov and Usvyatsov 2010).

The basic ideas of point-free geometry were firstly formulated by A. N. Whitehead in *An Inquiry Concerning the Principles of Natural Knowledge* (Whitehead 1919) and in *The concept of Nature* (Whitehead 1920), where he proposed as primitive notions *events* and *extension relation* between events (in geometrical terms, regions and inclusion relation). Later, in *Process and Reality* (Whitehead 1929), Whitehead proposed a different treatment, inspired by De Laguna (1922), in which the topological notion of 'connection' between two regions was assumed as primitive and the inclusion was defined (see Gerla 1994). Successively, in a series of papers, metric-based approaches to point-free geometry were proposed in which, apart the inclusion relation, distances and diameters are also considered (see Di Concilio and Gerla 2006; Gerla 1990; Gerla and Miranda 2004). The resulting notion of point-free pseudo-metric space is a promising base for a metric foundation of geometry in accordance with the ideas of L.M. Blumenthal (1970). Indeed, it is possible to associate every point-free pseudo-metric space with a metric space via a natural definition of point and distance.

C. Coppola (✉) • G. Gerla
Dipartimento di Matematica, Università degli Studi di Salerno, Salerno, Italy
e-mail: ccoppola@unisa.it; gerla@unisa.it

C. Calosi and P. Graziani (eds.), *Mereology and the Sciences*, Synthese Library 371,
DOI 10.1007/978-3-319-05356-1_5,

In this exploratory paper we suggest the possibility of applying the ideas of continuous logic to point-free geometry. This is done by assuming as primitives predicates, geometrical in nature, as 'to be included', 'to be small', 'to be close'. Indeed, since these predicates apply at different grades, we have to interpret them as fuzzy relations and therefore we have to refer to a first order multi-valued logic. Perhaps we can look the resulting formalisms as a way to modelize the passage from the original, naive, predicate based description of the geometrical space, qualitative in nature, to the modern real-number based approach to geometry, quantitative in nature.

More precisely, in Sect. 5.2 we propose the notion of inclusion space corresponding to some of the geometrical properties of the inclusion analyzed in Whitehead (1919, 1920). In Sect. 5.3 we give the notion of connection space corresponding to the analysis given in Whitehead (1929). Taking in account of the difficulties of the inclusion-based proposal in defining the notion of point, in Sects. 5.4 and 5.5 we reformulate it in the framework of multi-valued logic. This means that the inclusion is intended as a graded notion. We show that this enables us to overcome the observed difficulties. Finally, in Sect. 5.6 we reformulate the metric-based theory of point-free geometry into a theory in a multi-valued logic involving the graded predicates 'to be close' and 'to be small'.

5.2 Inclusion Spaces

We isolate the main properties considered by Whitehead (1919) and we transform them into a system of axioms. Indeed, we consider the following first order theory in a language $L_{\leq}$ containing only the binary predicate $\leq$. As usual, we write $x < y$ to denote the formula $(x \leq y) \wedge (\neg(x = y))$.

Definition 1. An *inclusion space* is a structure satisfying the following axioms:

I1 $\forall x\,(x \leq x)$ (reflexivity)
I2 $\forall x \forall y \forall z\,((x \leq z \wedge z \leq y) \Rightarrow x \leq y)$ (transitivity)
I3 $\forall x \forall y\,(x \leq y \wedge y \leq x \Rightarrow x = y)$ (anti-symmetry)
I4 $\forall z \exists x\,(x < z)$ (there is no minimal region)
I5 $\forall x \forall y\,(x < y \Rightarrow \exists z\,(x < z < y))$ (density)
I6 $\forall x \forall y\,(\forall x'(x' < x \Rightarrow x' < y) \Rightarrow x \leq y)$ (below approximation)
I7 $\forall x \forall y\,\exists z(x \leq z \wedge y \leq z)$ (upward-directed).

We call *regions* the elements of an inclusion space and *inclusion relation* the relation $\leq$. Then an inclusion space is an ordered set $(S, \leq)$ such that $\leq$ has not a minimum, it is dense and upward-directed. Moreover, in this set it is possible to approximate every region from below. To find a model for this theory, we refer to the notion of bounded closed regular subset of the Euclidean space $\mathbb{R}^n$. This is a natural candidate to represent the idea of region which is usually accepted in literature.

Definition 2. Given a topological space we call *closed regular* a subset which is a fixed point of the operator $creg$ defined by setting

$$creg(X) = cl(int(X))$$

where we denote by *cl* and int the closure and the interior operators, respectively.

We denote by $RC(\mathbb{R}^n)$ the class of all the closed regular subsets of $\mathbb{R}^n$. $RC(\mathbb{R}^n)$ is a very interesting example of complete atomic-free Boolean algebra. We are interested to the class $\mathcal{R}$ of the nonempty bounded elements of $RC(\mathbb{R}^n)$. It is easy to prove the following theorem.

Theorem 1. *The structure* $(\mathcal{R}, \subseteq)$ *is an inclusion space.*

We call *canonical inclusion space* the structure $(\mathcal{R}, \subseteq)$. Whitehead (1919) defines the points by the following basic notion.

Definition 3. Given an inclusion space $(S, \leq)$, we call *abstractive sequence* any sequence $(r_n)_{n\in\mathbb{N}}$ of regions such that

(i) $(r_n)_{n\in\mathbb{N}}$ is order-reversing with respect to the inclusion
(ii) There is no region which is contained in all the regions in $(r_n)_{n\in\mathbb{N}}$.

We denote by AS the class of abstractive sequences.

Whitehead's idea is that an abstractive sequence $(r_n)_{n\in\mathbb{N}}$ represents an 'abstract object' which is obtained as a 'limit' of $(r_n)_{n\in\mathbb{N}}$. On the other hand, it is possible that two different abstractive sequences define the same abstract object. Then, we introduce the following equivalence relation.

Definition 4. The *covering* relation $\leq_c$ is the relation in *AS* defined by setting, for every $(r_n)_{n\in\mathbb{N}}$ and $(s_n)_{n\in\mathbb{N}}$,

$$(r_n)_{n\in\mathbb{N}} \leq_c (s_n)_{n\in\mathbb{N}} \Leftrightarrow \forall n \exists m \; r_n \leq s_n.$$

The relation $\equiv$ is defined by setting

$$(r_n)_{n\in\mathbb{N}} \equiv (s_n)_{n\in\mathbb{N}} \Leftrightarrow (r_n)_{n\in\mathbb{N}} \leq_c (s_n)_{n\in\mathbb{N}} \text{ and } (s_n)_{n\in\mathbb{N}} \leq_c (r_n)_{n\in\mathbb{N}}.$$

It is possible to prove that $\leq_c$ is a pre-order and therefore that $\equiv$ is an equivalence. Then we can consider the quotient $AS/\equiv$ and an order relation in $AS/\equiv$ by setting

$$[(r_n)_{n\in\mathbb{N}}] \leq_c [(s_n)_{n\in\mathbb{N}}] \Leftrightarrow (r_n)_{n\in\mathbb{N}} \leq_c (s_n)_{n\in\mathbb{N}}.$$

The following definition remembers Euclid's definition of point.

Definition 5. We call *geometrical element* any element of the quotient $AS/\equiv$, i.e. any class of equivalence $[(r_n)_{n\in\mathbb{N}}]$ modulo $\equiv$. A *point* is a geometrical element which *has no part*, i.e. which is minimal in $(AS/\equiv, \leq_c)$.

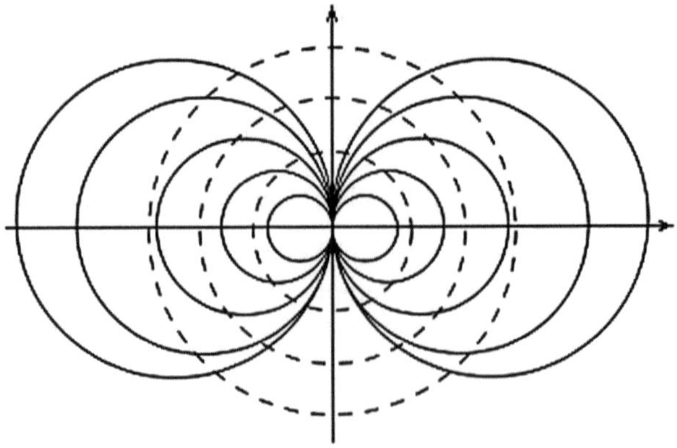

Fig. 5.1 Three different "points" in the origin

Unfortunately, in spite of the evident elegance of this definition of point, it is possible to prove the following theorem (see Gerla and Miranda 2004, 2008).

Theorem 2. *In a canonical inclusion space the definition of point is empty, i.e. there is no minimal element in* $(AS/\equiv, \leq_c)$.

Instead of an precise exposition of the proof of this theorem, we prefer to illustrate the idea which is on its basis by the following example. Consider in the Euclidean plane the abstractive sequence G defined by the sequence (B_n) of closed balls with center in the origin (0,0) and radius $r_n = 1/n$. From an intuitive point of view such an abstractive sequence represents a point. Unfortunately, we can consider the sequences G_1 and G_2 defined by the closed balls with radius r_n and centre in $(-1/n, 0)$ and $(1/n, 0)$, respectively (see Fig. 5.1). It is immediate that $[G] > G_1$, $[G] > G_2$ and that $[G]$, $[G_1]$ and $[G_2]$ are three different geometrical elements. This means that $[G]$ is not minimal and therefore that $[G]$ is not a point. Such an argument holds true if we start from any abstractive sequence. Perhaps Whitehead's passage from the inclusion-based approach to the connection-based approach was done to avoid such a counterintuitive behaviour. This theorem shows that the proposal of Whitehead of assuming as a primitive only the mereological notion of inclusion is unsatisfactory.

5.3 Connection Structures

Some years later the publication of Whitehead (1919, 1920), Whitehead (1929), proposed a different idea based on the primitive notion of *connection relation*. By isolating the main properties of the connection relation considered by Whitehead, we obtain the following theory. The considered language L_C has only a binary relation symbol C.

Definition 6. Denote by $x \leq y$ the formula $\forall z(zCx \Rightarrow zCy)$. Then we call *connection space theory* the following list of axioms.

C1 $\forall x \forall y\,(xCy \Rightarrow yCx)$ (symmetry)
C2 $\forall z \exists x \exists y\,((x \leq z) \wedge (y \leq z) \wedge (\neg xCy))$
C3 $\forall x \forall y \exists z\,(zCx \wedge zCy)$
C4 $\forall x(xCx)$
C5 $\forall x \forall y\,((x \leq y \wedge y \leq x) \Rightarrow x = y)$.

The intended interpretation is that the connection is either a surface contact or an overlap. It is easy to prove that in any connection space the relation $\leq$ is an order relation. We denote by xOy the formula$\exists z(z \leq x \wedge z \leq y)$ and we call *overlapping relation* the corresponding relation. Again we use the class $\mathscr{R}$ to find a model of this theory. We denote again by C the interpretation of the relation symbol C in $\mathscr{R}$.

Theorem 3. *Define in $\mathscr{R} \subseteq RC(\mathbb{R}^n)$ the relation C by setting*

$$XCY \Leftrightarrow X \cap Y \neq \emptyset.$$

Then $(\mathscr{R}, C)$ is a connection space in $\mathbb{R}^n$, whose associated order coincides with the set-theoretical inclusion.

We call *canonical connection space* a connection space defined in such a way. The observation of a canonical connection space makes evident way the connection relation is different from the overlapping relation. Indeed, while XCY means that there is a point belonging in both the regions, XOY means that there is a region contained in both the regions. To obtain an adequate definition of point, we need the notion of nontangential inclusion.

Definition 7. Given a connection space (S, C), we say that two regions have a *tangential connection* when

(i) They are connected,
(ii) They do not overlap.

We say that x is *non-tangentially included* in y and we write $x \prec y$ provided that

(j) x is included in y,
(jj) There is no region having a tangential connection with both x and y.

The following is a simple characterization of the non-tangential inclusion.

Proposition 1. *The non-tangential inclusion is the relation defined by the formula*

$$\forall z(zCx \Rightarrow zOy). \tag{5.1}$$

It is possible to prove that in a canonical connection space

$$X \prec Y \Leftrightarrow X \subseteq int(Y).$$

Definition 8. An *abstractive sequence* in a connection space is a sequence $(r_n)_{n \in \mathbb{N}}$ of regions such that

(j) $(r_n)_{n \in \mathbb{N}}$ is order-reversing with respect to the non-tangential inclusion,
(jj) There is no region which is contained in all the regions of $(r_n)_{n \in \mathbb{N}}$.

The notions of covering, equivalence, geometrical element, point are defined as in Sect. 5.2. Differently from the case of the inclusion spaces, in the canonical connection space $(\mathcal{R}, C)$ Whitehead's definition of point works well. Indeed the following theorem holds (see Coppola et al. 2010).

Theorem 4. *Consider the canonical space* $(\mathcal{R}, C)$ *and denote by* $B_n(p)$ *the closed ball centered in* p *and with radius* $1/n$*. Then the map associating every point* p *in* $\mathbb{R}^n$ *with the geometrical element* $[(B_n(p))_{n \in \mathbb{N}}]$ *is a one-to-one map from* $\mathbb{R}^n$ *to the set of points in* $(\mathcal{R}, C)$.

This theorem shows that connection space theory gives to point-free geometry a more suitable framework than the one of inclusion space theory. A further reason in furnished by the following theorem.

Theorem 5. *While in a canonical connection space* $(\mathcal{R}, C)$ *we can define the inclusion relation, in a canonical inclusion space* $(\mathcal{R}, \subseteq)$ *it is not possible to define the connection relation.*

The proof of this theorem is based on the fact that if a relation is definable in a structure, then it is invariant with respect to all the automorphisms of this structure. So, it is sufficient to exhibit an one-to-one map preserving the inclusion and not preserving the connection (for a complete proof see Gerla and Miranda 2004).

5.4 Multi-valued Logic for an Inclusion-Based Point-Free Geometry

As we have seen, there are some troubles in the inclusion-based point-free geometry. Indeed in rather natural models Whitehead's definition of point is empty, moreover the topological notion of connection cannot be defined. In spite of that, we claim that an inclusion-based approach it is possible provided that we consider the inclusion as a graded notion and therefore provided that we shift from classical logic to multi-valued logic. Namely, we refer to the first order logic associated with a continuous triangular norm $\otimes : [0, 1] \times [0, 1] \to [0, 1]$ (see for example Hájek 1998) and therefore to a first order language with:

- Two logical connectives $\wedge$ and $\Rightarrow$, interpreted by $\otimes$ and the related residuum $\to$,
- Two logical constant $\underline{0}$ and $\underline{1}$ to denote 0 and 1
- The quantifiers $\forall$ and $\exists$ interpreted by the operators inf and sup.

In addition, we consider a connective Ct we interpret by the function $ct : [0,1] \to [0,1]$ such that $ct(x) = 1$ if $x = 1$ and $ct(x) = 0$ otherwise. This means that the intended meaning of a formula as $Ct(\alpha)$ is '*α is completely true*'. To fix the ideas, we assume that $\otimes$ is the usual product and therefore that the implication is interpreted by the operation $\to$ such that $x \to y = 1$ if $x \leq y$ and $x \to y = y/x$ otherwise. Given a set D, an *n-ary fuzzy relation* in D is a map $r : D^n \to [0,1]$. We call *crisp* a fuzzy relation assuming only the values 0 and 1 and we identify a classical relation $R \subseteq D^n$ with the crisp relation $c_R : D^n \to [0,1]$ defined by setting $c_R(d_1, \ldots, d_n) = 1$ if $(d_1, \ldots, d_n) \in R$ and $c_R(d_1, \ldots, d_n) = 0$ otherwise. In other words, we identify R with its characteristic function c_R.

A *multi-valued interpretation* (D, I) is defined by a nonempty domain D and by a function I associating every constant c with an element $I(c) \in D$, every n-ary operation symbol with an n-ary operation in D and every n-ary relation symbol $\underline{r}$ with an n-ary fuzzy relation $r = I(\underline{r})$, i.e. a map $r : D^n \to [0,1]$. Given a multi-valued interpretation (D, I), the interpretation $I(t) : D^n \to D$ of a term t is defined as in classical logic. The valuation of the sentences is defined in a truth-functional way as follows.

Definition 9. Given a multi-valued interpretation (D, I), a formula α whose variables are among $x_1, \ldots, x_n$ and $d_1, \ldots, d_n$ in D, we define the value $Val(\alpha, d_1, \ldots, d_n)$ by recursion on the complexity of α, by the equations:

(i) $Val(\underline{r}(t_1, \ldots, t_p), d_1, \ldots, d_n) = I(\underline{r})(I(t_1)(d_1, \ldots, d_n), \ldots, I(t_p)(d_1, \ldots, d_n))$

(ii) $Val(\alpha_1 \underline{\diamond} \alpha_2, d_1, \ldots, d_n) = Val(\alpha_1, d_1, \ldots, d_n) \diamond Val(\alpha_q, d_1, \ldots, d_n)$

(iii) $Val(\underline{\bullet}\alpha, d_1, \ldots, d_n) = \bullet(Val(\alpha, d_1, \ldots, d_n))$

(iv) $Val(\forall x_h \beta, d_1, \ldots, d_n) = \inf(\{Val(\beta, d_1, \ldots, d_{h-1}, d, d_{h+1}, \ldots, d_n) : d \in D\})$

(v) $Val(\exists x_h \beta, d_1, \ldots, d_n) = \sup(\{Val(\beta, d_1, \ldots, d_{h-1}, d, d_{h+1}, \ldots, d_n) : d \in D\})$

where we denote by $\underline{\diamond}$ (by $\underline{\bullet}$) a binary (an unary) connective and by $\diamond$ (by $\bullet$) the corresponding interpretation.

We say that $d_1, \ldots, d_n$ *satisfy* α if $Val(\alpha, d_1, \ldots, d_n) = 1$. If α is a closed formula, then the value $Val(\alpha, d_1, \ldots, d_n)$ does not depend on the elements $d_1, .., d_n$ and we write $Val(\alpha)$ instead of $Val(\alpha, d_1, \ldots, d_n)$. In the case there are free variables in α, we write $Val(\alpha)$ to denote $Val(\forall x_1 \ldots \forall x_n(\alpha))$ where $\forall x_1 \ldots \forall x_n(\alpha)$ is the universal closure of α.

Definition 10. Given a theory T, we say that (D, I) is a *multi-valued model* of T if $Val(\alpha) = 1$ for every $\alpha \in T$.

The so defined multi-valued logic is rather expressive. For example, if $\underline{r}$ is an n-ary relation symbol then the formula

$$\forall x_1 \ldots \forall x_n(Ct(\underline{r}(x_1, \ldots, x_n)) \leftrightarrow \underline{r}(x_1, \ldots, x_n))$$

is satisfied if and only if $\underline{r}$ is interpreted by a crisp relation. Indeed it is sufficient to observe that this formula is satisfied if and only if $ct(r(d_1, \ldots, d_n)) = r(d_1, \ldots, d_n)$ for every $d_1, \ldots, d_n$ in D. In other words, 'to be crisp' is a first order property. This entails that all the classical notions which are definable in classical first order logic are definable in our multi-valued logic, too. In particular, the notion of order relation is definable.

Definition 11. Let (D, I) be a multi-valued interpretation and α be a formula whose free variables are among $x_1, \ldots, x_n$. Then the *extension* of α in (D, I) is the fuzzy relation $r_\alpha : D^n \to [0, 1]$ defined by setting $r_\alpha(d_1, \ldots, d_n) = Val(\alpha, d_1, \ldots, d_n)$ for every $d_1, \ldots, d_n$ in D. In such a case we say that r_α *is defined by* α. We call *crisp extension* of α the extension $r_{Ct(\alpha)}$ of $Ct(\alpha)$. In such a case we say that $r_{Ct(\alpha)}$ is the *crisp relation defined by* α.

Then the crisp relation defined by α is the (characteristic function of the) relation

$$\{(d_1, \ldots, d_n) \in D^n : \alpha \text{ is satisfied by } d_1, \ldots, d_n\}.$$

Coming back to point-free geometry, we consider the first order language with a binary relation symbol *Incl* and we write $x \leq y$ to denote the formula $Ct(Incl(x, y))$. An interpretation of such a language is defined by a pair $(S, incl)$ where S is a nonempty set and $incl : S \times S \to [0, 1]$ a fuzzy binary relation. The interpretation of $\leq$ is the (characteristic function of the) relation $\leq$ defined by setting

$$x \leq y \Leftrightarrow incl(x, y) = 1. \tag{5.2}$$

We call *the crisp inclusion* associated with *incl* this relation.

If $Sim(x, y)$ denotes the formula $Incl(x, y) \wedge Incl(y, x)$, then the interpretation of $Sim(x, y)$ is the fuzzy relation $sim : S \times S \to [0, 1]$ defined by setting

$$sim(x, y) = incl(x, y) \otimes incl(y, x). \tag{5.3}$$

We call *the graded identity associated with incl* this fuzzy relation.

Definition 12. A *graded preorder structure*, in brief *graded preorder*, is a multi-valued model $(S, incl)$ of the following theory:

A1 $\forall x(Incl(x, x))$
A2 $\forall x \forall y \forall z((Incl(x, z) \wedge Incl(z, y)) \to Incl(x, y))$.

Then a fuzzy relation *incl* is a graded preorder if and only if, for every $x, y, z \in S$,

a1 $incl(x, x) = 1$ (reflexivity)
a2 $incl(x, y) \otimes incl(y, z) \leq incl(x, z)$ (transitivity).

If the symmetry axiom is also satisfied then the fuzzy relation is called *fuzzy equivalence* or *similarity*. Then a similarity is a fuzzy relation $sim : S \times S \to [0, 1]$ such that

b1 $sim(x, x) = 1$ (reflexivity)
b2 $sim(x, y) \otimes sim(y, z) \leq sim(x, z)$ (transitivity)
b3 $sim(x, y) = sim(x, y)$ (symmetry).

This notion is a graded extension of the one of equivalence. It is easy to prove that the fuzzy relation sim defined by (5.3) is a similarity. A *fuzzy equality* is a similarity satisfying the following 'separation axiom'

b4 $sim(x, y) = 1 \Leftrightarrow x = y$.

To simulate Whitehead's definition of point, we define a notion of 'point-likeness' suggested by Euclid's definition of point as *minimal element*, i.e. an element x such that $x' \leq x$ entails $x' = x$.

Definition 13. We call *point-likeness* property the property expressed by the formula,

$$Pl(x) \equiv \forall x'(x' \leq x \rightarrow Sim(x, x')).$$

The extension of Pl is the fuzzy subset of regions pl defined by

$$pl(x) = \inf\{incl(x, x') : x' \leq x\}.$$

This means that all the regions are points at a suitable degree. The formula $Pl(x)$ enables us to express the next two axioms. The first axiom claims that if we apply the graded inclusion to regions which are (approximately) points, then such a relation is (approximately) symmetric .

A3 $Pl(x) \wedge Pl(y) \rightarrow (Incl(x, y) \rightarrow Incl(y, x))$.

This axiom is satisfied if and only if, for every x and y,

a3 $pl(x) \otimes pl(y) \leq (incl(x, y) \rightarrow incl(y, x))$.

The further axiom claims that every region x contains a point:

A4 $\forall x \exists x'((x' \leq x) \wedge Pl(x'))$.

Such an axiom is satisfied if and only if for every x,

a4 $\sup_{x' \leq x} pl(x') = 1$

i.e. if and only if for every x

$$\exists \epsilon > 0 \text{ there is } x' \leq x \text{ such that } pl(x') \geq 1 - \epsilon.$$

Definition 14. We call *graded inclusion space* every model of **A1–A4**.

The following notion enables us to emphasize the metrical nature of the graded inclusion spaces.

Definition 15. A *hemimetric space* is a structure (S, d) such that S is a nonempty set and $d : S \times S \to [0, \infty]$ is a mapping such that, for all $x, y, z \in S$,

d1 $d(x, x) = 0$
d2 $d(x, y) \leq d(x, z) + d(z, y)$.

Then, a metric space is a hemimetric space which is symmetric, i.e. such that $d(x, y) = d(y, x)$ for every $x, y \in S$, and such that $d(x, y) = 0$ only if $x = y$. Every hemimetric space is associated with a pre-order in the following way.

Proposition 2. *Let (S, d) be a hemimetric space, then the relation $\leq$ defined by setting:*

$$x \leq y \Leftrightarrow d(x, y) = 0$$

is a pre-order such that d is order-preserving with respect to the first variable and order-reversing with respect to the second variable.

In the case d is a metric, $\leq$ coincides with the identity relation. Given $x \in S$, we call *diameter* of x the number

$$\delta(x) = \sup\{d(x_1, x_2) : x_1 \leq x, x_2 \leq x\}.$$

Observe that this definition entails that

$$\delta(x) \geq d(y, x) \text{ for every } y \leq x. \tag{5.4}$$

In the case d is a metric, all the diameters are equal to 0.

The following proposition shows that the notion of hemimetric distance is 'dual' of the one of graded preorder. As usual, we put $10^{-\infty} = 0$ and $Log(0) = -\infty$.

Proposition 3. *Given a hemimetric space (S, d), the fuzzy relation incl defined by setting*

$$incl(x, y) = 10^{-d(x,y)}$$

is a graded preorder such that $pl(x) = 10^{-\delta(x)}$. Conversely, let $incl : S \times S \to [0, 1]$ be a graded preorder and let d be defined by setting

$$d(x, y) = -Log(incl(x, y)).$$

Then d is a hemimetric such that $\delta(x) = -Log(pl(x))$.

In the case d is a pseudo-metric the associated fuzzy relation *incl* is a similarity, obviously. We consider the following class of hemimetrics.

Definition 16. A *hemimetric space of regions* is a hemimetric space (S, d) such that for every x and y,

d3 $|d(x, y) - d(y, x)| \leq \delta(x) + \delta(y)$
d4 $\forall \epsilon > 0 \exists x' \leq x, \delta(x') \leq \epsilon.$

The following theorem shows a duality between the class of hemimetric spaces of regions and the one of the graded inclusion spaces (see also Di Concilio and Gerla 2006).

Theorem 6. *For every hemimetric space of regions* (S, d)*, the fuzzy relation incl defined by setting*

$$incl(x, y) = 10^{-d(x,y)}$$

defines a graded inclusion space of regions. Conversely, let $(S, incl)$ *be a graded inclusion space of regions and let* $d : S \times S \rightarrow [0, +\infty]$ *be defined by setting*

$$d(x, y) = -Log(incl(x, y)).$$

Then (S, d) *is a hemimetric space of regions.*

5.5 Canonical Graded Inclusion Spaces, Connection and Points

The most famous hemimetric is the excess measure used to define the Hausdorff distance.

Definition 17. Given a metric space (M, d) the *excess measure* is the map $e : \mathcal{P}(M) \times \mathcal{P}(M) \rightarrow [0, \infty]$ defined, for every pair X and Y of subsets of M, by setting

$$e(X, Y) = \sup_{p \in X} \inf_{q \in Y} d(p, q).$$

The following proposition is proved in Di Concilio and Gerla (2006).

Proposition 4. *Let* $\mathcal{R}$ *be the class nonempty, bounded, closed regular subsets of* (M, d)*. Then the excess measure defines in* $\mathcal{R}$ *a hemimetric space of regions. Consequently, by setting*

$$incl(X, Y) = 10^{-e(X,Y)} = inf_{p \in X} sup_{q \in Y} 10^{-d(p,q)}$$

we obtain a graded inclusion space. The induced order is the usual set theoretical inclusion and the point-likeness property is defined by

$$pl(X) = 10^{-|X|},$$

where $|x|$ is the usual diameter in a metric space.

We call *canonical graded inclusion space* the inclusion space obtained by such a proposition. Observe that if we consider a fuzzy equality $eq : M \times M \rightarrow [0,1]$, then by setting $d(x,y) = -Log(eq(x,y))$ we obtain a metric. Indeed, it is evident that $d(x,y) = 0$ if and only if $eq(x,y) = 1$ if and only if $x = y$. By applying Proposition 4, we obtain that

$$incl(X,Y) = inf_{p\in X} sup_{q\in Y} eq(p,q).$$

Assume that in the language there is a name Eq to denote eq. Then, in accordance with the usual interpretation of the quantifiers in multi-valued logic, we can interpret the value $incl(X,Y)$ as the interpretation of the formula $\forall p \in X \exists q \in Y(Eq(p,q))$, i.e. of the claim 'every point in X is (approximately) equal with a point in Y'.

We will show that, differently from Whitehead's inclusion spaces, in a graded inclusion space we can define the connection relation as the crisp extension of the formula expressing the overlapping relation in an inclusion space.

Theorem 7. *Consider in a canonical graded inclusion space $(\mathcal{R}, incl)$ the formula $O(x,y) \equiv \exists z(Incl(z,x) \wedge Incl(z,y))$. Then the connection relation C in the canonical connection space $(\mathcal{R}, C)$ is defined by the formula $Ct(O(x,y))$.*

In other words, we can define the connection between two regions by saying that the two regions completely overlaps (at degree 1).

The second question is to show that in a graded inclusion space it is possible to give a nonempty notion of point.

Definition 18. Given a graded inclusion space, we call *abstraction process* any sequence $< p_n >_{n\in\mathbb{N}}$ of regions which are order-reversing with respect to the order associated with the graded inclusion. We extend the point-likeness property to the abstraction processes by setting

$$pl(< p_n >_{n\in\mathbb{N}}) = lim_{n\rightarrow\infty} pl(p_n) = sup_n\, pl(p_n).$$

We say that $< p_n >_{n\in\mathbb{N}}$ *represents a point* if $pl(< p_n >_{n\in\mathbb{N}}) = 1$ and we denote by Pr the class of abstraction processes representing a point.

Observe that **A4** enables us to prove that every region of a graded inclusion space 'contains' an abstraction process representing a point and therefore that $Pr \neq \emptyset$. Indeed, in accordance with **a4**, for every region x there is $x' \leq x$ such that $pl(x') \geq 1 - 1/n$. Then we can consider the sequence $< p_n >_{n\in\mathbb{N}}$ defined by setting $p_1 = x$

and p_n equal to a region such that $p_n \leq p_{n-1}$ and $pl(p_n) \geq 1 - 1/n$. Obviously, $pl(< p_n >_{n \in \mathbb{N}}) = 1$.

The following theorem shows that in the class of abstraction processes representing points it is possible to define a pseudo-metric d. We give the proof in order to emphasize the role of **A3** and therefore of **d3** in proving the symmetry of d.

Theorem 8. *Let $(S, incl)$ be a graded inclusion space and d' be the associated hemimetric. Then the map $d : Pr \times Pr \rightarrow [0, \infty]$ obtained by setting*

$$d(< p_n >_{n \in \mathbb{N}}, < q_n >_{n \in \mathbb{N}}) = lim_{n \rightarrow \infty} d'(p_n, q_n),$$

defines a pseudo-metric space (Pr, d).

Proof. To prove the convergence of the sequence $< d'(p_n, q_n) >_{n \in \mathbb{N}}$, let n and k be natural numbers and assume that $n \geq k$. Then, since $d'(p_n, p_k) = 0$ and, by (5.4), $d'(q_k, q_n) \leq \delta(q_k)$ we have that,

$$d'(p_n, q_n) \leq d'(p_n, p_k) + d'(p_k, q_k) + d'(q_k, q_n) \leq d'(p_k, q_k) + \delta(q_k)$$

and therefore,

$$d'(p_n, q_n) - d'(p_k, q_k) \leq \delta(q_k).$$

Likewise, since $d'(q_n, q_k) = 0$ and $d'(p_k, p_n) \leq \delta(p_k)$,

$$d'(p_k, q_k) \leq d'(p_k, p_n) + d'(p_n, q_n) + d'(q_n, q_k) \leq d'(p_n, q_n) + \delta(p_k)$$

and therefore

$$d'(p_k, q_k) - d'(p_n, q_n) \leq \delta(p_k).$$

This entails

$$|d'(p_n, q_n) - d'(p_k, q_k)| \leq \max\{\delta(q_k), \delta(p_k)\}.$$

Obviously, in the case $n \leq k$

$$|d'(p_n, q_n) - d'(p_k, q_k)| \leq \max\{\delta(q_n), \delta(p_n)\}.$$

Thus

$$|d'(p_n, q_n) - d'(p_k, q_k)| \leq \max\{\delta(q_n), \delta(p_n), \delta(q_k), \delta(p_k)\}.$$

The convergence of the diameters entails that $< d'(p_n, q_n) >_{n \in \mathbb{N}}$ is a Cauchy sequence.

It is evident that $d(< p_n >_{n\in\mathbb{N}}, < p_n >_{n\in\mathbb{N}}) = 0$ and that d satisfies the triangular inequality.

To prove the symmetry, observe that, by **d3**, $|d'(p_n, q_n) - d'(q_n, p_n)| \leq \delta(p_n) + \delta(q_n)$ and therefore, since $lim_{n\to\infty}\delta(p_n) + \delta(q_n) = 0$, $lim_{n\to\infty}|d'(p_n, q_n) - d'(q_n, p_n)| = 0$. Since the sequences $< d'(p_n, q_n) >_{n\in\mathbb{N}}$ and $< d'(q_n, p_n) >_{n\in\mathbb{N}}$ are both convergent, $lim_{n\to\infty}d'(p_n, q_n) = lim_{n\to\infty}d'(q_n, p_n)$. Thus

$$d(< p_n > n \in \mathbb{N}, < q_n > n \in \mathbb{N}) = lim_{n\to\infty}d'(p_n, q_n) = lim_{n\to\infty}d'(q_n, p_n)$$
$$= d(< q_n >_{n\in\mathbb{N}}, < p_n >_{n\in\mathbb{N}}).\square$$

Such a proposition enables us to associate any graded inclusion space with a metric space. Indeed, recall that the *quotient* of a pseudo-metric space (X, d) is the metric space $(\underline{X}, \underline{d})$ defined by assuming that

- $\underline{X}$ is the quotient of X modulo the relation $\equiv$ defined by setting $x \equiv x'$ if and only if $d(x, x') = 0$,
- $\underline{d}([x], [y]) = d(x, y)$ for every $[x], [y] \in X'$.

Definition 19. We call *metric space associated with a graded inclusion space* $(S, incl)$ the quotient $(\underline{Pr}, \underline{d})$ of the pseudo-metric space (Pr, d). We call *point* any element in $\underline{Pr}$.

Then, the metric space $(\underline{Pr}, \underline{d})$ associated with a graded inclusion space $(S, incl)$ is obtained

- By starting from the class Pr of abstraction processes,
- By setting $\underline{Pr}$ equal to the quotient of Pr modulo the equivalence relation $\equiv$ defined by

$$< p_n >_{n\in\mathbb{N}} \equiv < q_n >_{n\in\mathbb{N}} \Leftrightarrow lim_{n\to\infty}incl(p_n, q_n) = 1,$$

- By defining $\underline{d} : \underline{Pr} \times \underline{Pr} \to [0, \infty]$ by the equation

$$\underline{d}(P, Q) = lim_{n\to\infty} - Log(incl(p_n, q_n))$$

  where $P = [< p_n >_{n\in\mathbb{N}}]$ and $Q = [< q_n >_{n\in\mathbb{N}}]$ are elements in $\underline{Pr}$.

5.6 To Be Closed and To Be Small

In a series of papers a metric approach to point-free geometry is proposed in which, in addition to the inclusion relation, the notions of diameter of a region and distance between two regions are assumed as primitives (see Gerla 1990).

Definition 20. A *point-free pseudo-metric space*, in short a *ppm-space*, is a structure $(S, \leq, \Delta, \delta)$, where $(S, \leq)$ is an ordered set, $\Delta : S \times S \rightarrow [0, \infty)$ is order-reversing, $\delta : S \rightarrow [0, \infty]$ is order-preserving and, for every $x, y, z \in S$:

D1 $\Delta(x, x) = 0$
D2 $\Delta(x, y) = \Delta(y, x)$
D3 $\Delta(x, y) \leq \Delta(x, z) + \Delta(z, y) + \delta(z)$.

The elements in S are called *regions*, the order $\leq$ *inclusion relation*, $\Delta(x, y)$ *distance between x and y*, $\delta(x)$ the *diameter* of x. Inequality $D3$ is a weak form of the triangular inequality taking in account the diameters of the regions. The notion of *ppm*-space extends the one of pseudo-metric space (and therefore of metric space). Indeed, if all the diameters are equal to zero, then $D3$ coincides with the triangular inequality and the *ppm*-space is a pseudo-metric space. More precisely, we can identify the pseudo-metric spaces with the *ppm*-spaces in which $\leq$ is the identity and all the diameters are equal to zero. We identify the metric spaces with the *ppm*-spaces satisfying these conditions and such that Δ is finite and $\Delta(x, y) = 0$ entails $x = y$.

Notice that we can also assume as primitive a function Δ satisfying $D1$ and $D2$ and define a diameter by setting $\delta(z) = \sup\{\Delta(x, y) - \Delta(x, z) - \Delta(z, y) : x, y \in S\}$. Indeed, to prove that the resulting structure $(S, \leq, \Delta, \delta)$ is a pmm-space, we observe that by setting $x = y = z$ we obtain that $\delta(z)$ is greater or equal to 0. It is evident that δ is order-preserving and that $D3$ holds true by definition. It is also possible to assume as primitive only the diameter function (see Gerla and Paolillo 2010; Pultr 1988).

The following proposition gives prototypic examples of *ppm*-space (see Gerla 1990).

Proposition 5. *Let* (M, d) *be a pseudo-metric space and let* C *be a nonempty class of bounded and nonempty subsets of* M*. Define* Δ *and* δ *by setting*

$$\Delta(X, Y) = \inf\{d(x, y) : x \in X, y \in Y\}$$

$$\delta(X) = \sup\{d(x, y) : x, y \in X\},$$

respectively. Then $(C, \subseteq, \Delta, \delta)$ *is a ppm-space.*

In particular, we call *canonical ppm-space* the space $(\mathscr{R}, \subseteq, \Delta, \delta)$. By referring to the just defined class of *ppm*-spaces, the meaning of the proposed axioms becomes evident. For example, the meaning of $D3$ is given by Fig. 5.2: Indeed, it is evident that in this case $\Delta(X, Y) > \Delta(X, Z) + \Delta(Z, Y)$ and therefore that the usual triangular inequality cannot be assumed. Instead, it is matter of routine to prove that $\Delta(X, Y) \leq \Delta(X, Z) + \Delta(Z, Y) + \delta(Z)$.

The notion of point is defined as in Sect. 5.5. Indeed, we call *abstraction process* any sequence $< p_n >_{n \in \mathbb{N}}$ of regions which are order-reversing and we call *distance* between two abstraction processes $< p_n >_{n \in \mathbb{N}}$ and $< q_n >_{n \in \mathbb{N}}$ the number:

$$d(< p_n >_{n \in \mathbb{N}}, < q_n >_{n \in \mathbb{N}}) = lim_{n \rightarrow \infty} \Delta(p_n, q_n)$$

Fig. 5.2 Approximate triangular inequality

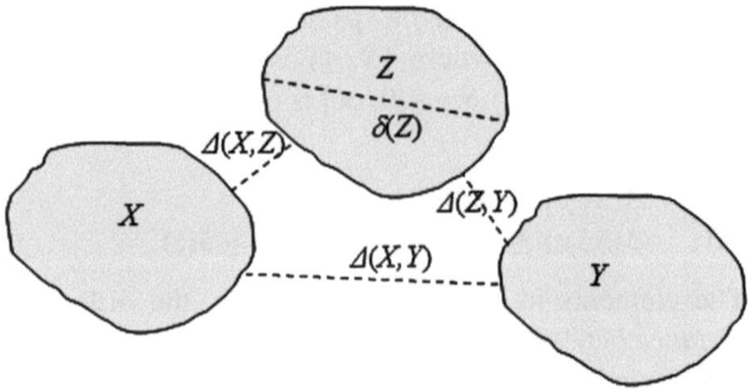

and *diameter* of an abstraction process $< p_n >_{n\in\mathbb{N}}$ the number

$$\delta(< p_n >_{n\in\mathbb{N}}) = lim_{n\to\infty}\delta(p_n).$$

Definition 21. We say that $< p_n >_{n\in\mathbb{N}}$ *represents a point* if its diameter is zero and we denote by *Pr* the class of abstraction processes representing a point.

It is matter of routine to prove that (Pr, d) is a pseudo-metric space.

Definition 22. A *point* is an element of the metric space $(\underline{Pr}, \underline{d})$ associated with (Pr, d).

The logical counterpart of the *ppm*-space is defined as follows. We refer to a first order language L_{CS} with the predicate symbols $\leq$, *Cl* and *Sm*. The intended meaning of $Cl(x, y)$ is 'x and y are close', the intended meaning of $Sm(x)$ is 'x is small'. We denote by $(S, \leq, cl, sm)$ an interpretation of L_{CS}.

Definition 23. A CS*-space* is an interpretation $(S, \leq, cl, sm)$ of L_{CS} such that $\leq$ is a crisp order relation and such that the following axioms are satisfied:

CS1 $\forall x\, Cl(x, x)$
CS2 $\forall x \forall y\, (Cl(x, y) \Rightarrow Cl(y, x))$
CS3 $\forall x \forall y \forall z\, (Cl(x, z) \wedge Cl(y, z) \wedge Sm(z) \Rightarrow Cl(x, y))$
CS4 $\forall y \forall x \forall x'\, (x \leq x' \Rightarrow (Cl(x, y) \Rightarrow Cl(x', y)))$
CS5 $\forall x \forall x'\, (x \leq x' \Rightarrow (Sm(x') \Rightarrow Sm(x)))$.

Observe that, as observed in Sect. 5.4, the logical connective *Ct* enables us to express the condition '$\leq$ is a crisp relation' by a first order formula in the multi-valued logic. Notice also that in Gerla (2008) the structures satisfying *CS1*, *CS2*, *CS3* are called *approximate similarity structures* and that they are proposed for a solution of Poincaré paradox.

The proof of the following proposition is obvious.

Proposition 6. *An interpretation* $(S, \leq, cl, sm)$ *of* L_{CS} *is a CS-space if and only if* $\leq$ *is an order relation, cl is order-preserving, sm is order-reversing and*

(i) $cl(x, x) = 1$,

(ii) $cl(x,y) = cl(y,x)$,
(iii) $cl(x,z) \otimes (y,z) \otimes (z) \leq cl(x,y)$.

The following theorem shows that there is a duality between the notions of *ppm*-space and the one of *CS*-space.

Theorem 9. *Let* $(S, \leq, \Delta, \delta)$ *be a ppm-space and define cl and sm by setting*

$$cl(x,y) = 10^{-\Delta(x,y)} \quad ; \quad sm(x) = 10^{-\delta(x)}.$$

Then $(S, \leq, cl, sm)$ *is a CS-space. Conversely, let* $(S, \leq, cl, sm)$ *be an approximate CS-space and set*

$$\Delta(x,y) = -Log(cl(x,y)) \quad ; \quad \delta(x) = -Log(sm(x)).$$

Then $(S, \leq, \Delta, \delta)$ *is a ppm-space.*

In particular, by starting from the canonical *ppm*-space $(\mathscr{R}, \subseteq, \Delta, \delta)$, we define the *canonical CS-space* $(\mathscr{R}, \subseteq, cl, sm)$ by setting $cl(X,Y) = 10^{-\Delta(X,Y)}$ and $sm(X) = 10^{-\delta(X)}$.

Theorem 10. *In the canonical CS space we can define the connection relation by the formula* $Ct(Cl(x,y))$.

Proof. It is sufficient to observe that $cl(X,Y) = 1$ if and only if $\Delta(X,Y) = 0$ if and only if $X \cap Y \neq \emptyset$.

References

Blumenthal, L. M. (1970). *Theory and applications of distance geometry*. New York: Chelsea.

Chang, C. C., & Keisler, H. J. (1966). *Continuous model theory*. Princeton: Princeton University Press.

Coppola, C., Gerla, G., & Miranda, A. (2010). Point-free foundation of geometry and multi-valued logic. *Notre Dame Journal of Formal Logic, 51*, 383–405.

De Laguna, T. (1922). Point, line and surface as sets of solids. *The Journal of Philosophy, 19*, 449–461.

Di Concilio, A., & Gerla, G. (2006). Quasi-metric spaces and point-free geometry. *Mathematical Structures in Computer Science, 16*, 115–137.

Gerla, G. (1990). Pointless metric spaces. *Journal of Symbolic Logic, 55*, 207–219.

Gerla, G. (1994). Pointless geometries. In F. Buekenhout (Ed.), *Handbook of incidence geometry* (pp. 1015–1031). Amsterdam/New York: Elsevier.

Gerla, G. (2008). Approximates similarities and Poincaré paradox. *Notre Dame Journal of Formal Logic, 49*, 203–226.

Gerla, G., & Miranda, A. (2004). Graded inclusion and point-free geometry. *International Journal of Pure and Applied Mathematics, 11*, 63–81.

Gerla, G., & Miranda, A. (2008). Mathematical features of Whitehead's pointfree geometry. In M. Weber & W. Desmond (Eds.), *Handbook of Whiteheadian process thought* (pp. 507–533). Frankfurt: Ontos.

Gerla, G., & Paolillo, B. (2010). Whitehead's point-free geometry and diametric posets. *Logic and Logical Philosophy, 19*, 289–308.
Hájek, P. (1998). *Metamathematics of fuzzy logic*. Dordrecht: Kluwer.
Pavelka, J. (1979). On fuzzy logic I many-valued rules of inference. *Zeitschrift für mathematische Logik und Grundlagen der Mathematik, 25*, 45–52.
Pultr, A. (1988). Diameters in locales: How bad they can be? *Commentationes Mathematicae Universitatis Carolinae, 4*, 731–742.
Whitehead, A. N. (1919). *An inquiry concerning the principles of natural knowledge*. Cambrige: Cambrige University Press.
Whitehead, A. N. (1920). *The concept of nature*. Cambrige: Cambrige University Press.
Whitehead, A. N. (1929). *Process and reality*. New York: Macmillan.
Yaacov, I. B., & Usvyatsov, A. (2010). Continuous first order logic and local stability. *Transactions of the American Mathematical Society, 362*, 5213–5259.

Chapter 6
The Relations of *Supremum* and *Mereological Sum* in Partially Ordered Sets

Rafał Gruszczyński and Andrzej Pietruszczak

6.1 Basic Axioms and Definitions

Let M be an arbitrary non-empty set and let $\sqsubseteq \subseteq M \times M$. We call $\sqsubseteq$ *the relation of being part of* and in case $x \sqsubseteq y$ we say that x *is part of* y, '$x \not\sqsubseteq y$' is to mean $\neg\, x \sqsubseteq y$. *Part of* is the only primitive concept of the theory we are going to present.

In the sequel we use standard logical constants: quantifiers $\exists$ and $\forall$, sentential operators $\neg$, $\wedge$, $\vee$, $\Rightarrow$ and $\Leftrightarrow$. For any set S, $\mathcal{P}(S)$ is its power set, while $\mathcal{P}_+(S) := \mathcal{P}(S) \setminus \{\varnothing\}$. Moreover, let $|S|$ be the cardinal number of S and id_S be the identity relation on S, i.e. $\mathrm{id}_S := \{\langle x, x\rangle : x \in S\}$.

A pair $\langle M, \sqsubseteq\rangle$ is a *degenerate structure* iff it consists of exactly one element, i.e. $|M| = 1$. We say that $\langle M, \sqsubseteq\rangle$ is a *partially ordered set* (*poset* for short) iff it satisfies the following three axioms of reflexivity, transitivity and antisymmetry:

$$\forall_{x \in M}\ x \sqsubseteq x\,, \qquad (\mathtt{r}_\sqsubseteq)$$

$$\forall_{x,y,z \in M}(x \sqsubseteq y \wedge y \sqsubseteq z \Longrightarrow x \sqsubseteq y), \qquad (\mathtt{t}_\sqsubseteq)$$

$$\forall_{x,y \in M}(x \sqsubseteq y \wedge y \sqsubseteq x \Longrightarrow x = y). \qquad (\mathtt{antis}_\sqsubseteq)$$

$\langle M, \sqsubseteq\rangle$ is a *quasi-partially ordered set* (*quasi-poset* for short) iff satisfies $(\mathtt{r}_\sqsubseteq)$ and $(\mathtt{t}_\sqsubseteq)$. Let **POS** and **QPOS** be respectively the class of all posets and the class all quasi-poset.

We introduce some standard relations definable by means of the only primitive relation and the identity relation:

$$x \sqsubset y \;:\Longleftrightarrow\; x \sqsubseteq y \wedge x \neq y\,, \qquad (\mathtt{df}\ \sqsubset)$$

R. Gruszczyński (✉) • A. Pietruszczak
Department of Logic, Nicolaus Copernicus University in Toruń, Toruń, Poland
e-mail: gruszka@umk.pl; pietrusz@umk.pl

C. Calosi and P. Graziani (eds.), *Mereology and the Sciences*, Synthese Library 371,
DOI 10.1007/978-3-319-05356-1_6,

$$x \bigcirc y \;:\Longleftrightarrow\; \exists_z (z \sqsubseteq x \wedge z \sqsubseteq y)\,, \tag{df$\bigcirc$}$$

$$x \wr y \;:\Longleftrightarrow\; \neg \exists_z (z \sqsubseteq x \wedge z \sqsubseteq y)\,. \tag{df$\wr$}$$

In case $x \sqsubset y$ (resp. $x \bigcirc y$, $x \wr y$) we say that x *is proper part of* y (resp. x *overlaps* y, x *is exterior to* y[1]). Only by definitions the relation $\sqsubset$ is irreflexive, the relations $\bigcirc$ and $\wr$ are symmetric, and $\wr$ is the (set theoretical) complement of $\bigcirc$, i.e.:

$$\forall_{x \in M}\; x \not\sqsubset x\,, \tag{irr$_\sqsubset$}$$

$$\forall_{x,y \in M} (x \bigcirc y \iff y \bigcirc x)\,, \tag{s$_\bigcirc$}$$

$$\forall_{x,y \in M} (x \wr y \iff y \wr x)\,, \tag{s$_\wr$}$$

$$\forall_{x,y \in M} (x \wr y \iff \neg\, x \bigcirc y). \tag{6.1}$$

If $\sqsubseteq$ satisfies (r$_\sqsubseteq$), then the relation $\bigcirc$ is reflexive and $\wr$ is irreflexive, i.e.:

$$\forall_{x \in M}\; x \bigcirc x\,, \tag{r$_\bigcirc$}$$

$$\forall_{x \in M}\; \neg\, x \wr x\,, \tag{irr$_\wr$}$$

If $\sqsubseteq$ satisfies (t$_\sqsubseteq$) and (antis$_\sqsubseteq$), then $\sqsubset$ is transitive, i.e.:

$$\forall_{x,y,z \in M} (x \sqsubset y \wedge y \sqsubset z \Longrightarrow x \sqsubset z). \tag{t$_\sqsubset$}$$

If $\sqsubseteq$ satisfies (antis$_\sqsubseteq$) then $\sqsubset$ is asymmetrical, i.e.:

$$\forall_{x,y \in M} (x \sqsubset y \Longrightarrow y \not\sqsubset x). \tag{as$_\sqsubset$}$$

Notice that from (df $\sqsubset$), (r$_\sqsubseteq$) and (antis$_\sqsubseteq$) we have that:

$$\forall_{x,y \in M} (x \sqsubset y \iff x \sqsubseteq y \wedge y \not\sqsubseteq x), \tag{6.2}$$

and from (df $\sqsubset$) and (r$_\sqsubseteq$) we get that:

$$\forall_{x,y \in M} (x \sqsubseteq y \iff x \sqsubset y \vee x = y). \tag{6.3}$$

To facilitate considerations in the sequel, we introduce three operations P, PP, O whose domain is M and co-domain $\mathcal{P}(M)$:

$$\mathsf{P}(x) := \{y \in M \mid y \sqsubseteq x\}\,, \tag{df P}$$

$$\mathsf{PP}(x) := \{y \in M \mid y \sqsubset x\}\,, \tag{df PP}$$

[1] Sometimes terms 'incompatible' or 'disjoint from' are used instead of the one used by us.

$$\mathsf{O}(x) := \{y \in M \mid y \circ x\}. \tag{df O}$$

Thus $\mathsf{P}(x)$ is the set of all parts of x, $\mathsf{PP}(x)$ the set of all its proper parts and $\mathsf{O}(x)$ the set of all these objects each of which has a common lower bound with x. Of course, the conjunction of $(\mathtt{r}_\sqsubseteq)$ and $(\mathtt{t}_\sqsubseteq)$ is equivalent to the following condition:

$$\forall_{x,y \in M}\big(x \sqsubseteq y \iff \mathsf{P}(x) \subseteq \mathsf{P}(y)\big). \tag{rt$_\sqsubseteq$}$$

Moreover, by $(\mathtt{r}_\sqsubseteq)$ we obtain:

$$\forall_{x \in M}\ \mathsf{P}(x) \subseteq \mathsf{O}(x) \tag{6.4}$$

and by $(\mathtt{t}_\sqsubseteq)$ we obtain:

$$\forall_{x,y \in M}\big(x \sqsubseteq y \Longrightarrow \mathsf{O}(x) \subseteq \mathsf{O}(y)\big), \tag{6.5}$$

$$\forall_{x,y \in M}\big(\mathsf{P}(x) \subseteq \mathsf{O}(y) \Longrightarrow \mathsf{O}(x) \subseteq \mathsf{O}(y)\big). \tag{6.6}$$

If **C** is a class of structures then any given sentence is said *to be true in this class* iff it is true in (satisfied by) every structure from this class. If φ is a formula expressing some property of the elements of **C**, then **C**+φ is the class of all these structures from **C** that satisfy φ.[2]

The symbol 'ι' is interpreted as the standard description operator, which we use to build the expression '$(\iota\, x)\ \varphi(x)$' being the individual constant 'the only object x such that $\varphi(x)$'. To use it, first we have to know that there exists exactly one object x such that $\varphi(x)$, i.e., the formula $\varphi(x)$ must fulfill the following two conditions:

$$\exists_x\ \varphi(x)\,,$$

$$\forall_{x,y}\big(\varphi(x) \wedge \varphi(y) \Rightarrow x = y\big).$$

In such case we also write: $\exists^1_x\ \varphi(x)$.

[2] All the notions such as *formula*, *sentence*, *satisfy*, *true* are imprecise here, since we do not present any formal theory – we have no alphabet, nor language specified. However this imprecision is intended here, since we do not want to get bogged down in formal details but rather would like to focus on semantical or model theoretical aspect of the problem. Yet it should be noticed, that with some effort the notions addressed in this footnote could be precised and formal theory could be built, similarly as it was for example done in Part B of Pietruszczak (2000). Then we would have some elementary language with suitable definitions of formulas and sentences for which the usual notion of model and satisfaction could be given. Then by a class of structures we would mean the class of all models of a given set of axioms. We use the notion of *class*, since the collections of structures considered are too big to be just sets.

6.2 The Supremum Relation for Posets

Let $\langle M, \sqsubseteq\rangle$ be a poset, $S \subseteq M$ and $x \in M$. Let us recall a couple of basic definitions.

We say that x is an *upper bound* (resp. a *lower bound*) of S iff $\forall_{y \in S}\ y \sqsubseteq x$ (resp. $\forall_{y \in S}\ x \sqsubseteq y$). We say that x is a *supremum* of the set S (with respect to $\sqsubseteq$) iff x is the least upper bound of S; formally:

$$x \mathsf{\ sup\ } S \ :\Longleftrightarrow\ \forall_{z \in S}\ z \sqsubseteq x \wedge \forall_{y \in M}(\forall_{z \in S}\ z \sqsubseteq y \Longrightarrow x \sqsubseteq y)\,. \qquad (\mathtt{df}\ \mathsf{sup})$$

Using the operation P we can give an alternative version of the definition:

$$x \mathsf{\ sup\ } S \ \Longleftrightarrow\ S \subseteq \mathsf{P}(x) \wedge \forall_{y \in M}(S \subseteq \mathsf{P}(y) \Longrightarrow x \sqsubseteq y)\,. \qquad (\mathtt{df}'\ \mathsf{sup})$$

The immediate consequences of $(\mathtt{df}\ \mathsf{sup})$ are stated in the following lemma.

Lemma 1. (i) *Only by its definition the relation* sup *is monotonic, i.e.*:

$$\forall_{S_1, S_2 \in \mathcal{P}(M)} \forall_{x,y \in M}(S_1 \subseteq S_2 \wedge x \mathsf{\ sup\ } S_1 \wedge y \mathsf{\ sup\ } S_2 \Longrightarrow x \sqsubseteq y). \qquad (\mathrm{M}_{\mathsf{sup}})$$

(ii) *From* $(\mathtt{r}_{\sqsubseteq})$ *it follows that*:

$$\forall_{x \in M}\ x \mathsf{\ sup\ } \{x\}\,, \qquad (6.7)$$

$$\forall_{x \in M}\ x \mathsf{\ sup\ } \mathsf{P}(x)\,. \qquad (6.8)$$

(iii) *From* $(\mathtt{antis}_{\sqsubseteq})$ *it follows that if a set has a supremum, then it is unique, i.e.*:

$$\forall_{S \in \mathcal{P}(M)} \forall_{x,y \in M}(x \mathsf{\ sup\ } S \wedge y \mathsf{\ sup\ } S \Longrightarrow x = y). \qquad (\mathrm{U}_{\mathsf{sup}})$$

(iv) *From* $(\mathtt{r}_{\sqsubseteq})$ *and* $(\mathtt{antis}_{\sqsubseteq})$ *it follows that*:

$$\forall_{x,y \in M}(y \mathsf{\ sup\ } \{x\} \Longrightarrow x = y). \qquad (\mathrm{S}_{\mathsf{sup}})$$

6.3 Definition and Basic Properties of Mereological Sum

Let $\langle M, \sqsubseteq\rangle$ be a poset, $S \subseteq M$ and $x \in M$.

We say that x is a *mereological sum of all elements* of S iff x is an upper bound of S and every part of x overlaps some element of S; formally:

$$x \mathsf{\ sum\ } S \ :\Longleftrightarrow\ \forall_{z \in S}\ z \sqsubseteq x \wedge \forall_{y \in M}(y \sqsubseteq x \Longrightarrow \exists_{z \in S}\ z \bigcirc y). \qquad (\mathtt{df}\ \mathsf{sum})$$

Using the operations P and O we can give an alternative version of the definition:

$$x \;\mathsf{sum}\; S \iff S \subseteq \mathsf{P}(x) \subseteq \bigcup \mathsf{O}[S]\,, \tag{$\mathtt{df}'\,\mathsf{sum}$}$$

where $\mathsf{O}[S]$ is the image of the set X under the operation O, i.e.:

$$\mathsf{O}[S] := \{\mathsf{O}(z) \mid z \in S\} \quad \text{and} \quad \bigcup \mathsf{O}[S] = \{y \in M \mid \exists_{z \in S}\, z \circ y\}\,.$$

By the definition we obtain that:

$$\forall_{x \in M}(x \;\mathsf{sum}\; \varnothing \iff \mathsf{P}(x) = \varnothing). \tag{6.9}$$

Moreover, by $(\mathtt{r}_{\sqsubseteq})$, for any $x \in M$ we obtain that $\varnothing \neq \{x\} \subseteq \mathsf{P}(x) \subseteq \mathsf{O}(x) = \bigcup \mathsf{O}[\{x\}] \subseteq \bigcup \mathsf{O}[\mathsf{P}(x)]$ and if $\mathsf{PP}(x) \neq \varnothing$, then $\mathsf{P}(x) \subseteq \bigcup \mathsf{O}[\mathsf{PP}(x)]$. Hence:

Lemma 2. *The following conditions are consequences of* $(\mathtt{r}_{\sqsubseteq})$:

$$\neg \exists_{x \in M}\; x \;\mathsf{sum}\; \varnothing\,, \tag{6.10}$$

$$\forall_{x \in M}\; x \;\mathsf{sum}\; \{x\}\,, \tag{6.11}$$

$$\forall_{x \in M}\; x \;\mathsf{sum}\; \mathsf{P}(x)\,, \tag{6.12}$$

$$\forall_{x \in M}\big(\mathsf{PP}(x) \neq \varnothing \Longrightarrow x \;\mathsf{sum}\; \mathsf{PP}(x)\big). \tag{6.13}$$

Now notice that:

Lemma 3 (Pietruszczak 2000). *The following condition is true in* **QPOS**:

$$\forall_{x,y \in M}(\mathsf{P}(x) \subseteq \mathsf{O}(y) \Longrightarrow x \;\mathsf{sum}\; \mathsf{P}(x) \cap \mathsf{P}(y)).$$

Proof. Suppose that $\mathsf{P}(x) \subseteq \mathsf{O}(y)$. Since $\mathsf{P}(x) \cap \mathsf{P}(y) \subseteq \mathsf{P}(x)$, we only need to prove that $\mathsf{P}(x) \subseteq \bigcup \mathsf{O}[\mathsf{P}(x) \cap \mathsf{P}(y)]$. To see this, notice that from the assumption it follows that if $z \sqsubseteq x$, then $z \circ y$. So, by $(\mathtt{df}\,\circ)$, for some z_0 we have that $z_0 \sqsubseteq z$ and $z_0 \sqsubseteq y$. By $(\mathtt{t}_{\sqsubseteq})$, $z_0 \sqsubseteq x$, so $z_0 \in \mathsf{P}(x) \cap \mathsf{P}(y) \cap \mathsf{P}(z)$. The more so $z_0 \circ z$, by $(\mathtt{r}_{\sqsubseteq})$, as required.

Since we are interested in mutual dependencies between sum and supremum, let, for brevity and reference reasons, (†) denote the condition that every sum is a supremum:

$$\mathsf{sum} \subseteq \mathsf{sup}\,, \tag{†}$$

and (‡) the reversed condition:

$$\mathsf{sup} \subseteq \mathsf{sum}\,. \tag{‡}$$

Fig. 6.1 A poset in which supremum counterparts of (6.10), (6.13), ($\mathrm{S_{sum}}$), ($\mathrm{U_{sum}}$) and ($\mathrm{M_{sum}}$) are not true

We will be interested as well in, weaker from (‡), the following sentence:

$$\forall_{x\in M}\forall_{S\in\mathcal{P}_+(M)}(x\ \mathsf{sup}\ S \Longrightarrow x\ \mathsf{sum}\ S) \tag{$\ddagger_+$}$$

6.4 Basic Differences Between the Relations sup and sum

Firstly, notice that the supremum counterparts of (6.10) and (6.13), i.e. $\neg\,\exists_{x\in M}\ x\ \mathsf{sup}\ \varnothing$ and $\forall_{x\in M}(\mathsf{PP}(x) \neq \varnothing \Longrightarrow x\ \mathsf{sup}\ \mathsf{PP}(x))$, are not true in **POS**. Indeed, let us consider a two-element poset with $M = \{0, 1\}$ and $\sqsubseteq\, = \mathrm{id}_M \cup \{\langle 0, 1\rangle\}$ (see Fig. 6.1). We have that $0\ \mathsf{sup}\ \varnothing$ and $\mathsf{PP}(1) = \{0\}$, but $\neg\ 1\ \mathsf{sup}\ \{0\}$, since $0\ \mathsf{sup}\ \{0\}$.

Secondly, notice that the mereological sum counterparts of ($\mathrm{S_{sup}}$), ($\mathrm{U_{sup}}$) and ($\mathrm{M_{sup}}$), i.e.:

$$\forall_{x,y\in M}(y\ \mathsf{sum}\ \{x\} \Longrightarrow x = y), \tag{$\mathrm{S_{sum}}$}$$

$$\forall_{S\in\mathcal{P}(M)}\forall_{x,y\in M}(x\ \mathsf{sum}\ S \wedge y\ \mathsf{sum}\ S \Longrightarrow x = y), \tag{$\mathrm{U_{sum}}$}$$

$$\forall_{S_1,S_2\in\mathcal{P}(M)}\forall_{x,y\in M}(S_1 \subseteq S_2 \wedge x\ \mathsf{sum}\ S_1 \wedge y\ \mathsf{sum}\ S_2 \Longrightarrow x \sqsubseteq y) \tag{$\mathrm{M_{sum}}$}$$

are not true in **POS**. Indeed, in the poset from Fig. 6.1, respectively by (6.11) and (6.13), we have that $0\ \mathsf{sum}\ \{0\}$ and $1\ \mathsf{sum}\ \{0\}$, but $1 \not\sqsubseteq 0$.

6.5 Basic Properties of ($\mathrm{S_{sum}}$), ($\mathrm{U_{sum}}$) and ($\mathrm{M_{sum}}$)

The lemma below is obvious.

Lemma 4. (i) *From* ($\mathtt{r}_\sqsubseteq$) *and* ($\mathrm{U_{sum}}$) *we obtain* ($\mathrm{S_{sum}}$). *Consequently* **QPOS**+($\mathrm{U_{sum}}$) $\subseteq$ **QPOS**+($\mathrm{S_{sum}}$).

(ii) *From* ($\mathtt{antis}_\sqsubseteq$) *and* ($\mathrm{M_{sum}}$) *we obtain* ($\mathrm{U_{sum}}$). *Consequently* **POS**+($\mathrm{M_{sum}}$) $\subseteq$ **POS**+($\mathrm{U_{sum}}$) $\subseteq$ **POS**+($\mathrm{S_{sum}}$).

Notice that enriching the axioms for posets with ($\mathrm{S_{sum}}$) (resp. ($\mathrm{U_{sum}}$)) does not entail uniqueness (resp. monotonicity) of sum. Indeed, we have:

Fact 1. (i) ($\mathrm{U_{sum}}$) *is not true in* **POS**+($\mathrm{S_{sum}}$). *Hence* **POS**+($\mathrm{U_{sum}}$) $\subsetneq$ **POS**+($\mathrm{S_{sum}}$).

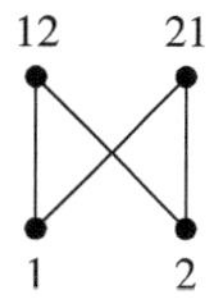

Fig. 6.2 A poset which satisfies $(\mathtt{S}_{\mathsf{sum}})$, but not $(\mathtt{U}_{\mathsf{sum}})$

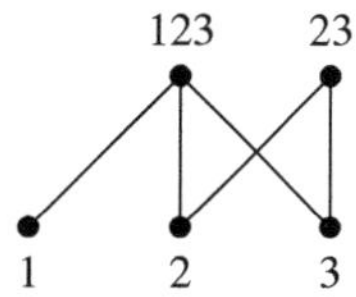

Fig. 6.3 A poset which satisfies $(\mathtt{U}_{\mathsf{sum}})$, but not $(\mathtt{M}_{\mathsf{sum}})$

(ii) $(\mathtt{M}_{\mathsf{sum}})$ *is not true in* **POS**+$(\mathtt{U}_{\mathsf{sum}})$. *Consequently* **POS**+$(\mathtt{M}_{\mathsf{sum}}) \subsetneq$ **POS**+$(\mathtt{U}_{\mathsf{sum}}) \subsetneq$ **POS**+$(\mathtt{S}_{\mathsf{sum}})$.

Proof. Ad (i): The poset from Fig. 6.2 with $M = \{1, 2, 12, 21\}$ and $\sqsubseteq \; = \mathrm{id}_M \cup \{\langle 1, 12\rangle, \langle 1, 21\rangle, \langle 2, 12\rangle, \langle 2, 21\rangle\}$ belongs to **POS**+$(\mathtt{S}_{\mathsf{sum}})$ and it shows that a set can have more than one mereological sum. Indeed, $(\mathtt{S}_{\mathsf{sum}})$ is satisfied in this poset, but $12 \;\mathsf{sum}\; \{1, 2\}$ and $21 \;\mathsf{sum}\; \{1, 2\}$.

Ad (ii): This time we consider the poset with $M = \{1, 2, 3, 12, 123\}$ and $\sqsubseteq \; = \mathrm{id}_M \cup \{\langle 1, 123\rangle, \langle 2, 123\rangle, \langle 3, 123\rangle, \langle 2, 23\rangle, \langle 3, 23\rangle\}$ (see Fig. 6.3). It satisfies $(\mathtt{U}_{\mathsf{sum}})$. On the other hand, we have that $23 \;\mathsf{sum}\; \{2, 3\}$ and $123 \;\mathsf{sum}\; \{1, 2, 3\}$, but $23 \not\sqsubseteq 123$.

Notice that:

Lemma 5. $(\mathtt{M}_{\mathsf{sum}})$ *entails* $(\mathtt{r}_{\sqsubseteq})$.

Proof. By $(\mathtt{M}_{\mathsf{sum}})$ we have (a): $\forall_{x \in M}(x \not\sqsubseteq x \Longrightarrow \neg \exists_{S \in \mathcal{P}(M)}\, x \;\mathsf{sum}\; S)$. By (a) we have (b): $\forall_{x \in M}(x \not\sqsubseteq x \Longrightarrow \exists_{y \in \mathsf{PP}(x)}\, y \not\sqsubseteq y)$. Indeed, if for any $y \in \mathsf{PP}(x)$ we have that $y \sqsubseteq y$, then $x \;\mathsf{sum}\; \mathsf{PP}(x)$. So $x \sqsubseteq x$ by (a).

By (b) we obtain (c): $\forall_{x \in M}(x \not\sqsubseteq x \Longrightarrow x \;\mathsf{sum}\; \mathsf{PP}(x))$. Indeed, let $x \not\sqsubseteq x$ and $y \sqsubseteq x$. Then $y \in \mathsf{PP}(x)$. Moreover, if $y \sqsubseteq y$, then $y \bigcirc y$. If $y \not\sqsubseteq y$, then by (b) there is $u \in \mathsf{PP}(y)$; so also $y \bigcirc y$. Hence in both cases there is $z \in \mathsf{PP}(x)$ such that $z \bigcirc y$.

By (a) and (c) we have that $(\mathtt{r}_{\sqsubseteq})$ holds.

We now point to some relationship between $(\mathtt{S}_{\mathsf{sum}})$ and the so-called *Weak Supplementation Principle*, used by Simons (1987):

$$\forall_{x,y \in M}\big(x \sqsubset y \Longrightarrow \exists_{z \in M}(z \sqsubset y \wedge z \wr x)\big), \tag{WSP}$$

which will let us obtain a connection between the relations sup and sum (see Theorem 2).

Theorem 1. (i) *From* (WSP) *we obtain* $(\mathtt{S}_{\mathsf{sum}})$.
(ii) *From* $(\mathtt{r}_{\sqsubseteq})$ *and* $(\mathtt{S}_{\mathsf{sum}})$ *we obtain* (WSP).
(iii) *From* $(\mathtt{r}_{\sqsubseteq})$ *and* $(\mathtt{U}_{\mathsf{sum}})$ *we obtain* (WSP).
In consequence, **QPOS**+(WSP) $=$ **QPOS**+$(\mathtt{S}_{\mathsf{sum}})$.

Proof. *Ad* (i): Suppose that y sum $\{x\}$ and $x \neq y$. Then $x \sqsubset y$. So, by (WSP), for some z we have: $z \sqsubset y$ and $z \wr x$. So we have a contradiction, since from $z \sqsubset y$ and y sum $\{x\}$ follows that $z \circ x$.

Ad (ii): Suppose that $x \sqsubset y$ and $\mathsf{PP}(y) \subseteq \mathsf{O}(x)$. Since by ($\mathtt{r}_{\sqsubseteq}$) we have that $y \circ x$, then y sum $\{x\}$. Thus $y = x$ by ($\mathtt{S}_{\mathsf{sum}}$), which is a contradiction.

(Hence y sum $\{x\}$ and, by ($\mathtt{r}_{\sqsubseteq}$), also $y \circ x$. So, by ($\mathtt{S}_{\mathsf{sum}}$), we obtain a contradiction: $x = y$.)

Ad (iii): We use (1) and Lemma 4(i).

Corollary 1. *The sentence* (WSP) *is true in the class* **POS**+($\mathtt{M}_{\mathsf{sum}}$).

Proof. By Lemma 4(ii), from ($\mathtt{antis}_{\sqsubseteq}$) and ($\mathtt{M}_{\mathsf{sum}}$) follows ($\mathtt{U}_{\mathsf{sum}}$). Moreover, by Theorem 1(iii), ($\mathtt{r}_{\sqsubseteq}$) and ($\mathtt{U}_{\mathsf{sum}}$) entail (WSP).

Now we prove that in every structure from **QPOS**+($\mathtt{S}_{\mathsf{sum}}$), if both sum and supremum exists, then they are equal.

Theorem 2 (Pietruszczak 2000). ($\mathtt{t}_{\sqsubseteq}$) *and* (WSP) *entail the following sentence*:

$$\forall_{S \in \mathcal{P}(M)} \forall_{x,y \in M} (x \text{ sup } S \wedge y \text{ sum } S \Longrightarrow x = y). \tag{6.14}$$

Proof. Let x sup S and y sum S. Then $S \subseteq \mathsf{P}(y)$, so $x \sqsubseteq y$. Suppose that $x \neq y$. Then $x \sqsubset y$. Hence, by (WSP), for some $z \in M$ we have that $z \sqsubset y$ and $z \wr x$. Hence, by (df sum), there are $u \in S$ and $v \in M$ such that $v \sqsubseteq u$ and $v \sqsubseteq z$. By the assumption, $u \sqsubseteq x$. Hence, by ($\mathtt{t}_{\sqsubseteq}$), also $v \sqsubseteq x$. So we have a contradiction: $z \circ x$.

Now we will prove an important lemma which will be useful a little bit further. Let us start with the following definition.

An object x is called *the zero element* of a poset $\langle M, \sqsubseteq \rangle$ iff every object from M is part of x, i.e. $\forall_{y \in M}\, x \sqsubseteq y$. The uniqueness of the zero element follows from antisymmetry of $\sqsubseteq$. Moreover, we immediately have that for any poset, if it has zero, then all objects overlap with each other:

$$\forall_{x,y \in M} (x \text{ is a zero} \wedge y \text{ is a zero} \Longrightarrow x = y), \tag{6.15}$$

$$\exists_{x \in M}\, x \text{ is a zero} \Longrightarrow \forall_{x,y \in M}\, x \circ y\,. \tag{6.16}$$

Lemma 6 (Pietruszczak 2000).

(i) *From* (WSP) *we obtain the following implication*:

$$|M| > 1 \Longrightarrow \exists_{x,y \in M}\, x \wr y\,.$$

(ii) *From* ($\mathtt{r}_{\sqsubseteq}$) *we obtain the following implication*:

$$\exists_{x,y \in M}\, x \wr y \Longrightarrow |M| > 1\,.$$

(iii) *From* (WSP) *and* ($\mathtt{r}_{\sqsubseteq}$) *we obtain the following equivalence*:

$$\exists_{x \in M} \forall_{y \in M}\ x \sqsubseteq y \iff |M| = 1\,.$$

Proof. *Ad* (i): Let $x_1, x_2 \in M$ be different: $x_1 \neq x_2$. Suppose that $\forall_{x,y \in M}\ x \bigcirc y$. So there is $u \in M$ such that $u \sqsubseteq x_1$ and $u \sqsubseteq x_2$. Moreover, either $u \sqsubset x_1$ or $u \sqsubset x_2$. In both cases, by (WSP) we obtain a contradiction: there is $z \in M$ such that $z \wr u$.

Ad (ii): By ($\mathtt{r}_{\sqsubseteq}$) we have ($\mathtt{irr}_{\wr}$); so if $x \wr y$, then $x \neq y$.

Ad (iii): "$\Rightarrow$" If $\exists_{x \in M} \forall_{y \in M}\ x \sqsubseteq y$ then $\forall_{x,y \in M}\ x \bigcirc y$, so we use (i). "$\Leftarrow$" Immediate, from ($\mathtt{r}_{\sqsubseteq}$).

By Corollary 1 we obtain:

Corollary 2. *The sentences from Lemma 6 are all true in the class* **POS**+($\mathsf{M_{sum}}$).

6.6 The Inclusions (†) and (‡) in the Class POS+($\mathsf{U_{sum}}$)

We show that neither (†) nor (‡) follows from the axioms for **POS** plus ($\mathsf{U_{sum}}$). In consequence none of them follows from the axioms for **POS** plus ($\mathsf{S_{sum}}$); see Lemma 4.

Fact 2. *None of the sentences* (†) *and* ($‡_{+}$) *is true in* **POS**+($\mathsf{U_{sum}}$).

Proof. In the poset from Fig. 6.3 we have: 23 $\mathsf{sum}\ \{2,3\}$, but $\neg$ 23 $\mathsf{sup}\ \{2,3\}$. So $\mathsf{sum} \not\subseteq \mathsf{sup}$. In the same poset we have: 123 $\mathsf{sup}\ \{1,2\}$, but $\neg$ 123 $\mathsf{sum}\ \{1,2\}$. Thus $\mathsf{sup} \not\subseteq \mathsf{sum}$ as well.

6.7 The Inclusions (†) and (‡) in the Class POS+($\mathsf{M_{sum}}$)

Firstly, we show that (‡) does not follow from the axioms for **POS** plus ($\mathsf{M_{sum}}$).[3]

Fact 3. *The sentence* ($‡_{+}$) *is not true in* **POS**+($\mathsf{M_{sum}}$).

Proof. We take the poset with $M = \{1, 2, 3, 123\}$ and $\sqsubseteq\ =\ \mathrm{id}_M \cup \{\langle 1, 123\rangle, \langle 2, 123\rangle, \langle 3, 123\rangle\}$ (see Fig. 6.4). Obviously, this poset satisfies ($\mathsf{M_{sum}}$) but not ($‡_{+}$), since e.g. 123 $\mathsf{sup}\ \{1,2\}$ while $\neg$ 123 $\mathsf{sum}\ \{1,2\}$.

Secondly, we can show that (†) is true in the class **POS**+($\mathsf{M_{sum}}$). Moreover we will demonstrate that for quasi-partially ordered sets the inclusion (†) is equivalent to the sentence ($\mathsf{M_{sum}}$). But earlier we need to prove some interesting facts.

[3] This, by Lemma 4, entails the case for (‡) in Fact 2.

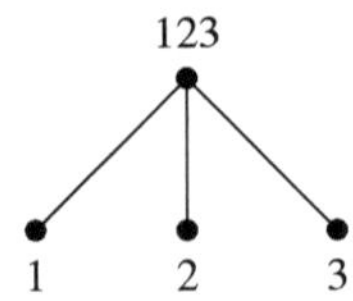

Fig. 6.4 A poset which satisfies $(\mathsf{M_{sum}})$, but not (‡)

Firstly, notice that to examine properties of the relation sum we will make use of the following condition which is related to $(\mathtt{df}'\,\mathsf{sum})$ and $(\mathsf{M_{sum}})$:

$$\forall_{S\in\mathcal{P}(M)}\forall_{x,y\in M}\big(\mathsf{P}(x)\subseteq\bigcup\mathsf{O}[S]\wedge S\subseteq\mathsf{P}(y)\Longrightarrow x\sqsubseteq y\big). \qquad (\mathsf{M'_{sum}})$$

Lemma 7 (Pietruszczak 2000). $(\mathsf{M'_{sum}})$ *entails* $(\mathsf{M_{sum}})$.

Proof. If $S_1\subseteq S_2$, $x\ \mathsf{sum}\ S_1$ and $y\ \mathsf{sum}\ S_2$, then $\mathsf{P}(x)\subseteq\bigcup\mathsf{O}[S_1]\subseteq\bigcup\mathsf{O}[S_2]$ and $S_2\subseteq\mathsf{P}(y)$. So $x\sqsubseteq y$, by $(\mathsf{M'_{sum}})$.

Lemma 8. *From* $(\mathsf{M_{sum}})$ *and* $(\mathtt{t}_\sqsubseteq)$ *we obtain* $(\mathsf{M'_{sum}})$.

Proof. By Lemma 5 we have $(\mathtt{r}_\sqsubseteq)$. If $\mathsf{P}(x)\subseteq\bigcup\mathsf{O}[S]$ and $S\subseteq\mathsf{P}(y)$, then $\mathsf{P}(x)\subseteq\bigcup_{z\in\mathsf{P}(y)}\mathsf{O}(z)$. Notice that by $(\mathtt{t}_\sqsubseteq)$ we have (6.5), so we obtain that $\bigcup_{z\in\mathsf{P}(y)}\mathsf{O}(z)\subseteq\mathsf{O}(y)$. Thus, $\mathsf{P}(x)\subseteq\mathsf{O}(y)$. Hence, by Lemma 3, $x\ \mathsf{sum}\ \mathsf{P}(x)\cap\mathsf{P}(y)$. Moreover, $y\ \mathsf{sum}\ \mathsf{P}(y)$, by $(\mathtt{r}_\sqsubseteq)$. Thus $x\sqsubseteq y$, by $(\mathsf{M_{sum}})$.

From Lemmas 7 and 8 we obtain:

Theorem 3 (Pietruszczak 2000). *The following sentence is true in* **QPOS**:

$$(\mathsf{M_{sum}})\iff(\mathsf{M'_{sum}}).$$

Thus, **QPOS**$+(\mathsf{M_{sum}})=$ **QPOS**$+(\mathsf{M'_{sum}})$.

Now we prove that:

Theorem 4. *The following sentence is true in* **QPOS**:

$$(\mathsf{M_{sum}})\iff(\dagger).$$

Thus, **QPOS**$+(\mathsf{M_{sum}})=$ **QPOS**$+(\dagger)$.

Proof. "$\Rightarrow$" Assume that $x\ \mathsf{sum}\ S$, i.e., $S\subseteq\mathsf{P}(x)\subseteq\bigcup\mathsf{O}[S]$. This gives us immediately the first conjunct of $(\mathtt{df}'\,\mathsf{sup})$. For the second one assume that $y\in M$ is such that $S\subseteq\mathsf{P}(y)$. Then $x\sqsubseteq y$, by $(\mathsf{M'_{sum}})$ and Theorem 3. So $x\ \mathsf{sup}\ S$.

"$\Leftarrow$" If $S_1\subseteq S_2$, $x\ \mathsf{sum}\ S_1$ and $y\ \mathsf{sum}\ S_2$, then by (†) it is the case that $x\ \mathsf{sup}\ S_1$ and $y\ \mathsf{sum}\ S_2$; so $x\sqsubseteq y$, by $(\mathsf{M_{sup}})$.

Remark 1. For structures from **POS**$+(\mathsf{M_{sum}})$ (= **POS**$+(\dagger)$, by Theorem 4) we have a simple proof of the sentence (6.14), i.e., if for a set has both sum and supremum, then they are equal. Indeed, if x sup S and y sum S, then also y sup S. Thus, since we can use (antis$_\sqsubseteq$), we obtain $x = y$, by (U$_{\mathsf{sup}}$).

6.8 Separative Partially Order Sets

Any quasi-poset which satisfies the following sentence:

$$\forall_{x,y\in M}\big(x \not\sqsubseteq y \Longrightarrow \exists_{z\in M}(z \sqsubseteq x \wedge z \wr y)\big), \qquad \text{(SSP)}$$

will be called *separative*. Let **SPOS** be the class of all separative posets.

In Simons (1987) the sentence (SSP) is called *Strong Supplementation Principle*. According to (SSP) if one object is not a part of another, than they can be distinguished by some object from the domain, but not only in the sense that this object is part of one but not the other element of the domain – it is exterior to the latter.

The sentence (SSP) can as well be expressed in the following, definitionally equivalent, way:

$$\forall_{x,y\in M}\big(\mathsf{P}(x) \subseteq \mathsf{O}(y) \Longrightarrow x \sqsubseteq y\big). \qquad \text{(SSP}^\circ\text{)}$$

Hence, by (6.4)–(6.6), we also obtain:

Fact 4. *The following sentence is true in all separative quasi-posets*:

$$\forall_{x,y\in M}\big(x \sqsubseteq y \iff \mathsf{O}(x) \subseteq \mathsf{O}(y)\big).$$

Now we will show that **QPOS**+(M$_{\mathsf{sum}}$) $=$ **QPOS**+(SSP). In the proof of the equality in question we will use the equality **QPOS**+(M$_{\mathsf{sum}}$) $=$ **QPOS**+(M$'_{\mathsf{sum}}$) from Theorem 3 together with the facts below.

Lemma 9 (Pietruszczak 2000, 2005). *From* (SSP) *and* (t$_\sqsubseteq$) *we obtain* (M$'_{\mathsf{sum}}$) *and* (M$_{\mathsf{sum}}$).[4]

Proof. For (M$'_{\mathsf{sum}}$): If $\mathsf{P}(x) \subseteq \bigcup \mathsf{O}[S]$ and $S \subseteq \mathsf{P}(y)$, then $\mathsf{P}(x) \subseteq \bigcup_{z\in\mathsf{P}(y)} \mathsf{O}(z)$. By (6.5) we obtain that $\bigcup_{z\in\mathsf{P}(y)} \mathsf{O}(z) \subseteq \mathsf{O}(y)$. Therefore $\mathsf{P}(x) \subseteq \mathsf{O}(y)$. So $x \sqsubseteq y$, by (SSP°). For (M$_{\mathsf{sum}}$): Use Lemma 7.

Lemma 10. (i) (M$'_{\mathsf{sum}}$) *entails* (SSP).
(ii) (M$_{\mathsf{sum}}$) *and* (t$_\sqsubseteq$) *entails* (SSP).

[4]Hence, by Lemma 5, we obtain that (SSP) and (t$_\sqsubseteq$) entail (r$_\sqsubseteq$). This fact was proven in Pietruszczak (2000, 2005). So separative posets can be defined by means of these three conditions: (t$_\sqsubseteq$), (antis$_\sqsubseteq$) and (SSP).

Proof. (i) Notice that, by Lemmas 5 and 7, we have ($\mathtt{r}_{\sqsubseteq}$). If $\mathsf{P}(x) \subseteq \mathsf{O}(y)$, then $\mathsf{P}(x) \subseteq \bigcup \mathsf{O}[\{y\}]$ and $\{y\} \subseteq \mathsf{P}(y)$ by ($\mathtt{r}_{\sqsubseteq}$). So $x \sqsubseteq y$, by ($\mathtt{M}'_{\mathsf{sum}}$).
(ii) By (i) and Lemma 8.

Thus, from the above lemmas and Theorem 4 we have:

Theorem 5. *The following sentence is true in* **QPOS**:

$$(\mathtt{SSP}) \iff (\mathtt{M}_{\mathsf{sum}}).$$

Thus, **QPOS**+($\mathtt{SSP}$) = **QPOS**+($\mathtt{M}'_{\mathsf{sum}}$) = **QPOS**+($\mathtt{M}_{\mathsf{sum}}$) = **QPOS**+(†).

Finally, we obtain:

Fact 5. (i) *The sentences* ($\mathtt{r}_{\sqsubseteq}$) *and* ($\mathtt{antis}_{\sqsubseteq}$) *entail the implication* ($\mathtt{SSP}$) $\Rightarrow$ ($\mathtt{WSP}$). *Consequently,* **SPOS** $\subseteq$ **POS**+($\mathtt{WSP}$).
(ii) **SPOS** $\subsetneq$ **POS**+($\mathtt{WSP}$).

Proof. (i): Let $x \sqsubset y$, i.e., $x \sqsubseteq y$ and $x \neq y$. Then $y \not\sqsubseteq x$, by ($\mathtt{antis}_{\sqsubseteq}$). Hence, by ($\mathtt{SSP}$), there is z such that $z \sqsubseteq y$ and $z \wr x$. We have that $z \neq y$, since $y \circ x$, by ($\mathtt{r}_{\sqsubseteq}$). So $z \sqsubset y$.
(ii): The poset from Fig. 6.2 satisfies ($\mathtt{WSP}$). It is the case that $12 \not\sqsubseteq 21$, but there is no z such that $z \sqsubseteq 12$ and $z \wr 21$. So ($\mathtt{SSP}$) is not true in the structure considered.

6.9 Mereological Structures

We now take into account the following axiom of existence of mereological sum:

$$\forall_{S \in \mathcal{P}_+(M)} \exists_{x \in M}\ x \ \mathsf{sum}\ S\,. \qquad (\exists\mathsf{sum})$$

Any separative poset which satisfies ($\exists\mathsf{sum}$) is called a (*classical*) *mereological structure*.[5] Let **MS** and **MS**$^+$ be respectively the class of all mereological structures and the class all non-degenerate mereological structures. Of course, **MS**$^+ \subsetneq$ **MS**.

By Lemma 4 the formula ($\mathtt{U}_{\mathsf{sum}}$) is true in **MS**. So the following sentence is also true in **MS**:

$$\forall_{S \in \mathcal{P}_+(M)} \exists^1_{x \in M}\ x \ \mathsf{sum}\ S\,. \qquad (\exists^1\mathsf{sum})$$

[5]In Tarski (1956) we find an equivalent axiomatization of mereological structures consisted of the following sentences: ($\mathtt{t}_{\sqsubseteq}$) and given below ($\exists^1\mathsf{sum}$) (which is equivalent to: ($\mathtt{t}_{\sqsubseteq}$), ($\mathtt{U}_{\mathsf{sum}}$) and ($\exists\mathsf{sum}$)). Various equivalent axiomatizations of mereological structures are presented e.g. in Pietruszczak (2000, 2005).

Hence in any mereological structure $\langle M, \sqsubseteq \rangle$ there is exactly one object x such that $x \ \mathsf{sum}\ M$. By (df sum), x is *the unity* in the sense that: $\forall_{y \in M}\ y \sqsubseteq x$. So in any mereological structure we put:

$$\mathbf{1} := (\iota x)\ x \ \mathsf{sum}\ M\,, \tag{df 1}$$

and by ($\mathtt{antis}_{\sqsubseteq}$) we have:

$$\mathbf{1} = (\iota x)\ \forall_{y \in M}\ y \sqsubseteq x\,. \tag{6.17}$$

Theorem 6 (Pietruszczak 2000, 2005). *The following sentences are true in* **MS**[6]*:*

$$\forall_{x \in M} \forall_{S \in \mathcal{P}(M)} (x \ \mathsf{sum}\ S \iff S \neq \varnothing \wedge x \ \mathsf{sup}\ S), \tag{sum-sup}$$

$$|M| > 1 \iff \mathsf{sup} \subseteq \mathsf{sum}\,. \tag{$\star$}$$

Proof. Ad (sum-sup): By Theorems 4 and 5 we have (†). So if $x \ \mathsf{sum}\ S$, then $x \ \mathsf{sup}\ S$ and $S \neq \varnothing$, by ($\mathtt{r}_{\sqsubseteq}$). Let now $S \neq \varnothing$ and $x \ \mathsf{sup}\ S$. Then, by ($\exists$sum), there is y such that $y \ \mathsf{sum}\ S$. So $x = y$, by (6.14); see Remark 1. Therefore $x \ \mathsf{sum}\ S$.

Ad ($\star$): Firstly, let $|M| > 1$ and $x \ \mathsf{sup}\ S$. Then $S \neq \varnothing$, since by Corollary 2, in M there is no zero element. Hence, $x \ \mathsf{sum}\ S$, by (sum-sup). Secondly, assume that M has only one element x. Then $x \ \mathsf{sup}\ \varnothing$. But $\neg x \ \mathsf{sum}\ \varnothing$, by ($\mathtt{r}_{\sqsubseteq}$). So $\mathsf{sup} \nsubseteq \mathsf{sum}$.

By the above theorem we get:

Corollary 3. *The equality* $\mathsf{sum} = \mathsf{sup}$ *holds in* $\mathbf{MS}^{+}$.

6.10 Weakening and Replacing the Sum Existence Axiom

Consider the following weakened versions of ($\exists$sum):

$$\forall_{S \in \mathcal{P}_{+}(M)} (\exists_{u \in M}\ S \subseteq \mathsf{P}(u) \Longrightarrow \exists_{x \in M}\ x \ \mathsf{sum}\ S), \tag{$\mathtt{W}_1\exists$sum}$$

$$\forall_{S \in \mathcal{P}_{+}(M)} \big(\forall_{y,z \in S} \exists_{u \in M}\ \{y, z\} \subseteq \mathsf{P}(u) \Longrightarrow \exists_{x \in M}\ x \ \mathsf{sum}\ S\big). \tag{$\mathtt{W}_2\exists$sum}$$

The first one says that every non-empty set which is bounded from above has its mereological sum. The second (stronger than the first one) says that if every subset $\{y, z\}$ of S is bounded in M, then S has its sum.[7]

[6] The first one to prove (sum-sup), in original language of Leśniewski's mereology, was A. Tarski (see Leśniewski (1930), p. 87).

[7] This not exactly upward directedness of S. A subset S in a poset $\langle M, \sqsubseteq \rangle$ is *upward directed* iff $\forall_{y,z \in S} \exists_{u \in S} (y \sqsubseteq u \wedge z \sqsubseteq u)$, while we require the existence of upper bound in M. Consequently,

We have the following fact.

Fact 6. *The sentence* (sum-sup) *is true in both* **QPOS**+(SSP)+($W_1\exists$sum) *and in* **QPOS**+(SSP)+($W_2\exists$sum).

Proof. By Theorem 4 we have (†); so if x sum S, then x sup S and $S \neq \varnothing$, by ($r_\sqsubseteq$). Moreover, let $S \neq \varnothing$ and x sup S. Then $S \subseteq \mathsf{P}(x)$. Hence, there is y such that y sum S, by ($W_1\exists$sum) or ($W_2\exists$sum). So $x = y$, by (6.14); see Remark 1. Therefore x sum S.

The above fact shows that we can weaken the sum existence axiom to the forms presented yet keep the equality between sum and supremum. Of course, this does not solve the problem of characterization of structures from classes **SPOS**+($W_1\exists$sum) and **SPOS**+($W_2\exists$sum). In our opinion further study concerning their properties seems to be interesting from the following, a bit philosophical, point of view. The unrestricted sum axiom ($\exists$sum) is often objected as counterintuitive in case of some so called ontological interpretations of mereology.[8] It is argued for example, that the Moon and a cup of coffee standing in front of you are parts of the world, yet it is hard to find anything that could be their sum. Axioms ($W_1\exists$sum) and ($W_2\exists$sum) could be interpreted (at least in a way) as saying that only these objects which have something in common (in the sense that they are both parts of something bigger) have their mereological sums.

No we consider the following principle, intimately connected with those analyzed by us in previous sections:

$$\forall_{x,y \in M}\big(x \not\sqsubseteq y \Longrightarrow \exists_z(z \sqsubseteq x \wedge z \wr y \wedge \forall_u(u \sqsubseteq x \wedge u \wr y \Longrightarrow u \sqsubseteq z))\big) \quad \text{(SSP+)}$$

which we will call *the super strong supplementation principle* or "*SSP plus*". What it intuitively says is that if x is not part of y, then we can not only find some z being part of x which is external to y, but we can also find an element of the structure in question satisfying the aforementioned property and being the largest such object in the structure. The sentence (SSP+) is assumed as an axiom in Grzegorczyk's system of mereology from Grzegorczyk (1955).

Theorem 7. *The sentence* (SSP+) *is true in the class* **QPOS**+(SSP)+($W_1\exists$sum).[9]

both axioms are equivalent in posets with the unity, since antecedents of ($W_1\exists$sum) and ($W_2\exists$sum) are both true in the presence of one.

[8]In our opinion these objections are not properly addressed and they result from a twisted perspective, as we can see it. Nothing is wrong with ($\exists$sum) and no one should demand the world to behave according to it in all its aspects. Yet there are such applications of mereology in which it is very useful, as in building point-free systems of geometry for example, where elements of the domain are treated as regions of space. For details see Tarski (1956), Gruszczyński and Pietruszczak (2008, 2009, 2010), and Grzegorczyk (1960).

[9]Of course by this theorem (SSP+) is true in the class **MS** as well.

Proof. If $x \not\sqsubseteq y$, then by (SSP) the set $S_0 := \{z \in M \mid z \sqsubseteq x \wedge z \wr y\}$ is not empty and $S_0 \subseteq \mathsf{P}(x)$. Hence, by ($\mathtt{W}_1\exists\mathsf{sum}$), for some z_0 we have that z_0 sum S_0.

Firstly, notice that $z_0 \sqsubseteq x$. Indeed, suppose towards contradiction that $z_0 \not\sqsubseteq x$. Then, by (SSP), there is u such that $u \sqsubseteq z_0$ and $u \wr x$. Hence, by (df sum), there are $v \in S_0$ and $w \in M$ such that $w \sqsubseteq v \sqsubseteq x$ and $w \sqsubseteq u$. So $u \bigcirc x$, by ($\mathtt{t}_\sqsubseteq$); a contradiction.

Secondly, notice that $z_0 \wr y$. Indeed, suppose towards contradiction that $z_0 \bigcirc y$. Then there is u such that $u \sqsubseteq z_0$ and $u \sqsubseteq y$. Hence, by (df sum), there are $v \in S_0$ and $w \in M$ such that $w \sqsubseteq v \wr y$ and $w \sqsubseteq u \sqsubseteq y$. So $w \wr y$ and $w \sqsubseteq y$, by ($\mathtt{t}_\sqsubseteq$). Moreover, $w \bigcirc y$, by ($\mathtt{r}_\sqsubseteq$); a contradiction again.

Thirdly, if $u \sqsubseteq x$ and $u \wr y$, then $u \in S_0$. So $u \sqsubseteq z_0$, by (df sum).

There is one more issue that can be addressed with respect to axioms and mutual relationship between sum and supremum – *in what effect results replacing* ($\exists\mathsf{sum}$) *with the following version of completeness*:

$$\forall_{S \in \mathcal{P}_+(M)} \exists_{x \in M}\, x \ \mathsf{sup}\ S\,. \tag{$\exists\mathsf{sup}$}$$

Algebraically speaking we consider the class **SPOS**+($\exists\mathsf{sup}$) elements of which are separative posets being complete join-semilattices. The following fact answers the question.

Fact 7 (Pietruszczak 2005). *The sentence* ($\ddagger_+$) *is not true in* **SPOS**+($\exists\mathsf{sup}$). *Therefore the counterpart of Theorem 6 does not hold for* **SPOS**+($\exists\mathsf{sup}$).

Proof. The structure from Fig. 6.4 belongs to **SPOS**+($\exists\mathsf{sup}$) and does not satisfy the sentence in question, since e.g. 123 sup $\{1,2\}$, but $\neg$ 123 sum $\{1,2\}$.

6.11 Mereological Posets

Any structure from the class **SPOS**+($\ddagger_+$) will be called a *mereological poset* (*mereoposet* for short). Let **MPOS** be the class of all mereoposets. By Fact 3 and Theorem 5 we have that **MPOS** $\subsetneq$ **SPOS**. By Theorem 5, the sentences ($\mathtt{M}_{\mathsf{sum}}$), ($\mathtt{M}'_{\mathsf{sum}}$) and ($\dagger$) are true in **MPOS**. Moreover, by Fact 5 (or Corollary 1) the sentence (WSP) is true in **MPOS** as well.

We will also be interested in the class **MPOS**$^+$:= **SPOS**+($\ddagger$). By the definition, **MPOS**$^+$ $\subseteq$ **MPOS**. Below we show that **MPOS**$^+$ is the class of all non-degenerate mereoposets. So **MPOS**$^+$ $\subsetneq$ **MPOS**, which is a result of the following lemma.

Lemma 11. *No poset from* **POS**+($\ddagger$) *has a zero element. Consequently, it is a non-degenerate structure.*

Proof. If a poset $\langle M, \sqsubseteq\rangle$ has the zero element $\mathbf{0}$, then $\mathbf{0}$ sup $\varnothing$. But $\neg\exists_{x \in M}\, x$ sum $\varnothing$, by ($\mathtt{r}_\sqsubseteq$). So $\mathsf{sup} \not\subseteq \mathsf{sum}$.

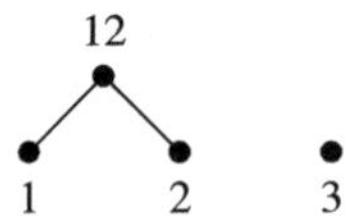

Fig. 6.5 The non-degenerate mereoposet without unity

From the above lemma, Corollary 2 and Theorem 5 we obtain:

Corollary 4. *No poset from* $\mathbf{MPOS}^+$ *has a zero element. Consequently, every structure from* $\mathbf{MPOS}^+$ *is non-degenerate and has at least two elements which are external two each other.*

Remark 2. Non-existence of zero element in the class $\mathbf{MPOS}^+$ and both supplementation principles are considered to be distinctive and fundamental features of structures that are examined within the field known as *mereology*.

Let $\langle M, \sqsubseteq \rangle$ be a mereoposet. We say that x is *isolated* in this structure iff x is a proper part of no element of M and no element of M is proper part of x. Let **is** be the set of all isolated elements, i.e.:

$$\mathbf{is} := \{x \in M \mid \neg \exists_{y \in M} (y \sqsubset x \vee x \sqsubset y\} . \qquad (\mathtt{df}\,\mathbf{is})$$

The simplest example of non-degenerate mereoposet is a pair $\langle M, \sqsubseteq \rangle$ with $M := \{1, 2\}$ and $\sqsubseteq := \mathrm{id}_M$. So this is a structure that consists of two isolated objects. Less trivial example is a four element structure $\langle \{1, 2, 12, 3\}, \sqsubseteq \rangle$, where $\sqsubseteq := \mathrm{id} \cup \{\langle 1, 12\rangle, \langle 2, 12\rangle\}$ and 3 is isolated (see Fig. 6.5).

The above model shows that the existence of unity is not a consequence of the axioms for mereological posets. However, neither is its non-existence, since every non-degenerate mereological structure is a mereoposet. So we have the following corollary.

Corollary 5. *Existence of unity is independent from axioms for mereoposets.*

Since the equality $\mathsf{sum} = \mathsf{sup}$ is true in $\mathbf{MS}^+$ (see Corollary 3) and by the structure from Fig. 6.5, we obtain:

Corollary 6. *Every non-degenerate mereological structure is a mereoposet, but not every mereposet is a mereological structure. So* $\mathbf{MS}^+ \subsetneq \mathbf{MPOS}^+$.

On the other hand we have the following interesting result about mereoposets.

Fact 8. *The sentence* (SSP+) *is not true in the class* $\mathbf{MPOS}^+$.

Proof. We consider the following non-degenerate mereoposet $\langle M, \sqsubseteq \rangle$ with $M := \{-1, 1, 2, 3, -11, -12, 13, 23, -112, 123\}$ and for $x, y \in M$: $x \sqsubseteq y$ iff $\#x$ is part of $\#y$, where $\#x$ is the numeral of x (see Fig 6.6).

We have that $123 \not\sqsubseteq 3$, but only 1 and 2 are such that $1 \sqsubseteq 123$ and $1 \wr 3$, $2 \sqsubseteq 123$ and $2 \wr 3$. Notice that the set $\{1, 2\}$ does not have supremum, since $\{1, 2\} \subseteq \mathsf{P}(-112)$ and $\{1, 2\} \subseteq \mathsf{P}(123)$.

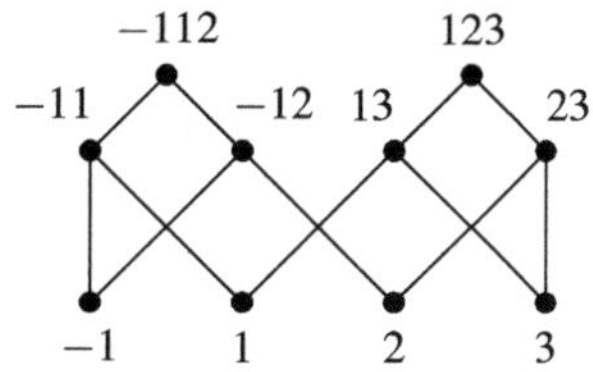

Fig. 6.6 The non-degenerate mereoposet without (SSP+)

Finally, we prove that:

Theorem 8 (Pietruszczak 2000).

(i) *The sentences* (sum-sup) *and* ($\star$) *are true in the class* **POS**+(SSP+). *So* **POS**+(SSP+) $\subsetneq$ **MPOS**.

(ii) *The equality* sum $=$ sup *is true in all non-degenerate posets which satisfy* (SSP+).

Proof. Ad the part "$\Rightarrow$" of (sum-sup): By Lemma 2 we have (6.10). By Theorem 5 we have (†).

For ($\ddagger_+$): Suppose towards contradiction that (a) x sup S, $S \neq \varnothing$ and (b) $\neg\, x$ sum S. Hence there is u_0 such that (c) $u_0 \sqsubseteq x$ and (d) $\forall_{z \in S}\, z \wr u_0$.

We notice that $u_0 \neq x$. Indeed, if $u_0 = x$ then, by (a), (d) and ($\mathtt{r}_\sqsubseteq$), for some $z_0 \in S$ we have a contradiction: $z_0 \sqsubseteq x$ and $z_0 \wr x$,

Thus $x \not\sqsubseteq u_0$, by (c) and ($\mathtt{antis}_\sqsubseteq$). Hence, by (SSP+), there is y_0 such that (e) $y_0 \sqsubseteq x$, (f) $y_0 \wr u_0$ and (g) for any $v \in M$: if $v \sqsubseteq x$ and $v \wr u_0$, then $v \sqsubseteq y_0$. From (a) and (d) we obtain that $\forall_{z \in S}(z \sqsubseteq x \wedge z \wr u_0)$. Hence, by (g), we have that $\forall_{z \in S}\, z \sqsubseteq y_0$. So $x \sqsubseteq y_0$, by (a). Hence $x = y_0$, by (e) and ($\mathtt{antis}_\sqsubseteq$). Thus, by (c), (f) and ($\mathtt{r}_\sqsubseteq$), we obtain a contradiction: $u_0 \sqsubseteq y_0$ and $y_0 \wr u_0$.

Since ($\ddagger_+$) is true in **POS**+(SSP+), then by Fact 8 we have: **POS**+(SSP+) $\subsetneq$ **MPOS**.

Ad ($\star$): Firstly, let $|M| > 1$ and x sup S. Then $S \neq \varnothing$, since by Theorem 5 and Corollary 2, in M there is no zero element. Hence, x sum S by (sum-sup). Secondly, assume that M has only one element x. Then x sup $\varnothing$. But $\neg\, x$ sum $\varnothing$, by ($\mathtt{r}_\sqsubseteq$). So sup $\not\subseteq$ sum.

Ad (ii): By (i).

References

Gruszczyński, R., & Pietruszczak, A. (2008). Full development of Tarski's geometry of solids. *Bulletin of Symbolic Logic, 14*, 481–540.

Gruszczyński, R., & Pietruszczak, A. (2009). Space, points and mereology. On foundations of point-free Euclidean geometry. *Logic and Logical Philosophy, 18*, 145–188.

Gruszczyński, R., & Pietruszczak, A. (2010). How to define a collective (mereological) set. *Logic and Logical Philosophy, 19*, 309–328.

Grzegorczyk, A. (1955). The systems of Leśniewski in relation to contemporaty logical research. *Studia Logica, 3*, 77–97.

Grzegorczyk, A. (1960). Axiomatizability of geometry without points. *Synthese, 12*(2–3), 228–235.

Leśniewski, S. (1930). O podstawach matematyki. *Przeglad Filozoficzny, XXXIII*, 77–105.

Pietruszczak, A. (2000). Metamereologia (in Polish). Toruń: Nicolaus Copernicus University Press.

Pietruszczak, A. (2005). Pieces of mereology. *Logic and Logical Philosophy, 14*(2), 211–234.

Simons, P. (1987). *Parts: A study in ontology*. Oxford: Clarendon.

Tarski, A. (1956). Les fondements de la géometrié de corps. In *Księga Pamiątkowa Pierwszego Polskiego Zjazdu Matematycznego*, suplement to *Annales de la Societé Polonaise de Mathématique* (pp. 29–33). English translation: Fundations of the geometry of solid. In *Logic, semantics, metamathematics: Papers from 1923 to 1938* (pp. 24–29). Oxford: Clarendon.

Chapter 7
Natural Mereology and Classical Mereology

Paul Hovda

The main goal of this paper is to sharpen our understanding of what is at stake between two opposing philosophical views, or orientations, on certain issues within and related to mereology. On the one hand, there is a view that reality includes a great deal of *natural mereological structure*, which must be discovered (at least partly) by empirical means, and for which there is no a priori reason to think that it will fit any neat formal pattern. Crudely, we may take this first view to be the view that x is part of y if and only if y is an organic or natural union which x partakes in. *Perhaps* the parthood relation has some neat formal properties like transitivity and anti-symmetry, perhaps not; investigation is required. Moreover, it is far from evident than every arbitrary collection of objects constitutes a natural unity, so there probably are many collections for which there is nothing that deserves to be called the "mereological sum" of this collection of objects. Broadly, we should leave it to empirical (natural) science to settle which natural units there are, and what the overall structure of the parthood relation "looks like."

On the other hand, there is a view that there is an a priori[1] science of mereology whose truths reveal a great deal about the overall *pattern* of part-whole connections in the universe. Crudely, we may take this view to be that Classical Mereology (or some similar formal theory) gives the one true theory of the part-whole relation. Very broadly, while the first view might be associated with Aristotle, the second might be associated with more modern figures like Quine and Lewis (though anticipations of it can be found in Descartes and Hume, and elsewhere in the early modern period). As Lewis writes: "I myself take [Classical Mereology] to

[1]If we reject a sharp distinction between a priori and not, in favor a graduated distinction, then we may substitute "very close to as a priori as it gets" for "a priori" here.

P. Hovda (✉)
Reed College, Portland, OR, USA
e-mail: paul.hovda@reed.edu

C. Calosi and P. Graziani (eds.), *Mereology and the Sciences*, Synthese Library 371,
DOI 10.1007/978-3-319-05356-1_7,

be perfectly understood, unproblematic, and certain."[2] Let us call this second type of view "formalistic."

Modern proponents of the first type of view—let us call it "naturalistic"—include van Inwagen (according to Van Inwagen (1990), there are partless simples and mereological fusions of partless simples that are jointly caught up in a life; there is nothing else) and Koslicki (according to Koslicki (2008), whether some material things have a fusion turns on whether they realize a structure); Kit Fine might also be suggested, but is harder to place (see Fine (1999, 2010), and elsewhere).

Imagine now a third party to a dispute between a proponent of a naturalistic view and a proponent of a formalistic view, who wishes to make a kind of peace between them by arguing that their differences are not ultimately as great as might at first appear. The general strategy the third party employs is to try to show that for each of the two disputants, the third party can find, within the things and structure the disputant believes in, a kind of "image" of the things and structure the other disputant believes in. It may be that, after "looking at the world from each other's point of view" the disputants find that the differences between them are negligible; or, perhaps more likely, that the exact nature of the disagreement between them is made sharper by getting clearer on why the differences, despite the existence of the "images," are not negligible.

7.1 Informal Presentation

7.1.1 A Simplification: Sets as Natural Kinds

A comparison to a somewhat simpler dispute will help make clear what I have in mind. Consider a dispute between two philosophers, the first of whom, in "naturalistic" fashion, holds that some but not all classes of material objects correspond to *natural kinds* (e.g., the class of all dogs corresponds to a natural kind, but the class of all dogs that are in a country whose name begins with "E" does not). The naturalistic philosopher believes in arbitrary classes of things, and *in addition*, a few *kinds* of things. The second, "formalistic," philosopher is skeptical of the existence of *kinds* above and beyond the classes themselves. Now imagine a third party who gets *both* philosophers to agree that every class (of material objects) corresponds to one and only one *set* of objects (perhaps they take a class to itself be a set, or perhaps they take a class to be a mere plurality and a set to be a single thing). The third party then proposes that the naturalistic philosopher might see the formalistic one as simply concerned to deny that there are any further entities that "collect" material objects, above and beyond the sets, so that if there are natural kinds, they are just sets. Meanwhile, the formalistic philosopher might

[2]Lewis (1991, p. 75).

see the naturalistic one as holding that among all the many sets of material objects, some are special, and deserve to be singled out as "natural."

If the formalistic philosopher agrees that some sets are especially natural, and the naturalistic one does not think that an *ontology* of kinds is necessary, in addition to the distinction between natural and unnatural sets, then it is unclear that the two really disagree on anything that matters. The situation can be compared to the dispute about universals, between David Armstrong and David Lewis, as portrayed by Lewis in (1983). Lewis (in the role corresponding to our formalistic philosopher) at first wants to deny that there are universals, in addition to arbitrary collections of possibility. But he then comes to recognize that the whole system, advocated by Armstrong, of a sparse ontology of universals,[3] together with certain features of them (their direct relations to laws of nature, to objective similarity, etc.) has great theoretical utility. But instead of adding a superstratum of universals to his arbitrary collections, Lewis proposes that all of the theoretical work that universals need to do can be done by the collections *together with* a crucial distinction between *perfectly natural* collections and other collections. One might say that his ontology is formalistic, but his ideology is naturalistic.

The disagreement between our two philosophers thus might be merely superficial: they might ultimately agree in ontology (sets alone, no other ontological type required) and theoretical ideology (there is an extremely important natural/non-natural distinction among sets). The disagreement might instead be deep, perhaps because the naturalistic philosopher takes himself to have good reasons to believe that kinds are not certain special sets, or perhaps because the formalistic philosopher takes the distinction between natural and unnatural to be unacceptable, either in general, or in its application to sets. Or again, perhaps both philosophers agree that sets do not change their members, and the first philosopher holds that natural kinds do change their members: e.g., the kind *dog* loses a member each time a dog dies. Then there appears to be a good reason to think that the set is intrinsically, hopelessly, unsuited to play the theoretical role required of the kind.

For our purposes, it is worth dwelling on this story just a little longer. While it is not implausible that sets do not change their members, while kinds do, it is also not implausible to think that this is a mere appearance of difference, resulting from typical ways of talking, rather than the natures of the things themselves. For it may be agreeable to both philosophers that a set has its members "eternally," so that the set of all dogs that ever exist currently has members that do not presently exist. Set membership, on this view, doesn't occur, or relate a member to a set, at one time rather than another; instead, it happens timelessly. Yet if this is the case there is still a reasonable notion of a set s losing a member x at a time t: x might be a member of s such that [x exists over a long span of time up to t, and x does not exist after t]. Once it is recognized that both (1) set membership is an eternal affair (so that $x \in s$ either once-and-for-all or never); and (2) nonetheless, there is a reasonable derivative notion of membership-at-a-time ($x \in_t s$ iff [$x \in s$ and x exists at t]) it is less clear

[3]"Sparse," because not every arbitrary collection corresponds to a universal.

that the fact that we tend to think of the natural kind *dog* as subject to membership-change while we tend to think of sets as membership-stable is a good reason to think that sets and natural kinds are different types of things. For it may be that when we think of the changeable membership relation on natural kinds we are really just thinking of the derivative changeable membership relation on (natural) sets.

The philosophers debating on natural kinds might continue to disagree. The naturalistic one might say that sets are ineligible to be kinds for another reason: they have the wrong *spatial* properties. The natural kind *dog* might be something that exists on earth, while the set of dogs exists nowhere or everywhere. But again, there is reason to wonder if this is a genuine difference rather than an appearance. For we may certainly define a notion of location for a set at a time that will behave, one might think, much like the notion of location for a kind does: the set will be located, at a time, wherever its members that exist at that time are. More precisely, we will need to say something like: the location of s at t is the union of the regions occupied at t by the members of s that exist at t. It is worth noting that part of what makes this particular definition work is that it is relatively uncontroversial that *regions* amalgamate in a natural way: for any collection of regions, there is the union of those regions, basically a region that you partly occupy if and only if you partly occupy any of the regions in the collection.

Now, it is possible to insist that such a notion of the "location" of a set is somehow second-rate ("unnatural" or "fake" or "merely derivative," etc.), while the notion of the "location" of a kind is first-rate, not second-rate. But it is unclear how such an asymmetric ranking of the two notions of location can be justified.

Similarly, if the naturalistic philosopher protests that kinds are made of matter, while sets are not, we might wonder why a well-defined notion of the material content of a set (if we can find one) is second-rate. Say that a set is "perfectly materialistic" if it is non-empty and every one of its members is made of matter. If we may suppose that for any bits of matter, there is some matter that functions as the "union" of those bits, in much the way that for any collection of regions of space there is a union of the regions, then we may say that a perfectly materialistic set is "made of" exactly the union of the bits of matter that make up its members.

It is not obvious how far such strategies can actually work to remove apparent differences between the set of dogs and the natural kind *dog*. But the basic point should now be clear enough: that the dispute between the two philosophers who seem to disagree about natural kinds might well turn out to be a shallow or merely verbal dispute, since it may turn out that each philosopher believes in a system of items and features of those items, a system that plays the theoretical role that the whole system of natural kinds is supposed to play.

7.1.2 Sets as Things

Now to return to the main theme: the suggestion of this paper is that the dispute between the "natural unities" mereologist and the "mathematical pattern" mereologist may turn out to be similarly largely shallow or verbal. In particular, the suggestion will be that if the naturalistic mereologist agrees to the existence of arbitrary *sets* of the material objects he or she already embraces, while the formalistic mereologist agrees to a crucial distinction between natural and non-natural objects and sets of objects, the two may equally regard the whole of reality to consist of a formally well-behaved pattern of objects and sets of objects (a pattern whose global properties are what the formalistic mereologist was always emphasizing), together with an important, formally unpredictable, natural/non-natural distinction among the nodes in this pattern (which the naturalistic mereologist was always emphasizing).

To illustrate a little: where the formalistic mereologist takes there to be a fusion of all objects which are either cats or dogs, the naturalistic one takes there to be the set of all things that are either cats or dogs. Now, both agree that the set exists, and we take it that it is negotiable that the set might inherit a *location*, and other minimal physical properties, from its members. But then how different is the set, as seen from the point of view of the naturalist, from the fusion, as seen from the point of view of the formalist? By downplaying the differences, we hope to make good on our suggestion that the mutually acceptable set can play the role of the fusion. Assuming that this works for this particular object (the fusion), our main task is to show how to coordinate things so as to make an entire formalistic network of objects, and part-whole relations among them, mutually agreeable. The mutually accepted network will have exactly the formal character that the formalistic mereologist emphasized; yet the naturalistic philosopher will still maintain that there is a special *natural* sub-network of the larger, formally well-behaved one, with natural objects as nodes, linked by a natural sub-relation of the larger part-whole relation.

The rest of this paper is concerned with some technical details involved with fleshing out this suggestion, particularly from the point of view of the naturalistic mereology. The main project at hand is of this form: assuming nothing formally about the most basic, *given* system of objects D and primitive "natural" part-whole relation $\mathbf{N}_0$ on them, what needs to be done, using nothing more than set theory together with the given objects and relation, to construct on and around it a formally "well-behaved" system of objects H and defined part-whole relation $\leq$ on them? We wish to "preserve" as much structure as possible, with D a subset of H and $\mathbf{N}_0$ a sub-relation of $\leq$, and such that the relation $\leq$, when restricted to its sub-domain D, should be identical with, or at least very closely related to, $\mathbf{N}_0$. To make this project more exact, we will take the notion of being "well-behaved" to be the notion of "obeying the laws of Classical Mereology," so that what we are up to is finding a transformation Ψ, that could in principle be applied to any relational structure $\langle X, \mathbf{R}\rangle$, so that

$$\Psi(\langle X, \mathbf{R}\rangle) = \langle X', \mathbf{R}'\rangle$$

has exactly the formal structure that Classical Mereology requires; that is, $\langle X', \mathbf{R}'\rangle$ is guaranteed to be a model of Classical Mereology, no matter what X and $\mathbf{R}$ are.

What makes the project formally non-trivial is that there are basically two sorts of formal task here, that tend to work against one another, but must be executed simultaneously. The first task is this: given a "natural" part-whole relation $\mathbf{N}_0$ and its "natural" domain D, *extend* the relation—that is, add relational "links" to $\mathbf{N}_0$, among things already present in D—in such a way that the resulting relation is formally well-behaved in the sense of possessing such features as reflexivity, transitivity, and obeying the *strong supplementation*[4] constraint of Classical Mereology. The second task is to add objects to the "natural" domain D (together with relational links) so as to provide mereological fusions for arbitrary non-empty subsets of this domain.

We would be on our way to executing the second task, if we were to imitate in a straightforward way what we considered saying about natural kinds above: "let us add to D every non-empty subset of D, and count members as parts." Thus we would get a candidate for the mereological fusion of all dogs: the set of all dogs would now be counted as a material object, alongside the dogs, and each dog would count as a part of it. Many objections to so counting the set can be met with, as discussed above. But this way of executing the second task has made it harder to execute the first task. For example, our new relation will not be transitive on its domain, since a given dog's foot will not be counted as a part of the fusion of all dogs. Moreover, we may have "too many things" in some cases, playing the same formal role: for example, if p is the set of parts (in the original, given sense of part) of a dog d, then both d and p are suited to play the formal role of being the mereological fusion of the members of p.

Thus the non-trivial formal difficulty is in executing both tasks simultaneously. But it can be done, in a fairly natural way. While the formal device explored here is, it is hoped, sufficient to give a "proof of concept" for the more general philosophical idea, it is really only a first step, as there are a number of questions one might raise about it that we will not have the space to discuss. A couple will be touched on briefly at the end of the paper, once the device is in view.

7.1.3 Overview of the Formal Device

Here is a brief informal sketch of the technique. We begin with some *natural objects* (to be thought of as concrete *natural units* on the model of the naturalistic mereology) and a given part-whole relation on them; call the set of these objects the *natural domain* and the relation the *natural* part-whole relation. Then we take the reflexive and transitive closure of the natural part-whole relation; next we *extend the domain* by adding non-empty, non-singleton *sets* of the members of the natural domain. We then extend the relation further, reaching a relation on the extended domain that is logically guaranteed to almost satisfy CM. *Almost,* because, in a very clear sense, the *only* possible failing is that the resulting relation might not be anti-symmetric. In the final stage, we *restrict the domain and relation* that resulted from the composite of our previous transformations, basically choosing (in a principled

[4]See below for a formally exact statement of this constraint.

way) one "representative" from each cluster of items that contravene anti-symmetry, thus guaranteeing that we move from *almost* satisfying CM to actually satisfying it.

An interesting feature of the general transformation is this: if we start with a domain and relation that satisfies CM, the construction winds up exactly where it started: the combined effect of our sequence of transformations will be nothing at all. CM is, structurally, a "fixed-point" of the construction.

7.2 Formalization

We turn to the technical details of the transformation; the discussion assumes only an elementary acquaintance with logic and set theory, and should be accessible to anyone interested in the formal details of Classical (and other) mereologies.

We will be discussing various transformations on *relational structures*, that is, ordered pairs $\langle X, \mathbf{R}\rangle$, where X is a set and $\mathbf{R}$ is a relation on that set (the *carrier set*). Relations are simply sets of ordered pairs, and what it means that $\mathbf{R}$ is a relation *on* X is just that for every ordered pair $\langle x, y\rangle$ in the relation, $x \in X$ and $y \in X$, or, to put it another way, $\mathbf{R} \subseteq X \times X$. We will often write '$x$ $\mathbf{R}$ y' for '$\langle x, y\rangle \in \mathbf{R}$'; we will also say '$x$ bears $\mathbf{R}$ to y' for this. Another notion we will want is the notion of the *restriction* of a relation to a given set: if $\mathbf{R}$ is a relation and Y is a set, then $\mathbf{R} \upharpoonright Y$ is the relation $\{\langle x, y\rangle : x \mathbf{R} y \text{ and } x, y \in Y\}$.

We will focus on "part-like" relations and structures, and a particular sequence of transformations on them. But the transformations we consider can be defined in a general way, independent of their application here; we will consider the general definitions as well as the application.

The first transformation, Φ_r, is simply to take the reflexive closure of a relation (on the carrier set):

$$\Phi_r(\langle X, \mathbf{R}\rangle) = \langle X, \mathbf{R} \cup \{\langle x, x\rangle : x \in X\}\rangle.$$

Clearly, if $\mathbf{R}$ is itself reflexive, then $\Phi_r(\langle X, \mathbf{R}\rangle) = \langle X, \mathbf{R}\rangle$. So Φ_r is *self-fixing:* for any relational structure $\mathfrak{B}$, $\Phi_r(\Phi_r(\mathfrak{B})) = \Phi_r(\mathfrak{B})$.

The next transformation, Φ_t, takes the transitive closure of the given relation. Given $\langle X, \mathbf{R}\rangle$, say that $\mathbf{S}$ *transitively extends* $\mathbf{R}$ *within* X if $\mathbf{S} \subseteq X \times X$, $\mathbf{R} \subseteq \mathbf{S}$, and $\mathbf{S}$ is transitive. $\mathbf{R}^t$ is then $\bigcap\{\mathbf{S} : \mathbf{S} \text{ transitively extends } \mathbf{R} \text{ within } X\}$, and we define Φ_t so that

$$\Phi_t(\langle X, \mathbf{R}\rangle) = \langle X, \mathbf{R}^t\rangle.$$

The transitive closure of a relation is itself transitive, since the intersection of a set of transitive relations is itself transitive. If $\mathbf{R}$ is itself transitive, then $\Phi_t(\langle X, \mathbf{R}\rangle) = \langle X, \mathbf{R}\rangle$. So Φ_t is also self-fixing. Further, $\Phi_t(\Phi_r(\mathfrak{B})) = \Phi_r(\Phi_t(\mathfrak{B}))$.[5]

[5]One can get an especially clear view of the effect of Φ_t by considering how it can be *built up* from iterated application of a simpler transformation. Define Φ_{t0} so that

$$\Phi_{t0}(\langle X, \mathbf{R}\rangle) = \langle X, \mathbf{R} \cup \{\langle x, z\rangle (\in X \times X) : \exists y\, (x \mathbf{R} y \wedge y \mathbf{R} z)\}\rangle.$$

To begin our discussion of the application, let D be the set of natural objects. (We assume that they form a set.) We may allow that there are many specific part-whole relations on D; let us define

$x\ \mathbf{N}_0\ y$

so that for $x, y \in D$, $x\ \mathbf{N}_0\ y$ just in case x bears one of these relations to y. $\mathbf{N}_0$ is the resulting generalized *natural* part-whole relation.

Formally, we make no assumptions whatever about $\mathbf{N}_0$: $\langle D, \mathbf{N}_0 \rangle$ is an arbitrary non-empty relational structure (a non-empty set with a relation on it). Informally, we will use natural examples like John's foot being part of John.

Now let $\langle D, \mathbf{N} \rangle$ be $\Phi_r(\Phi_t(\langle D, \mathbf{N}_0 \rangle))$, so that $\mathbf{N}$ is the relation that arises from taking the transitive closure of $\mathbf{N}_0$ and adding reflexivity.

Our next general transformation Φ_1 is somewhat complicated. Say that a set is *suitable* if it has two or more members. Given $\langle X, \mathbf{R} \rangle$, let A be the set of all suitable subsets of X, and let $B = X \cup A$. Let $\mathbf{S}$ be the relation on B that holds of x and y just in case

$x\mathbf{R}y$, or $x \in y$, or $x \subseteq y$.

Then let $\Phi_1(\langle X, \mathbf{R} \rangle) = \langle B, \mathbf{S} \rangle$. Clearly Φ_1 is not self-fixing; in fact, almost the opposite: provided the carrier set X itself is suitable, $\Phi_1(\langle X, \mathbf{R} \rangle) \neq \langle X, \mathbf{R} \rangle$.

Let $\langle E, \mathbf{P}_0 \rangle$ be $\Phi_1(\langle D, \mathbf{N} \rangle)$, i.e., $\Phi_1(\Phi_r(\Phi_t(\langle D, \mathbf{N}_0 \rangle)))$. Then we can show that $x\ \mathbf{P}_0\ y$ if and only if one of the following holds:

$x\ \mathbf{N}\ y$, or $x \in y$, or $x \subseteq y$.

Each of the three disjuncts excludes the other two.

Let E° be the set of suitable subsets of D, so that $E = D \cup E^\circ$ and $D \cap E^\circ = \emptyset$. Let $\langle E, \mathbf{P} \rangle$ be $\Phi_t(\langle E, \mathbf{P}_o \rangle)$. Then one can confirm that $x\ \mathbf{P}\ y$ just in case either:

$x\ \mathbf{P}_0\ y$ or

$x \in D$ and $y \in E^\circ$, and there is some $b \in y$ such that $x\ \mathbf{P}\ b$.

To show this, consider what was added when we applied Φ_t to $\langle E, \mathbf{P}_0 \rangle$ (show the easy Lemmas 1 and 2 below first). This shows that to define $\mathbf{P}$, we could have used these clauses instead of Φ_t, in our particular application. Also, instead of applying Φ_r and Φ_t to get $\mathbf{N}$ from $\mathbf{N}_0$ first, we could have applied Φ_1 directly to $\langle D, \mathbf{N}_0 \rangle$ and then applied Φ_r and Φ_t (or the above clauses); the result would be the same.

Let us observe some more features of $\mathbf{P}$. First, some informal examples: let *foot* be John's foot and *hand* be John's hand. Then

foot **P** John

John **P** { John, the Eiffel Tower }

(and hence) *foot* **P** { John, the Eiffel Tower }.

But

it is not the case that { *hand, foot* } **P** John.

$\Phi_{t_0}(\mathfrak{B})$ is a first approximation of $\Phi_t(\mathfrak{B})$; a second approximation is $\Phi_{t_0}(\Phi_{t_0}(\mathfrak{B}))$. One can show that $\Phi_t(\mathfrak{B})$ is the "limit" of the approximations. More precisely: let $\mathfrak{B}^0$ be $\mathfrak{B} = \langle X, \mathbf{R} \rangle$ and let $\mathfrak{B}^{i+1}$ be $\Phi_{t_0}(\mathfrak{B}^i)$. Let $\mathbf{R}^i$ be the relation in $\mathfrak{B}^i$. Then $\mathbf{R}^t$, the relation of $\Phi_t(\mathfrak{B})$, is the relation

$\{\langle x, y \rangle (\in X \times X) : \exists i \in \mathbb{N}\ \langle x, y \rangle \in \mathbf{R}^i\}$.

Second, some structural features. **P** has a "top" element, namely D: every member of E bears **P** to D. So everything in the wider domain is "part of" the set of all objects (the narrow domain). Clearly, **P** is reflexive and transitive (on E). A very important feature we will use later is this: if some $b \in D$ bears **P** to some $i \in E^\circ$, then b bears **P** to some $c \in i$ (in fact, $b\ \mathbf{N}\ c$). That is,

Lemma 1. $(b \in D \wedge i \in E^\circ) \rightarrow (b\ \mathbf{P}\ i \rightarrow \exists c \in D(c \in i \wedge b\ \mathbf{P}\ c))$.

Also note

Lemma 2. $(i \in E^\circ \wedge j \in E^\circ) \rightarrow (i\ \mathbf{P}\ j \leftrightarrow i \subseteq j)$ *and* $(i \in E^\circ \wedge b \in D) \rightarrow \neg i\ \mathbf{P}\ b$.

P is in the direction of the Classical Mereologist's part-whole relation: the set of some objects from D is playing something like the role of the mereological fusion of its members, since every part (in the sense of $\mathbf{N}_0$) of every member bears **P** to the set. But this approximation, to the "fusion" of a set of things that happen to be parts of something x, may not bear **P** to x, so we are not there yet. For example, if x is the set of John's parts, $x \in E^\circ$ (assuming John has more than one part) and it is not the case that x **P** John.

7.2.1 *Minimal Upper Bounds and Complements*

The next transformation takes us much closer. Given any structure $\langle X, \mathbf{R}\rangle$, define the relation $\circ_\mathbf{R}$ (**R**-overlap) on X as:

$(\forall\, x, y \in X)\ (x \circ_\mathbf{R} y\ \leftrightarrow\ \exists z(z\ \mathbf{R}\ x \wedge z\ \mathbf{R}\ y))$.

Then define **S** as: $x\ \mathbf{S}\ y$ iff $\forall z\ (z \circ_\mathbf{R} x \rightarrow z \circ_\mathbf{R} y)$. Finally, define Φ_o so that

$\Phi_o(\langle X, \mathbf{R}\rangle) = \langle X, \mathbf{S}\rangle$.

Let $\langle E, \sqsubseteq\rangle$ be $\Phi_o(\langle E, \mathbf{P}\rangle)$ i.e., $\Phi_o(\Phi_t(\Phi_1(\Phi_r(\Phi_t(\langle D, \mathbf{N}_0\rangle)))))$. Let us notate the relation of **P**-overlap as ∇. Consider again { *hand, foot* }; temporarily call it i. Given $x \in E$, if $x\ \nabla\ i$, then there is a $w \in E$ that bears **P** to x and to i. We argue now that there is a $b \in D$ such that b bears **P** to w and either to *hand* or to *foot*. If $w \in D$ then let $b = w$ (see Lemma 1). If $w \in E^\circ$, $w \subseteq i$, so $w = i$ (since i is a doubleton), and let b= *hand*. But b then bears **P** to John; and b bears **P** to x (since $b\ \mathbf{P}\ w$ and $w\ \mathbf{P}\ x$); thus, x **P**-overlaps John. This all shows that

{ *hand, foot* } $\sqsubseteq$ John.

Let us now consider the structural features of $\sqsubseteq$. It is easy to see from its definition (without even knowing what ∇ means) that $\sqsubseteq$ is reflexive and transitive. We also have

Lemma 3. *If* $\Phi_o(\langle X, \mathbf{R}\rangle) = \langle X, \mathbf{S}\rangle$, *then, provided that* **R** *is transitive,* $(\forall x, y \in X)\ (x\ \mathbf{R}\ y\ \rightarrow\ x\ \mathbf{S}\ y)$.

In particular, $(\forall x, y \in E)\ (x\ \mathbf{P}\ y\ \rightarrow\ x \sqsubseteq y)$.

We now are much closer to the behavior of fusions, since we have

$\{\, x : x$ **P** John $\} \sqsubseteq$ John.

To show how close we are will require some work. First, we will define a sum-like notion. Given a non-empty $X \subseteq E$, let

$\sigma'(X) = \{b \in D : (\exists y \in X)\ b\ \mathbf{P}\ y\}$.

$\sigma'(X)$ is obviously non-empty. It is a singleton if and only if X is a singleton of a **P**-atom (a member of D that nothing else bears **P** to); and then $\sigma'(X) = X$. In this case, $\sigma'(X) \notin E$; otherwise $\sigma'(X) \in E$. Accordingly, let

$\sigma(X) = \sigma'(X)$ if $\sigma'(X) \in E$; otherwise, let $\sigma(X)$ be the one member of $\sigma'(X)$.

We will prove that $\sigma(X)$ is a minimal upper bound on X: every member of X bears $\sqsubseteq$ to it, and it bears $\sqsubseteq$ to any such thing.

Lemma 4. $(\forall x, y \in E)\ (x \bigtriangledown y \rightarrow (\exists b \in D)\ (b\ \mathbf{P}\ x \wedge b\ \mathbf{P}\ y))$

Lemma 5. $(\forall b \in D)(\forall i \in E^{\circ})\ (b \bigtriangledown i \rightarrow (\exists c \in i)\ b \bigtriangledown c)$

Both of these Lemmas are easy to confirm from Lemmas 1 and 2.

Lemma 6. $(\forall X \subseteq E) :\ X \neq \emptyset\ \rightarrow\ (\forall x \in X)\ x \sqsubseteq \sigma(X)$

Proof. Let $x \in X$. Then, if $y \bigtriangledown x$, by Lemma 4, we have a $b \in D$ with $b\ \mathbf{P}\ x$ and $b\ \mathbf{P}\ y$. By the definition of $\sigma(X)$, $b \underline{\in} \sigma(X)$ (i.e., either $b = \sigma(X)$ or $b \in \sigma(X)$); so $b\ \mathbf{P}\ \sigma(X)$. So $y \bigtriangledown \sigma(X)$.

Lemma 7. $(\forall y \in E)(\ ((\forall x \in X)\ x \sqsubseteq y)\ \rightarrow \sigma(X) \sqsubseteq y)$

Proof. Suppose $(\forall x \in X)\ x \sqsubseteq y$. Suppose $w \bigtriangledown \sigma(X)$. Then, by Lemma 4, we have a $b \in D$ such that $b\ \mathbf{P}\ w$ and $b\ \mathbf{P}\ \sigma(X)$. By Lemma 1, there must be a $c \in D$ with $c \underline{\in} \sigma(X)$ such that $b\ \mathbf{P}\ c$. Since $c \underline{\in} \sigma(X)$, for some $x' \in X\ \ c\ \mathbf{P}\ x'$; by Lemma 3 and our original supposition, $c \sqsubseteq y$. $w \bigtriangledown c$, hence $w \bigtriangledown y$, and we are done.

Lemmas 6 and 7 together say that $\sigma(X)$ is a minimal upper bound for X, with respect to the $\sqsubseteq$ relation. Formally, define: *y is a $\sqsubseteq$-minimal upper bound on X* if and only if

$(\forall x \in X)\ x \sqsubseteq y\ \ \wedge\ \forall z(\ ((\forall x \in X)\ x \sqsubseteq z)\ \rightarrow y \sqsubseteq z)$.

We have now shown

Theorem 1. *For every non-empty $X \subseteq E$, X has a $\sqsubseteq$-minimal upper bound.*

$\sigma(X)$ plays this role, so $\sigma(X)$ is an approximation of the fusion of X.

7.2.2 Complements

The $\sqsubseteq$ relation on E has even more in common with the classical mereologist's part-whole relation, since it includes what we may call *complements*. Roughly, for almost any object in E, there is a another object that represents "everything else" in E: the complement is "disjoint" from the original, but everything "overlaps" one or the other. The only objects without complements are objects of which everything is already a "part."

Define the $\sqsubseteq$-overlap relation (symbolized with $\Box$) as

$x \Box y \;\leftrightarrow\; \exists z\,(z \sqsubseteq x \wedge z \sqsubseteq y)$

Lemma 8. $(\forall x, y \in E)\;(x \Box y \;\leftrightarrow\; x \nabla y)$

Proof. The right-to-left direction is straightforward from Lemma 3. For the left-to-right direction, we give a visual proof. Straight lines represent holdings of the **P** relation from lower to higher, and squiggly lines represent holdings of the $\sqsubseteq$ relation from lower to higher.

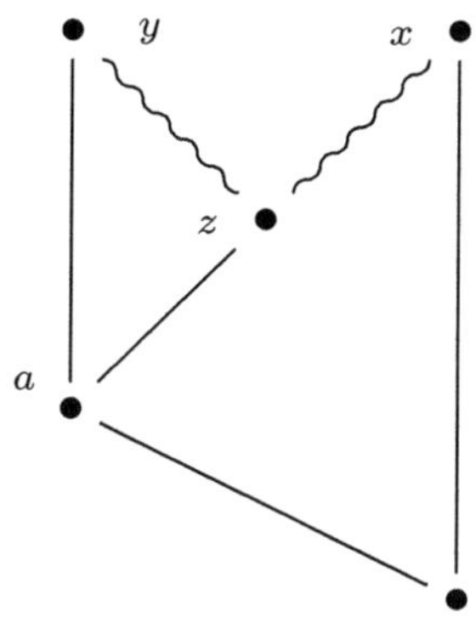

a has to exist, since $z \nabla z$ and $z \sqsubseteq y$; but then $a \nabla x$ as well.

In view of Lemma 8, we can interchange $\Box$ and ∇ as we please.

Lemma 9. $(\forall x, y \in E)(\,(\forall z \in E)(z \sqsubseteq x \rightarrow z \Box y) \rightarrow x \sqsubseteq y)$

Proof. Suppose the antecedent and that $w \nabla x$, and let z **P** w and z **P** x. By Lemma 3 and the antecedent, $z \Box y$. By Lemma 8, $z \nabla y$, so $w \nabla y$.

We will also want the notions of **P**-disjointness and $\sqsubseteq$-disjointness, where each is non-overlap of the relevant sort. Given Lemma 8 these relations are interchangeable. For notation, set

$x \wr y \;\leftrightarrow\; \neg\, x \nabla y$ (or equivalently)

$x \wr y \;\leftrightarrow\; \neg\, x \Box y$

Now we find, for almost any member of E, an object that will play the role of its complement. Given $x \in E$, *if there is a* $y \in E$ *with* $y \not\sqsubseteq x$, *then* define

$x^\star \;=\; \sigma\{y \in E : y \wr x\}$

We can use Lemma 9 to show that $\{y \in E : y \wr x\}$ is non-empty: so $x^\star$ exists.

Lemma 10. $x \wr x^\star$

Proof. Suppose for *reductio* $x \nabla x^\star$. Then either $x^\star \in D$ (in which case $\{y \in E : y \wr x\}$ was $\{x^\star\}$ and it is clear from the definition that $x^\star \wr x$) or get a $b \in D$ with

b **P** x and b **P** $x^\star$; since b **P** $x^\star$, get (by Lemma 1) a $c \in x^\star$ with b **P** c. Using the def. of $x^\star$, confirm that $c \wr x$. But b **P** c and b **P** x, so $c \ \nabla \ x$; contradiction.

Lemma 11. $y \wr x \ \rightarrow \ y \sqsubseteq x^\star$

Proof. Suppose $y \wr x$ and $w \ \nabla \ y$. Get (by Lemma 4) $b \in D$ with b **P** w and b **P** y . Now if $b \ \nabla \ x$ then $y \ \nabla \ x$; we supposed not, so $b \wr x$. So $b \ \underline{\in} \ x^\star$. So b **P** $x^\star$, so $w \ \nabla \ x^\star$.

Lemma 12. $y \wr x^\star \ \rightarrow \ y \sqsubseteq x$

Proof. Suppose $y \wr x^\star$ and $w \ \nabla \ y$. Get $b \in D$ with b **P** w and b **P** y. Get that $b \wr x^\star$, so $b \ \underline{\notin} \ x^\star$, so it is not the case that $b \wr x$, so $b \ \nabla \ x$, and hence $w \ \nabla \ x$.

Putting the last three lemmas together, we have that everything that is not all-inclusive has a "$\sqsubseteq$-complement" where we define: *y is a $\sqsubseteq$-complement of x* if and only if

$y \wr x$ and
$\forall z((z \wr x \rightarrow z \sqsubseteq y)$ and $(z \wr y \rightarrow z \sqsubseteq x))$

Theorem 2. *For all $x \in E$, if $\exists y(y \not\sqsubseteq x)$ then x has a $\sqsubseteq$-complement.*

For all, except the all-inclusive $x \in E$, x has at least one complement, and $x^\star$ is one.

7.2.3 *Anti-symmetry*

The relation $\sqsubseteq$ on E is formally very much like the Classical Mereologist's part-whole relation. For we have shown that $\sqsubseteq$ and E are a relation **R** on a set X such that

(2) **R** is transitive.
(3) All non-empty subsets of X have an **R**-minimal upper bound.
(4) For any member of X, if not everything bears **R** to it, then it has a complement.

If a relation **R** on a domain X satisfies (2)–(4), then the structure $\langle X, \mathbf{R} \rangle$ satisfies the axioms of Classical Mereology, provided it has two further features: (1) **R** is anti-symmetric; and (5) either there is only one member of X or there is no member of X that bears **R** to every member of X.[6]

The members of E fall into "clusters" of things that bear $\sqsubseteq$ to one another. These are like the equivalence classes of an equivalence relation, except that members of different clusters may (anti-symmetrically) bear $\sqsubseteq$ to one another. If a member of a cluster k bears $\sqsubseteq$ to a member of some other cluster l, then every member of k bears

[6]See Sect. 4 of Hovda (2009); the five conditions here correspond to the five axioms in the last of the five axiom-sets given there.

$\sqsubseteq$ to every member of l, and no member of l bears $\sqsubseteq$ to any member of k. There is a simple way to transform the structure $\langle E, \sqsubseteq \rangle$ into a Classical Mereology. We simply treat each "cluster" of things that bear $\sqsubseteq$ to each other as a single element, and let the clusters inherit the other aspects of the $\sqsubseteq$ relation. Formally, for each $x \in E$, define

$$[x] = \{y \in E : x \sqsubseteq y \wedge y \sqsubseteq x\}$$

Let F be $\{y : \exists x \in E\ y = [x]\}$. For $[x], [y] \in F$, with $x, y \in E$, define

$$[x] \leq^F [y] \text{ if and only if } \exists z \in [x]\ \exists w \in [y]\ z \sqsubseteq w.$$

We can think of this as an instance of a general transformation Φ_a taking us from $\langle E, \sqsubseteq \rangle$ to $\langle F, \leq^F \rangle$; the definition is confined to a footnote.[7]

Lemma 13. *$\leq^F$ on F is reflexive, anti-symmetric, and transitive.*

Suppose $X \subseteq F$ is non-empty. Let z be $\{c \in E : [c] \in X\}$. Theorem 1 tells us that z has at least one $\sqsubseteq$-minimal upper bound d. Let $\bigvee X$ be $[d]$.

Lemma 14. *$\bigvee X$ is a least upper bound for X (in F).*

That is, for every $x \in X$, $x \leq^F \bigvee X$, and, for any $y \in F$, if every $x \in X \leq^F y$, then $\bigvee X \leq^F y$. This is straightforward to show. (We call this a "least" upper bound since, because of anti-symmetry, it is unique.)

Lemma 15. *If F has more than one element, then there is no $x \in F$ such that $\forall y \in F, x \leq^F y$.*

Proof. It is clear that $\forall x \in E, x \leq \sigma(E)$, and so $[x] \leq^F [\sigma(E)]$. Now consider any $[x] \in F$ such that $[\sigma(E)] \nleq^F [x]$. Apply Theorem 2 and get $x^\star$ with $x^\star \wr x$; hence $x \not\sqsubseteq x^\star$. Thus $[x] \nleq^F [x^\star]$.

Finally, suppose that for a given $x \in F$, there is a $y \in F$ with $y \nleq^F x$. Then there is a $\leq^F$-complement for x (uniquely so, because of transitivity). Define $x, y \in F$ *are $\leq^F$-disjoint* (symbolized $\wr^F$) as

$$x \wr^F y \text{ if and only if } \neg \exists z \in F(z \leq^F x \wedge z \leq^F y)$$

For $x, y \in F$, define x *is a complement of* y as

$$x \wr^F y \text{ and}$$
$$\forall z \in F(\ (z \wr^F x \rightarrow z \leq^F y) \text{ and } (z \wr^F y \rightarrow z \leq^F x)\ )$$

Lemma 16. *For every $x \in F$, if $\exists y(y \nleq^F x)$, then x has a $\leq^F$-complement.*

Proof. Suppose we have an x as in the antecedent. Then pick some $a \in x$ and consider $[a^\star]$.

By the last four Lemmas, we have

---

[7]Given any structure $\langle X, \mathbf{R} \rangle$, let $A = \mathcal{P}(X)$. Given any $x \in X$, let $[x] = \{y \in X : x\ \mathbf{R}\ y \wedge y\ \mathbf{R}\ x\}$. Let B be $\{e \in A : \exists x \in X \wedge e = [x]\}$. Let $\mathbf{S}$ be the relation on B defined as follows: for any e and f in B,

$$e\ \mathbf{S}\ f \text{ if and only if } (\exists z \in e)(\exists w \in f)\ z\ \mathbf{R}\ w.$$

Then define $\Phi_a(\langle X, \mathbf{R} \rangle) = \langle B, \mathbf{S} \rangle$. In general, this transformation is much more natural when combined with prior application of Φ_r and Φ_t; the composite $\Phi_a \circ \Phi_r \circ \Phi_t$ transforms any relational structure into a partial ordering.

Theorem 3. *$\langle F, \leq^F \rangle$ is a Classical Mereology.*

Now, we can "project" the structure of $\leq^F$ into E by mapping each $f \in F$ to some representative member of it. The set of representatives would be a subset of E, and the restriction of $\sqsubseteq$ to this subset would be isomorphic to $\leq^F$.

There are at least two fairly natural ways to choose representatives. The first is this: for each $[x] \in F$, we pick $\sigma([x])$. To see that this works, we need to show

Lemma 17. $(\forall x \in E)\, \sigma([x]) \in [x]$

Proof. Suppose $y \in [x]$. Then, by Lemma 6, $y \sqsubseteq \sigma([x])$. And for all $z \in [x], z \sqsubseteq y$. Hence, by Lemma 7, $\sigma([x]) \sqsubseteq y$.

So now let G be $\{x : x = \sigma(f)$ for some $f \in F\}$. Then $G \subseteq E$, and we let $\leq^G$ be $\sqsubseteq \upharpoonright G$. Then $\langle G, \leq^G \rangle$ is isomorphic to $\langle F, \leq^F \rangle$: σ is a one-one map from F onto G, and $f \leq^F g$ iff $\sigma(f) \leq^G \sigma(g)$.

The second, preferred, way to choose representatives that we will consider is to choose the "smallest" representative, if there is one; otherwise choose the "largest," namely $\sigma([x])$. For each $x \in E$: if $x \cap D = \{b\}$ for some b, then let $\rho([x]) = b$; otherwise, let $\rho([x]) = \sigma([x])$. Let H be $\{x : x = \rho(f)$ for some $f \in F\}$. Then $H \subseteq E$, and we let $\leq$ be $\sqsubseteq \upharpoonright H$. Clearly, $\langle H, \leq \rangle$ is also isomorphic to $\langle F, \leq^F \rangle$. So we have:

Theorem 4. *$\langle G, \leq^G \rangle$ and $\langle H, \leq \rangle$ are Classical Mereologies, and each is isomorphic to $\langle F, \leq^F \rangle$.*

We may think of the composite of the operations of going "up" from $\langle E, \sqsubseteq \rangle$ to $\langle F, \leq^F \rangle$ and "down" to $\langle H, \leq \rangle$ as a single operation that is applied to $\langle E, \sqsubseteq \rangle$ to yield $\langle H, \leq \rangle$; this is more natural for our application, but harder to define in general. It can be done, however, yielding the generally defined transformation Φ_ρ.[8]

7.2.4 Overview of the Construction

The construction of $\langle H, \leq \rangle$ from $\langle D, \mathbf{N}_0 \rangle$ proceeded by five steps.

Given

$$\langle D, \mathbf{N}_0 \rangle$$

[8]For a fully general definition, we need some way to tell apart the members of a cluster that are of lower *rank* from the others; in our application, these were members of D rather than of E°. Assuming that our set theory provides a natural way to rank everything in the universe (as does Zermelo-Fraenkel set theory with ur-elements, choice, and foundation) a general transformation Φ_ρ on arbitrary $\langle X, \mathbf{R} \rangle$ may be defined by first applying Φ_t, then, taking a cluster to be a maximal set of members of X that bear $\mathbf{R}^t$ to one another, for each cluster, choosing its single lowest ranked member, if there is one, and the union of all its lowest-ranked sets, otherwise. Φ_ρ is then defined by taking the "chosen" items as carrier set and taking the "inherited" relation.

take a reflexive and transitive closure:

$$\Phi_r(\Phi_t(\langle D, \mathbf{N}_0 \rangle)) = \langle D, \mathbf{N} \rangle$$

add suitable sets of given objects, along with part-like relations ($\in$, $\subseteq$) between them and the given objects and on them:

$$\Phi_1(\langle D, \mathbf{N} \rangle) = \langle E, \mathbf{P}_0 \rangle$$

take a transitive closure:

$$\Phi_t(\langle E, \mathbf{P}_0 \rangle) = \langle E, \mathbf{P} \rangle$$

take the overlap-implication:

$$\Phi_o(\langle E, \mathbf{P} \rangle) = \langle E, \sqsubseteq \rangle$$

and then choose "leasts or sums" as representatives:

$$\Phi_\rho(\langle E, \sqsubseteq \rangle) = \langle H, \leq \rangle.$$

Each of these steps preserves important aspects of the structures involved, and there are a couple of senses in which the structure of Classical Mereology is a natural "fixed-point" for this sequence of transformations.

7.2.5 From Classical Mereology to Itself

Suppose that $\langle D, \mathbf{N}_0 \rangle$ is itself a Complete Classical Mereology (CCM).[9] Then F is related back to $\langle D, \mathbf{N}_0 \rangle$ as follows. For any $[x], [y] \in F$, with $x, y \in E$, if $x, y \in D$, then $[x] \leq^F [y]$ iff x $\mathbf{N}_0$ y; if $x, y \in E^\circ$, then there is a unique $b \in D$ with $b \in x$, and a unique $c \in D$ with $c \in y$, and ($[x] \leq^F [y]$ iff b $\mathbf{N}_0$ c)—in fact, for each other $z \in [x]$, b is the $\mathbf{N}_0$-fusion of the members of z, and similarly for c. The map that takes us from $[x]$ to its representative in D (x or b) as in the last sentence is our ρ. In fact, we have

Theorem 5. *(Variation 1) If $\langle D, \mathbf{N}_0 \rangle$ is a CCM, then $\langle F, \leq^F \rangle$ is isomorphic to it. More precisely: let the transformation Ψ_1 be*
$\Phi_a \circ \Phi_o \circ \Phi_t \circ \Phi_1 \circ \Phi_t \circ \Phi_r$.

[9]A structure is a Complete Classical Mereology if it satisfies any standard set of axioms for Classical Mereology *with the fusion axiom given set-theoretically*. That is, the fusion axiom is a single axiom given with the use of set-theory, rather than an axiom scheme; see Sect. 1.2 of Hovda (2009).

Then if $\mathfrak{B}$ *is a CCM,* $\Psi_1(\mathfrak{B})$ *is isomorphic with* $\mathfrak{B}$.

(Variation 2): If $\langle D, \mathbf{N}_0 \rangle$ *is a CCM, then* $\langle H, \leq \rangle$ *is identical with it. More precisely: let* Ψ_2 *be*

$\Phi_\rho \circ \Phi_o \circ \Phi_t \circ \Phi_1 \circ \Phi_t \circ \Phi_r$.

Then if $\mathfrak{B}$ *is a CCM,* $\Psi_2(\mathfrak{B}) = \mathfrak{B}$.

To prove this, the main key is Lemma 20 below. Before turning to the proof, observe that, given the above analysis, for each $b \in D$: if b is a Mereological *atom* in $\langle D, \mathbf{N}_0 \rangle$ (i.e., there is no $c \in D$ with $c \neq b$ and $c\ \mathbf{N}_0\ b$) then $[b] = \{b\}$ and $\sigma([b]) = \rho([b]) = b$. Otherwise, $\rho([b]) = b$ and $\sigma([b])$ is the set of b's $\mathbf{N}_0$ parts.

Now, of the transformations that we used along the way, three of them involve changing the relation only, and do not alter the carrying set: taking the reflexive or transitive closure (Φ_t and Φ_r), and taking the "overlap inclusion" (Φ_o). These transformations do not alter any structure that is a CCM. This is obvious for Φ_r and Φ_t, since a CCM is already reflexive and transitive. For Φ_o we may use the following Lemma. (The object-language version of this Lemma is called the "strong supplementation" theorem (or, as it may be, axiom) in Classical Mereology).

Lemma 18. *If* $\langle X, R \rangle$ *is a CCM, then* $(\forall x, y \in X)$, *if* $(\forall z \in X)(z\ \mathbf{R}\ x \rightarrow z \circ_{\mathbf{R}} y)$ *then* $x\ \mathbf{R}\ y$.

We now prove Theorem 5. Let $\langle D, \mathbf{N}_0 \rangle$ be a CCM, and let $\langle E, \sqsubseteq \rangle$ arise from it as described above, by applying $\Phi_o \circ \Phi_t \circ \Phi_1 \circ \Phi_t \circ \Phi_r$.

Lemma 19. $(\forall\, b, c \in D)\ (b \sqsubseteq c \leftrightarrow b\ \mathbf{P}\ c \leftrightarrow b\ \mathbf{N}_0\ c)$.

Proof. Clearly, $\mathbf{P} \restriction D$ is just $\mathbf{N}_0$, so we need only show that the step from $\mathbf{P}$ to $\sqsubseteq$ does not add anything: $\sqsubseteq \restriction D$ is the same relation. We get this from Lemma 18.

Last, we need a lemma telling us that for each $i \in E^\circ$, that there is a unique "small representative" $b \in D$; b is the $\mathbf{N}_0$-fusion of i. It is a theorem of CCM that if for each non-empty subset i of the domain, there is a unique fusion of it in this sense: a thing $f(i)$ such that for all y, y overlaps $f(i)$ if and only if it overlaps a member of i. Given i in E°, let $f(i) \in D$ be its $\langle D, \mathbf{N}_0 \rangle$-fusion.

Lemma 20. $(\forall i \in E^\circ)\,(\forall b \in D)\,(\,(b \sqsubseteq i \wedge i \sqsubseteq b) \leftrightarrow b = f(i)\,)$

Proof. That $f(i) \sqsubseteq i$ is clear from the fusion properties of $f(i)$; that $i \sqsubseteq f(i)$ is clear from those properties and Lemmas 1 and 4. Uniqueness follows basically from those properties with Lemmas 5, 18, and 19, and the anti-symmetry of $\mathbf{N}_0$.

This suffices to show Theorem 5.

7.2.6 Final Reflections

We also note a couple results that help to show under what conditions our constructions "leave intact" the structure of $\mathbf{N}_0$. Consider the "Strong Supplementation" axiom of Classical Mereology as applied to $\mathbf{N}$:

$(\forall x, y \in D)((\forall b \in D)(b \mathbf{N} x \to b \circ_{\mathbf{N}} y) \to x \mathbf{N} y)$
One result is that this holds if and only if $\mathbf{N}{=}\sqsubseteq\restriction D$. A further easy result is that $\mathbf{N}$ is anti-symmetric iff $\mathbf{P}$ is. Moreover, $\mathbf{N}$ is anti-symmetric iff there are no "proper cycles" (in D) under $\mathbf{N}_0$, where a proper cycle is a finite sequence $a_1, \ldots, a_n$ with $n > 2$, with $a_1 = a_n$, and for each $i \leq n$, $a_i \neq a_{i+1}$ and $a_i \mathbf{N}_0 a_{i+1}$.

Further, if $\langle D, \mathbf{N}_0 \rangle$ is structurally "well-behaved" in that it features no proper cycles and the resulting $\langle D, \mathbf{N} \rangle$ obeys Strong Supplementation, then $\mathbf{N}{=}\leq\restriction D$, since for no $x \in D$ will there be a $y \in D$ such that $[x] = [y]$. Thus, if the naturalistic philosopher's original part-whole structure is "well-behaved" in this sense, our composite transformation Ψ_2 does fairly little, if any, "damage" to the relation $\mathbf{N}_0$ over its original domain: the restriction of $\leq$ to that domain is just the transitive and reflexive closure of $\mathbf{N}_0$.

So if the original part-whole structure is so "well-behaved" that its relation $\mathbf{N}_0$ is also already reflexive and transitive (hence identical to $\mathbf{N}$), then the restriction of $\leq$ to the original domain D is identical with the original relation $\mathbf{N}_0$: Ψ_2 has then done nothing but "filled in the gaps," with objects and relational links, so as to provide mereological fusions for arbitrary subsets of the domain, without adding to, or subtracting from, the original links, on the original objects.

Even if the original $\mathbf{N}_0$ is not already reflexive and transitive, it may be that $\mathbf{N}_0$ can be recovered from $\mathbf{N}$ in an interesting way. For example, if $\mathbf{N}_0$ is irreflexive, but transitive, then $\mathbf{N}_0$ is just $\mathbf{N}$ but with all self-links removed. And even if $\mathbf{N}_0$ is not transitive, it might still be formally "well-behaved" in this sense: for all $x, y \in D$, $x \mathbf{N}_0 y$ iff ($x \mathbf{N} y$ and there is no $z \in D$ such that $x \mathbf{N} z$ and $z \mathbf{N} y$); i.e., a part in the most basic sense is an immediate part in the transitive closure of the most basic sense. It is natural to think that this condition might hold in the non-classical mereological systems considered by Koslicki in (2008) and in Fine (1999). The system(s) considered in Fine (2010), or some important sub-class of them, might also satisfy this condition; the notion of *component* in Fine (2010) might be taken as a candidate for our $\mathbf{N}_0$.

These remarks should give a taste for the sort of refinements of the results we might reach by further exploration of the kind of technique explored in this paper. A broad statement of the general idea is that if the naturalistic mereologist's part-whole relation on its given domain obeys some apparently very weak formal constraints, it will be possible to define out of it, assuming constructions with set theory, a closely related structure which obeys much more stringent formal constraints that might be favored by the formalistic philosopher, such as those of Classical Mereology, in such a way that the original structure can be recovered as a sub-structure. In this way, the naturalistic mereologist might make peace with the formalistic one, provided the formalistic one is prepared to grant a special status (e.g., being *natural*, or *carving at the joints,* to use a metaphor favored by Sider in 2011) belonging uniquely to that particular sub-structure—to its objects and part-whole relation. Or, put another way, their original dispute might turn out to be merely verbal, the two simply using the words "part" and "object" in different, but ultimately mutually recognizable, ways.

7.2.7 Coda: Quick Response to Some Concerns About Sets

As we discussed briefly above, the project will only succeed if sets, or some replacement for sets, are granted the sorts of properties the formalistic philosopher ascribes to typical objects, e.g., being located. There are three points about this feature of our project that we may briefly address in closing.

First, it might be thought that if we grant sets location, then we will have a great many co-located sets, e.g., the set d whose members are all the dogs (and nothing else), the set $\{d\}$, the set $\{d, \{d\}\}$, and so forth. If d inherits location from its members, why shouldn't these other sets? Call two sets "materially equivalent" if the transitive closure of the one's membership is identical with the transitive closure of the other's. The reply to this concern would begin by suggesting that if two distinct sets are materially equivalent, then they are *qualitatively* indiscernible: they have the same basic physical properties. The next step would be to argue that it is acceptable for many purposes to pretend that qualitatively indiscernible sets are identical. The expectation would then be that the output of our Ψ_2 transformation captures exactly the right level of distinction among sets: two sets that are materially equivalent with the same element of the output of Ψ_2 are not, for many purposes, different; and every set is materially equivalent with a unique set-or-object in the output of Ψ_2.

Second, as an alternative to arguing for treating materially equivalent sets as the same (in some contexts), we could find a replacement for sets throughout the entire construction of Ψ_2. Interestingly, we could use plural quantification over the originally given domain, so that, for example, the role played by a doubleton $\{x, y\} \in E^\circ$ (with $x, y \in D$ and $x \neq y$) would now be played by those things such that: x is one of them, y is one of them, and nothing else is one of them. Arguably, there should be even less resistance to treating pluralities as having properties like location, and there is no problem about there being "too many of them" constructible out of the basic, given, objects.

Third, there is a concern that, given that sets do not change their members over time, they remain unsuited to play the roles of objects. There is much to say about this concern, and here we can only note that it seems worth exploring the possibility that considerations about time will only complicate the story, but not fundamentally change it. For example, if it can be agreed by both the naturalistic and formalistic philosopher that parthood may adequately be treated as a three-placed relation, so that we say "x is part of y at time t" instead of the bare "x is part of y," then we should consider how all of our formalization might be re-cast accordingly. Or perhaps we may take objects to have temporal parts, or, more non-traditionally, take (some) sets to change their members.[10]

[10]See Hovda (2013) for a discussion of the interaction of formalistic mereology with time and tense. The main idea pursued in Hovda (2013) is to re-conceive formalistic mereology while taking tense (or metaphysical modality) seriously, and allowing objects (including fusions) to change their parts. To wed, in a natural way, the approach in Hovda (2013) with the idea in this paper would seem to require a set theory in which sets can change their members. Such a set theory should

References

Fine, K. (1999). Things and their parts. *Midwest Studies in Philosophy, 23*, 61–74.
Fine, K. (2010). Towards a theory of part. *Journal of Philosophy, 107*(11), 559–589.
Hovda, P. (2013). Tensed mereology. *Journal of Philosophical Logic, 42*(2), 241–283. doi:10.1007/s10992-011-9220-4.
Hovda, P. (2009). What is classical mereology? *Journal of Philosophical Logic, 38*(1), 55–82. http://dx.doi.org/10.1007/s10992-008-9092-4.
Koslicki, K. (2008). *The structure of objects*. Oxford/New York: Oxford University Press.
Lewis, D. (1983). New work for a theory of universals. *Australasian Journal of Philosophical, 61*, 343–377.
Lewis, D. (1991). *Parts of classes*. Oxford/Cambridge: Blackwell.
Sider, T. (2011). *Writing the book of the world*. Oxford/New York: Oxford University Press.
Van Inwagen, P. (1990). *Material beings*. Ithaca: Cornell University Press.

be buildable by modifying untensed set theory in something like the manner that Hovda (2013) modifies untensed Classical Mereology to yield a tensed mereology.

Here is the barest sketch of how this would go. Naïve Set Theory consists of the axiom of Extensionality (sets x and y are the same iff x and y have the same members) together with the Naïve Comprehension scheme for set existence. The scheme is this (for any predicate $\phi(x)$ in which x occurs free and y does not, an instance of the scheme is): there exists at least one set y such that: for all x, $x \in y$ iff $\phi(x)$. Let Tensed Naïve Set Theory be Tensed Extensionality (sets x and y are the same iff *it is always the case that* x and y have the same members) together with a tensed correlate of Comprehension: there exists at least one set y such that *it is always the case that* for all x ($x \in y$ iff $\phi(x)$). An instance of this scheme thus implies that there is a set y such that at every time, for every x, $x \in y$ at that time iff x is (at that time) a dog. This set would have no members when there are no dogs, and its membership would wax and wane with the existence of dogs. Of course the tensed scheme inherits the inconsistency of the Naïve Comprehension scheme; to find a reasonable, consistent tensed set theory, one would modify ZFC with ur-elements, or the like.

Part III
Natural Sciences

Introduction to Part III: Mereology and Natural Sciences

It is difficult to imagine a contemporary scientific practice that trade more in mereological notions than the so called natural sciences, such as chemistry or biology. Their objects of study are in fact crucially composite objects such as atoms (which, despite the name are not *mereological atoms*), molecules, cells, living organisms and so on. The chemical and biological hierarchies, from atoms to molecules, from basic units of biological organization such as genes to more complex formations such as cells, tissues and living organisms are structured according to parthood relations on the one hand and probably several other relations such as functional relations, causal relations, dependence relations on the other. It is a substantive question how these relations interact, both from an empirical and a metaphysical perspective. Chemical, biological, even medical practices and theorizing may even call into question some common-sense intuitions about parthood. For example, in what sense are the reagents parts of the product of a particular chemical reaction? Are they really mereological parts of it? Note that arguably this question traces back to the early days of Greek philosophy. Mereological considerations, in particular the endorsement of gunk mereologies, were pivotal in the stoic solution of the problem of mixture, as argued in Nolan (2006). And this is just an example of what Lewis (2010) calls folk chemistry which involves crucially mereological reflection. Sharvy (1983) contains other examples. Consider then biological cases of parthood. We seem to have a solid intuition that my heart is a mereological part of me. But what about the human flora living in my body? And what if bacteria were found in my bloodstream? Would they count as a mereological part of me?

The papers in this section deal with similar questions, with a particular focus on chemistry, biology and medicine.

In the first one (*Developing the Mereology of Chemistry*) Llored and Harré contend that different mereological theories that are present in the literature are overly simplistic to fit chemical practices in that they do not consider different *ways* in which chemical substances are parts of other substances. This inadequacy of current mereological theories to give a satisfactory account of chemical phenomena leads them to a rather pragmatic development of new a mereological approach where on the one hand temporal and modal parameters such as affordance and dependence become crucial, and on the other hand traditional mereological principles such as transitivity loses their relevance.

In the second paper of this part (*Crisp Island in Vague Sea*) Jansen and Schulz address the crucial question we were asking in this brief introduction, i.e. the possibility of identifying precise conditions to ascribe parthood between pairs of biological entities. They make a convincing case drawing from a variety of different concrete biological examples that is extremely difficult to draw the inference from inclusion to parthood. Then they put forward different axioms specifying conditions that allow us to distinguish three crucial cases, namely (i) the case in which an *inclusion* statement is refined into a *parthood* statement, (ii) the case in which an *inclusion* statement is refined into a *containment* statement and (iii) the case in which an *inclusion* statement cannot be refined further. These axioms, together with some other criteria, in particular a *functional* criterion, help us deciding pragmatically whether a biological organism is part of another given that the latter spatially includes the former.

This part contains examples and discussions taken from entire scientific disciplines such as chemistry or biology, rather than particular theories. We therefore recommend general introductions to the philosophical aspects of those disciplines, referring the reader to the extensive bibliographies therein. Baird et al. (2007) is a good example of an introduction to the philosophy of chemistry. A recent one is Llored (2013). Among the many introductions to the philosophy of biology we point out the following recent ones: Hull and Ruse (2007); Rosenberg and McShea (2008). Finally, when it comes to philosophy of medicine, let us mention Sadegh-Zadeh (2012).

References

Baird, D., Scerri, E., & McIntyre, L. (2007). *Philosophy of chemistry. synthesis of a new discipline*. Boston: Springer.

Hull, D. L., & Ruse, M. (Eds.) (2007). *The Cambridge companion to the philosophy of biology*. Cambridge: Cambridge University Press.

Lewis, E. (2010). Diogenes laertius and the stoic theories of mixture. *Bullettin of the Institute of Classical Studies*, *35*(1), 84–90.

Llored, J. P. (Ed.) (2013). *The philosophy of chemistry. Practices, methodologies and concepts*. Newcastle Upon Tyne: Cambridge Scholars.

Nolan, D. (2006). Stoic gunk. *Phronesis*, *51*(2), 162–183.

Rosenberg, A., & McShea, D. W. (2008). *Philosophy of biology. A contemporary introduction*. London: Routledge.

Sadegh-Zadeh, K. (2012). *Handbook of analytic philosophy of medicine*, Dordrecht/New York: Springer.

Sharvy, R. (1983). Mixtures. *Philosophy and Phenomenological Research*, *44*, 227–239.

Chapter 8
Crisp Islands in Vague Seas: Cases of Determinate Parthood Relations in Biological Objects

Ludger Jansen and Stefan Schulz

8.1 Introduction

Parts are important for describing (types of) biological organisms: Mammals have lungs, fishes do not. Most cells have nuclei, red blood cells do not. Parthood is generally considered a fundamental relation both in formal ontology and in knowledge representation. As a complement to a taxonomic view on organizing things by categories, an (orthogonal) mereological view often constitutes an additional organizational principle for the representation of biological entities. Comprehensive representations of mereological hierarchies can be found in many biomedical terminology systems. Parthood relations are represented, e.g., in the anatomy branch of the MeSH thesaurus (Medical Subject Headings National Library of Medicine 2013), but there they are not formally distinguished from taxonomical relations. They are represented formally in the Foundational Model of Anatomy (Rosse and Mejino 2008), the Gene Ontology (Ashburner et al. 2000), ChEBI (Chemical Entities of Biological Interests; Hastings et al. 2013) and other OBO Foundry (Smith et al. 2007) ontologies that describe biological structures. In the huge clinical ontology SNOMED CT (International Health Terminology Standards Development Organisation (IHTSDO) 2013), an extensive mereological hierarchy describing anatomy classes is represented as a taxonomy of reified parthood relations, according to the structure-entity-part (SEP) triple architecture proposed by Schulz and Hahn (2005), where auxiliary classes like, e.g., *Heart structure* are introduced to model a partonomic statement like "The myocardium is part of the

L. Jansen (✉)
Department of Philosophy, University of Münster, Münster, Germany
e-mail: ludger.jansen@uni-muenster.de

S. Schulz
Institute for Medical Informatics, Statistics and Documentation, Medical University of Graz, Graz, Austria
e-mail: stefan.schulz@medunigraz.at

C. Calosi and P. Graziani (eds.), *Mereology and the Sciences*, Synthese Library 371, DOI 10.1007/978-3-319-05356-1_8,

heart" as a subclass relation between the class *Myocardium* and the class *Heart structure*.

Philosophical accounts of parthood (Simons 1987; Varzi 2014) have attracted much attention in the analysis of ontological problems in biology and medicine; and numerous use cases have driven the development of domain ontologies that depend on formal descriptions of biological structures. Controversial issues include the transitivity of part-of and the so-called propagation of attributes via partonomic hierarchies (Horrocks et al. 1996; Schulz and Hahn 2005; Schulz et al. 2006). These issues are crucial wherever mereology hierarchies are expected to support generalisable inference mechanisms. For example, a disease or injury being part of an anatomical entity could imply a disease or injury of the whole: A disorder of the retina is a disorder of the eye and a fracture of an elbow is a fracture of an upper extremity. On the other hand, if a part is lacking, this does not imply that the whole is lacking. Such patterns could be used in medical decision support systems or in other intelligent applications in life sciences and health care where we have to account for the broad range between normal and abnormal constitution, shape and function, as well as for the developmental stages that characterize the life cycle of organisms (Schulz and Hahn 2007; Schulz and Johansson 2007). They could also be used to test whether we deal with parthood or not. In both cases, of course, the inference patterns in question need to be valid ones. In this paper, we will discuss a number of candidates for such patterns and assess their scope.

When it comes to its application to biomedical domains, mereology opens up a multifaceted problem space, which has been repeatedly addressed both in philosophy and biomedical informatics. In order to prepare the ground for our discussion in this paper, we will discuss some of these dimensions in the remainder of this introduction.

Domain thesauri, as well as ontology-like artefacts derived from them, often represent taxonomies together with *parthood statements on class level* as opposed to particular things. These statements on class level are expressed in seemingly straightforward formulations like:

$$\textit{part-of}(\textit{Finger}, \textit{Hand}) \tag{8.1}$$

Although this statement closely resembles natural language sentences like "A finger is part of a hand", its precise semantic is a matter of debate. Several competing interpretations for such statements have been suggested. Some read them as universal statements (Smith et al. 2005), others read them as set-theoretical statements (Schulz and Hahn 2002) and yet others – less explicitly – as prototypical statements (Rosse and Mejino 2008). Out of these different possibilities, the first one has been integrated into the Relation Ontology (RO) as a suggested standard for biomedical ontologies (Smith et al. 2005). It takes the time-indexed instance-level relation **part-of** (a, b, t) as a primitive and defines class-level parthood as follows:

$$\begin{aligned} \textit{part-of}(A, B) =_{def} \forall a, t(\textbf{instance-of}(a, A, t) \rightarrow \\ \exists b(\textbf{instance-of}(b, B, t) \wedge \textbf{part-of}(a, b, t))) \end{aligned} \tag{8.2}$$

Of course, this leaves open the interpretation of instance-level parthood relations. It has, indeed, been debated whether standard mereology is appropriate for modelling this relation within the biomedical domain. To start with, the standard mereological **part-of** relation is the reflexive, antisymmetric and transitive proper-or-improper parthood relation (Simons 1987), whereas in biomedicine there is an *implicit restriction to proper parthood*, which is irreflexive and asymmetric. Biological and medical terms containing "partial" are, indeed, generally opposed to terms containing "total". E.g., SNOMED CT contains 804 preferred terms with the modifier "partial" in their name, like "partial larynx" or "partial agenesis of pericardium". In all of these cases, "partial" is opposed to and explicitly excludes "total" or "complete". For instance, "total mastectomy" is never seen as a special case of "partial mastectomy". To generalise, there is no use of the word "part" in biological or medical discourse that would contradict irreflexivity (Schulz et al. 2006). However, this is only a terminological issue: It is an empirical linguistic fact that the scientific language of biomedicine uses "part of" to denote an irreflexive relation, whereas from an logical point of view it is a mere matter of convention whether to use the reflexive "(proper or improper) part" or the irreflexive "proper part" as a primitive.

More substantial is an attack on the transitivity of the standard mereological proper-or-improper parthood: Some authors have argued for the *non-transitivity of parthood* in biology (Johansson 2004; Varzi 2006). E.g., the sheep is part of the flock, the sheep's stomach is part of the sheep, but the sheep's stomach is not part of the flock. However, if we leave aside the use of "part" in ordinary discourse, which must not be considered to be an exact fit to the formal-mereological parthood relation, there are no convincing arguments against the transitivity of the most general parthood relation, for the alleged counterexample can be dissolved if we analyse "part of" in the first premise as a linguistic substitute for "member of", which is a more specialised mereological relation that is not transitive. Below, we will discuss in more detail intransitive subrelations of **has-part**, such as **has-granular-part** or **has-component** (Beisswanger et al. 2008).

More of a problem is the domain-specific application of the parthood predicate. Anatomy, for example, needs to talk about *immaterial parts of material objects*, like cavities and one- or two-dimensional boundaries (Schulz et al. 2006). Biology also talks about *disconnected parts of non-connected wholes*, such as the molecules in a volume of gas, the sheep in a herd or a component of an organ system such as the thyroid gland as part of the endocrine system. Though not physically connected, these entities are considered parts of their wholes as long as they fulfil certain other criteria. We will analyse this later in more detail. General discourse about parts also includes *removed or lost parts*, e.g. a lost hair, a blood sample taken from the body, or a fallen apple, as well as *future parts*, such as car parts. In cases of self-connected wholes, the loss of connection is a clear criterion for the loss of parthood, whereas this is less clear with regard to, e.g. a sheep that has moved away from its herd or a gas molecule that dissipates from the remainder of a volume of gas.

There are a number of phenomena that make the decision about biological parthood a non-trivial problem. The first of these is the *continuous mereological change of biological objects over time*, without which life would not even be possible. My body remains numerically the same even after my hairs have been cut and probably all of a body's molecules are being replaced during life. Mereological essentialism (Simons 1987) postulates that if an object loses a part, it is no longer the same entity. This may hold, e.g., for information entities; a word from which you remove a letter is, indeed, no longer the same word. But mereological essentialism is counterintuitive when it comes to living entities: The loss of a hair does not make me a different individual and metabolic processes in organisms are not feasible without the exchange of matter. Such questions of identity and mereological change lead up to well-known paradoxes, like the infamous Ship of Theseus or the Paradox of Tibbles and Tib (Rea 1995); however, these do not surface in current endeavours to develop biomedical ontologies.

Another issue is the *indeterminacy in the spatial demarcation of biological objects*, with their surfaces exhibiting the most diverse shapes of cavities, tunnels etc. An object located, e.g., in a microscopic cavity of a biological surface structure like the intestinal mucosa, is therefore located within this surface structure. If the cavity were not part of the wall, then the object would be outside the wall and, as a consequence, outside the object that "hosts" it. Especially the condition of some object being "in" a biological object often reflects a situation of being located in some not fully enclosed immaterial part (Schulz and Johansson 2007). Moreover, biomedicine combines the analysis of *different granular partitions* where parthood across different such partitions is often difficult to determine, as in parthood predications with regard to countable objects vs. homogeneous collections of particles of heterogeneous "chunks of stuff" (Jansen and Schulz 2011; Schulz et al. 2006).

All this raises the hypothesis that parthood statements are complicated both by ontological vagueness and by epistemic indeterminacy. In Table 8.1, we list a number of examples of objects that are spatially included within other objects and for which common sense arguments in favour and against the assertion of parthood are collected. For spatial inclusion without parthood we will introduce a new relation, *containment*, which will be formally described below. This allows us to harvest intuitions that we will analyse with more scrutiny in the following sections. However, it also shows that our intuitions are not unambiguous.

8.1.1 Vagueness and Indeterminacy of Biological Parthood

Important inference patterns are sensitive to the interpretation of the parthood predicate. Therefore, the question whether something is part of, e.g., a cell or an organism, or merely contained by it, has noteworthy implications: Radiologists identify a certain structure in a medical image and ask themselves whether this depicts a modified structure of the body or rather something contained in it as a foreign body. Also, the distinction between "own" and "alien" is fundamental to the functioning of an organism's immune system. It can even be of legal and ethical

Table 8.1 Biological problem cases, with arguments in favour and against parthood, given a spatial inclusion relation between objects

Case	Arguments in favour of parthood	Arguments against parthood
(A) My body presently includes a certain portion of urine. Is this portion of urine a part of my body?	(i) The portion of urine is the result of a biological process that occurred in my body, (ii) It contains organic material with my body's genetic identity, (iii) It is completely surrounded by body parts.	(iv) The urine does not contribute to the functioning of the body, (v) Its presence in the body is short-lived, (vi) Most of the urine volume is of an inorganic nature (H_2O), (vii) A longer presence in the body will negatively affect the functioning of the body.
(B) A cell contains a certain volume of water. Is this water part of the cell?	(i) The cell would not be functional if this water volume were removed. Furthermore, it would probably be completely destroyed.	(ii) The water volume is inorganic.
(C) This H_2O molecule is now included in one of my cells. Is it part of this cell?	(i) The molecule contributes (together with many others) to the functioning of the cell.	(ii) H_2O is inorganic and does not require a biological process to be synthesised, (iii) It does not necessarily come into being in the cell, (iv) Its presence therein may be very short.
(D) A brain now includes this brain metastasis. Is this (ill-formed; Schulz and Hahn 2007) object a part of this brain?	A metastasis derives from malignant tissue that originates elsewhere. (i) It shares the genetic identity with the brain, (ii) It is tightly connected with the brain, (iii) It exchanges matter with it.	(iv) It did not originate there, (v) It is a non-canonical structure, body part, (vi) it negatively affects the function of the brain.
(E) My body now includes this E. coli population. Is it a body part?	(i) Intestinal bacteria contribute to canonical body functions, (ii) Every organism hosts a bacteria population.	(iii) The bacteria are genetically different, (iv) They are not attached to the organism.
(F) This single bacterium is included in my body. Is it a part of my body?	(i) The bacterium is part of a larger bacterial population bearing important body functions.	(ii) The bacterium may not have started its life within my body and may cease to exist outside (cf. (E).(i)).

relevance whether something is part of a human organism or merely located within it. If we subscribe to the principle that a defect of a part is also a defect of the whole, we will have to accept that a defect of an artificial heart valve implies a defective heart if this artificial valve is considered to be a part of the heart. On the other hand,

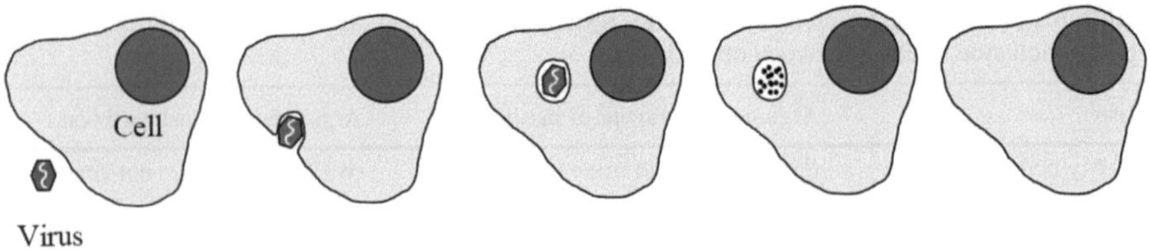

Fig. 8.1 A virus is ingested by a cell. When do virus components become part of the cell? (Schulz and Johansson 2007)

it would be counterintuitive to interpret a structure abnormality of an embryo as an abnormality of the maternal organism.

As opposed to parthood, containment is a matter of mere spatial location, like cookies are contained in a box without being a part of it. Later in this section, we will introduce a formal definition of containment. To be sure, containment is itself a matter of vagueness and it is not at all clear where parthood starts and containment ends (Schulz and Johansson 2007).

One example from biology is phagocytosis, i.e. the gradual incorporation and re-use of foreign material by a cell, such as depicted in Fig. 8.1. Similar problems arise with ingestion and digestion processes in general: We could trace the fate of a single carbon atom which is taken in into the body as part of some piece of food and later becomes part of an animal's tissue (Donnelly 2009). At which time does it become part of the animal's body?

In the following, our discussion of such phenomena will be based on a number of assumptions. Of some of these assumptions we are convinced that they are true, although we will not argue for them in this paper. Foremost, we subscribe to a notion of biological objects, according to which they are independent continuants (Grenon and Smith 2004) that either are living three-dimensional material beings or material parts thereof (Schark 2005). In addition, we will make some methodological restrictions for the sake of simplicity. E.g., this paper will only consider objects at the times they are included in or constituting some living organism. That is, we will not discuss, e.g., dead bodies, fossils, tissue samples and processed biological material like food items, leather, cotton, timber etc., as long as they are not included in living biological systems. Moreover, we will assume that every biological object occupies at any time an exactly one spatial region. That is, we will not consider vague boundaries here, although they are typical for many biological surfaces as mentioned above.

We take into consideration, however, the possibility that biological objects may have "fiat boundaries", i.e. that they may be arbitrarily delineated by human fiat (Smith and Varzi 2000; Vogt et al. 2012). Material objects may also have immaterial objects as parts, especially cavities, in line with the current specification of BFO version 2 (Smith et al. 2013). We will, however, not consider zero to two-dimensional boundaries (i.e. points, lines and planes), which could also be considered to be parts of their hosts (Donnelly 2011).

We also assume that when asserting parthood between two objects, this assertion refers to the current state of affairs, despite the common extension of the word "part" to denote past and future parthood, such as in calling something a car part because it is being produced with the intention to become part of a car (intended part), or calling something a body part because it has been severed from the body (past part). Our fundamental assumption in this paper is that, in accordance with (Bittner 2004), spatial inclusion is a necessary condition for material parthood. Standard material parthood, that is, implies spatial inclusion:

$$\textbf{has-part}(I, i, t) \rightarrow \textbf{includes}(I, i, t) \tag{8.3}$$

with "I" ("includer") and "i" ("includee") denoting individuals.

We need to say more about spatial inclusion, however. Any three-dimensional entity can be included in any other three-dimensional entity, be they material objects, immaterial spatial objects or spatial regions. We use the term "inclusion" instead of the more common term "location" because the latter is normally used for the relation of an object to a spatial region only and not, as we need, to another material object. The conception of inclusion is thus broader than the standard notion of spatial location (Casati and Varzi 1999) where the range is restricted to spatial regions. This extension of the concept of location is in line with the ongoing BFO 2 specification (Smith et al. 2013), as well as with the BioTop upper domain ontology (Beisswanger et al. 2008).

The inclusion predicate can formally be defined in terms of overlapping regions, where the region of a thing is, in turn, the set of points in space occupied by this thing. If we follow this strategy, talk about an object I including another object i is a simplified way of stating that, at a given time, all of i's parts are located in I's region or, even more simple, i is located within I's region.

We can give a formal semantics for the inclusion relation using notions from mathematical topology. On this account, each physical object c occupies exactly one spatial region at a time t, namely **region**(c, t). This spatial region is, by definition, exactly occupied by c at time t. Using **has-part** as a primitive, we can then define the relation of material inclusion and containment as follows:

$$\textbf{includes}(I, i, t) =_{def} \textbf{point-subset-of}(\textbf{region}(i, t), \textbf{region}(I, t)) \tag{8.4}$$

$$\textbf{contains}(I, i, t) =_{def} \textbf{includes}(I, i, t) \wedge \neg\textbf{has-part}(I, i, t) \tag{8.5}$$

$$\textbf{part-of}(i, I, t) =_{def} \textbf{has-part}(I, i, t) \tag{8.6}$$

The relation **includes** is defined in terms of topologic (point-set) inclusion, whereas containment is defined as spatial inclusion without parthood. As **has-part** has been introduced as a primitive term, we cannot expect any sufficient criteria for parthood with general applicability. The question, however, is, whether we can state necessary conditions or conditions that are sufficient for certain subdomains. All of these three relations (i.e. **has-part**, **includes** and **contains**) are considered to be transitive (Bittner 2004), but we will later discuss intransitive sub-relations of **has-part**.

On this basis, the main question of the deliberations in the remainder of this paper can be phrased as follows: What distinguishes full-blown parthood from mere containment? Are there criteria that can be added to inclusion in order to get parthood? And are there criteria that can be added to inclusion to get containment?

8.2 Soft Criteria: Function, Origin and Genetic Identity

In this section, we discuss an extended list of criteria and explore whether they support either parthood or containment. The assessment is, above all, guided by cognitive adequacy. In Sect. 8.3, we strive for clearly delineated, ontological criteria, aware, however, that we can, at most, identify "islands" of ontological certainty within an ocean of vagueness. These criteria will be mostly categorial criteria, i.e. criteria that refer to the kind of beings includer and includee belong to. In this section, however, we will explore a number of prima facie criteria referring to origin, function and genetic identity.

We will now start our quest for such criteria by revisiting a decision algorithm that has been suggested by Schulz et al. (2005). This algorithm makes use of a set of simple decision rules (see Table 8.2): (i) Is the includee an artefact? (ii) Is the function of the includee relevant for the integrity of the includer? (iii) Do includee and includer have the same genetic origin? (iv) Has the includer hitherto been located in the includer or in a part of the includer?

Of these, the first criterion appears to be a monadic criterion, asking for a property of the includee alone. It is based, however, on the assumptions that the includer is a biological entity and that no artefact can be part of a biological entity. However, the distinction between biological and artificial entities is not as clear-cut as would be necessary for successfully applying this criterion (Jansen 2013). Engineered cells and genetically modified mice are both artefacts and biological entities. Moreover, the question whether something of artificial origin can be part of a biological entity will be answered differently with respect to different levels of granularity. Were there artificially produced lipid molecules, we see no reason why these should not become part of, say, a membrane that is part of a living cell. We will now look at the remaining criteria, discussing in turn (i) genetic identity, (ii) functionality and (iii) spatiotemporal origin. First, we can check whether includer and includee have the same genetic identity. For criminalistic purposes, the 'genetic fingerprint' is normally a reliable guide to decide whether detached organic materials derive from a certain individual organism. This criterion can, however, not be generalised: On the one hand, the cells of monozygotic twins have exactly the same genetic identity. On the other hand, mosaic organisms occur where different organs can have cells of different genetic identity. Further problem cases for genetic criteria are colonial organisms (e.g. microorganisms in a biofilm or the zooids constituting the jellyfish-like *Portuguese man o' war*), composed of a multitude of single organisms which cannot survive when separated. The boundary between a colony and a single organism is vague. Slime moulds normally live as colonies

Table 8.2 Decision algorithm proposed in Schulz et al. (2005)

```
If located-in (c, d, t) then
  If Artifact (c) then
    contained-in (c, d, t)
  Else
    If function-integrity-relevant (c, d, t) then
      part-of (c, d, t)
    Else
      If (not same-genetic-origin (c, d, t)) or
        (instance-of (c, MaterialObject) and instance-of (d, ImmaterialObject)) then
          contained-in (c, d, t)
      Else
        If hitherto-located-in (c, d, t) or (hitherto-located-in (c, m, t) and part-of (m, d, t)) then
          part-of (c, d, t)
        Else
          contained-in (c, d, t)
        End If
      End If
    End If
  End If
End if
```

of single cells, but can aggregate to form a single body-like whole. Composite organisms like lichens aggregate fungal and prokaryotic cells. Talking about genetic identity or similarity is, therefore, not trivial. These problem cases notwithstanding, the following cases are typically distinguished:

- Two entities are **autogenic** if and only if they share the genetic identity of the same individual. Example: a saphenous vein autograft used in coronary artery bypass surgery.
- Two entities are **allogenic** if and only if they share the genetic identity of the same species, but not of the same individual. Example: typical organ transplants (kidney, liver, heart) from one human to another.
- Two entities are **xenogenic** if and only if they do not share the genetic identity of the same species. Example: a baboon heart transplanted to a human baby.

All three predicates can only be applied in a meaningful way if both includer (I) and includee (i) carry genetic information, and they are difficult to apply in the case of chimeras that contain body parts of different genetic identity. On first sight, we have reason to assert parthood for autogenic includees and to assert containment for allogenic and xenogenic includees. But it is difficult to derive clear-cut criteria for parthood or containment from this. If, for example, xenogenic entities play a vital role in the organism, it could be justified to consider them as parts of the organism.

This points us to the next feature that could be used in parthood criteria, i.e. the *functionality* of the includee for the includer:

- We have reason to assume that the includee i is part of the includer I if i is **part-for** a certain type I_{TYPE}, of which I is an instance. The relation **part-for** holds between an individual includee i and a type, if and only if i has a (token) function f, which can only be realized in case i is included in some instance of I_{TYPE}. A typical example is an ion channel protein which is synthesised in a ribosome, but becomes functional only once it is included in a cell membrane. Similarly, despite having its own independent genome, a mitochondrion can be considered to be part of some eukaryotic cell because mitochondria completely depend upon their host cell for survival and functioning, viz. ATP supply. This contrasts with an intestinal *E. coli* bacterium (or any other cell from a human organism's microbiome), which is able to live and function outside its host.
- In contradistinction, we have reason to assume that the includee i is only contained in the includer I if i has a function that cannot be realised while it is included in some instance of I. Examples: the ion channel as long as it is still within the cytoplasm; a surgically removed heart valve in a cooling container. The heart valve is only functional when included in some biological heart, i.e. it is a **part-for** the type heart. It cannot be functional when isolated from a heart in the cooling container.

Conversely, the function of the includer depends on an includee. In this case, we can say that the includer is a "whole for" the includee, i.e. the part is necessary for the functioning of the whole. This allows us to formulate two additional criteria:

- We have reason to assume that the includee i is *part of* the includer I if I has a function that can only be realised if some instance of a certain type i_{TYPE} is included in it. Examples: My digestive tract requires a stomach for its complete functioning, or a cell cannot live without mitochondria.
- We have reason to assume that the includee i is *contained* in the includer I if the latter has a function that cannot be realised if some instance of a certain type i_{TYPE} is included. An example is the complex of a receptor and an antagonist molecule. In such a complex, a small molecule a of a certain type is bound to a receptor and blocks its function. In this situation, we have reason to deny that a is a part of the receptor.

Functional criteria can, however, be in conflict with criteria referring to the *spatiotemporal origin* of the includee in the includer and the *physical connection* (Schulz and Johansson 2007) of includer and includee.

- We have reason to assume that the includee is part of the includer if the includee originated within the includer. Example: The heart and the brain of a human originate within that human.
- We have reason to assume that the includee is contained in the includer if the includee originated outside the includer. Example: A brain metastasis of a breast cancer is included within the brain, but it originated in the breast tissue.
- We have reason to assume that the includee is part of the includer if the includee is physically connected to the includer (it cannot be severed from the

includer without physical damage). This criterion would allow the rejection of the parthood hypotheses for containees like urine or faeces, which are not tightly connected to the surrounding body structures. In contrast, a metastasis is tightly connected to the surrounding tissue (blood supply), whereas a sequestrum (piece of dead bone) is much less connected.

These criteria can be combined with additional conditions, such as:

- We have reason to assume that the includee is part of the includer if the includee originated together with the includer. Example: The endothelium of my aorta originated together with my aorta.
- We have reason to assume that the includee is part of the includer if the spatiotemporal inclusion is permanent. Examples: all the organs in the development of an embryo, a primary tumour of the brain.

Criteria that draw on the origin of the includee could be disputed because they draw on (causally irrelevant) historical properties of the includee. This problem is especially pressing because within one and the same includer, instances of the same type may fulfil different spatiotemporal criteria. Consider, for example, a cell in a thyroid gland containing several molecules of the hormone L-thyroxin. Some of these molecules may have been synthesised within this cell and never left it, others were synthesised within the cell, left it and returned, others were synthesised in a neighbour cell, whereas still others could have been synthesised in a lab and entered the body as a drug. The problem is here, that within a snapshot view of the cell, there is no difference between these molecules. If the fact that a certain molecule never left the cell were a sufficient criterion to consider it a part of it, why should we withhold this status to the other L-thyroxin molecules just because of their deviant history? The criterion of connection (for the discussion of several strengths of connection, see Schulz and Johansson 2007), however, would make a difference also in a snapshot view. If i originated within I and is tightly connected to it (and cannot just go in and out like many molecules in cells), then this seems to be a good criterion for granting parthood.

These criteria capture some of the intuitions underlying our judgements about biological parthood (cf. Table 8.1). At times, however, their results are unsatisfactory and can also contradict each other. Artificial heart valves are, by all means, necessary for the functioning of the including organism, but they do not originate from this body, nor do they have a genetic identity. Similarly, medically indicated allografts and xenografts are normally necessary for the functioning of the receiving organism, but they do not originate in these organisms, nor do they share the same genetic identity. The bacteria on the mucous membranes of a human body are canonical contributors to its proper functioning, they may have originated within this body, but they do not share its genetic identity. As they allow for exceptions, these criteria can only be considered ‘soft criteria’; they are more or less reliable rules of thumb.

8.3 Hard Criteria: Containers, Grains and Components

8.3.1 *Topological Descriptions of Material and Immaterial Objects*

Our question was: how and when can we infer parthood from spatial inclusion? This question is only meaningful if we can characterise spatial inclusion in a non-circular way, i.e. if we can describe the makeup of physical objects in terms of their spatial arrangement without any reference to parthood. The following example illustrates the challenges of such an approach.

Imagine a simple toy object o, consisting of a wooden box b and a dice d (Fig. 8.2 i, ii), which is located inside b, i.e. it fills the hollow space h (Fig. 8.2 iii) enclosed by b. A standard mereological approach would describe o as a mereological sum of its two component parts b and d, like in Fig. 8.2 iv. Note that b *contains* d, which is, therefore, not a part of b, but both would then form the sum o.

How can we describe this scenario without any reference to parthood and mereological sums? Let us assume that we can find out for each point in space whether it is occupied by some material entity. This allows us to define the toy object o as all physical matter that exactly coincides with the points that make up the regions of the box and the dice. We can generalise on this by defining what we may call a "topological sum", or "t-sum" for short, which is solely based on point-set theoretical considerations. By this, we avoid the use of the mereological part-of relation.

Definition 1. Let $s_1, \ldots, s_n$ be material or immaterial objects. S is the t-sum of the summands $s_1, \ldots, s_n$ if and only if S is the (self-connected or scattered) object that comprises any spatial region occupied by some summand and the totality of matter that is located at any point in space occupied by some of its summands.

According to this definition, the complete toy object o would then be constituted by the totality of material entities in the joint region occupied by b and d (Fig. 8.2 iv). Objects like b in our example may act as a container, i.e. as material entities that host a hollow space like h (Fig. 8.2 iii) which, in turn, can be filled by another object which is not itself part of the host. Hollow spaces exist, whether they are empty or not, i.e. whether they contain material objects such as d, a marble or a bug. Using Definition 1 we could try to define the container c as everything that is within the regions occupied by the material object b and the hollow space h (Fig. 8.2 vi). But then we have the problem of how to distinguish between c in case it is filled by d on the one hand and the whole object o on the other hand, for the dice d included in h is located within the region of c and, according to our definition, it will inevitably be included in the t-sum, as it lies inside its spatial region. A filled container would, therefore, always be a different kind of thing than an empty container. Moreover, we were not able to distinguish between the container and the container plus its content. That is, according to our example (as d is inside h) the following holds:

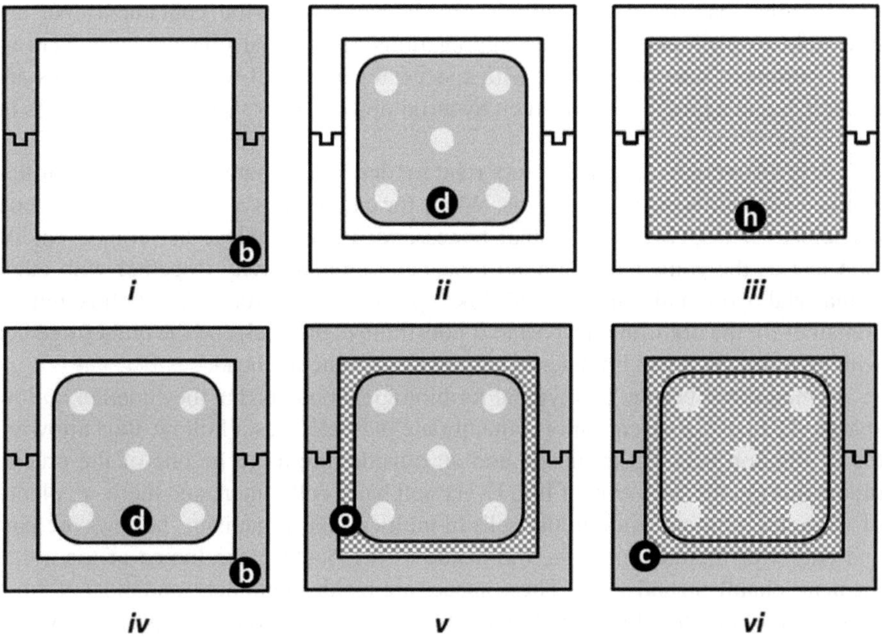

Fig. 8.2 The figure shows different (partly overlapping) objects that can be composed out of a box b (**i**) and a dice d (**ii**). The immaterial hollow h (**iii**) is the space enclosed in b. These three elements can be combined in different ways: (**iv**) shows the t-sum of the material parts b and d; (**v**) shows the t-sum of the material parts b and d and the immaterial h; (**vi**) shows the container c as the ti-sum of b and h that excludes d even if it is included in b. The *grey mark* signifies that both the spatial region and the matter situated in it belong to the object in question, while the chequered area signifies that only the spatial region belongs to the object, but not the matter contained in this region

$$\textbf{t-sum}(b,h) = \textbf{t-sum}(b,h,d) \tag{8.7}$$

This shows that the t-sum does not satisfy our purpose because it always adds the included entity to the container. We can avoid this by defining a sum that excludes the material includees of immaterial includers (ti-sum):

Definition 2. Let $s_1, \ldots, s_n$ be material or immaterial objects. S is the ti-sum of the summands $s_1, \ldots, s_n$ if and only if S comprises any space occupied by some of the summands and the totality of matter that is located at this space except the matter located inside of immaterial summands.

In contrast to proposition 8.7 above that holds for the t-sum, the following hold for the ti-sum:

$$\textbf{ti-sum}(b,h) \neq \textbf{ti-sum}(b,h,d) \tag{8.8}$$

$$c = \textbf{ti-sum}(b,h) \tag{8.9}$$

This means that the ti-sum allows to adequately describe containers. We are now able to distinguish between aggregations of material objects, containers as aggregations of material objects with spaces, regardless of whether the spaces are occupied, and aggregations between material objects, their spaces and the fillers of the spaces.

These distinctions are hugely important for describing material biological entities from cell components to organisms. What, for example, is a skull? Just as the box *b* encloses *h* (Fig. 8.2), the cranial bones and tissues enclose the cranial cavity. And just as the container *c*, cranial bones and cranial cavity (together with other immaterial parts) make up the skull. Like *c* as a container for *d*, the skull is, thus, a container for the brain, and just *c* and *d* constitute *o*, the skull and the brain (together with other material and immaterial entities) constitute the head.

As mentioned before, it is typical for biological objects that they contain hollow spaces. In fact, the invention of a membrane that encloses a hollow, thus allowing the distinction between an inside and an outside, seems to be one of the crucial innovations in the evolution of life. Every cell has a cell lumen, and there are plenty of vescicles with an inside in the cell. In multicellular organisms, hollows are part and parcel of the anatomy, like the inside of the bladder, the lumen of the aorta, the heart chambers and so on. These spaces are inside containers, and most of these spaces are in fact filled with material entities. These material entities are, however, not part of the container in question: what is inside the skull is not part the skull, the content of the bladder is not part of the bladder, and the content of the aorta is not part of the aorta. To generalise, things included in a container are not part of the container but contained in it.

There are many more examples that suggest themselves as clear cases of mere containment as opposed to parthood:

- If Mary swallows a glass marble, this marble does not become a part of her stomach, but is only contained in it because the marble existed before swallowed, it is expected to be expelled, it is not a biological object, it is not physically connected to her stomach and it does not have any function in her stomach. Even if it remains jammed in some body structure, it is not part of it. Also, if a milk tooth were swallowed instead of the marble, it would not be considered to be a part of the stomach either, although the tooth would be an object of biological origin, originating even in the same body.
- Similarly, a cotton wool pad in my mouth during a dental treatment is contained in my head, but not part of it because the pad existed before, the pad remains only for a short time in the head, it is not physically connected to my head and although it has a specific function while located in my oral cavity during the treatment, this is not a physiological function.
- My oral cavity does now contain a certain amount of saliva, which is not a part of it. This is because a cavity is an immaterial object, whereas saliva is a material object and an immaterial object cannot possibly have a material part.

We have thus identified inclusion in the immaterial hollow of a container as a sufficient condition for containment. To be sure, this is a very local criterion. But as a

Table 8.3 Types of complex entities

	mono-sortal	multi-sortal
flexible re. number	"flexible collectives" (e.g., a portion of water)	"flexible compounds" (e.g., a hand)
strict re. number	"strict collectives" (e.g., a pair of kidneys)	"strict compounds" (e.g., a water molecule)

sufficient condition for containment it is also a sufficient condition against parthood. It should, however, be noted that we can only somewhat tautologically infer non-parthood with respect to the container itself. It is possible that the container is both a body part and containing other body parts. It is only excluded that the container contains container parts, i.e. parts of itself. The blood included in the aorta, for example, is not part of the aorta. It is, however, part of the body of which the aorta is also a part.

8.3.2 Collections and Compounds

More parthood relations can be drawn from subrelations of **part-of**. In Jansen and Schulz (2011), we distinguished different kinds of complex entities based on the parthood relation (Table 8.3). A collective is the sum of one or more grains that are all instances of the same type, while a compound is the sum of one or more components that may be of different types. A portion of water can be regarded as a collection of grains of the same type, namely H_2O molecules, while a H_2O molecule is a compound of two hydrogen atoms and one oxygen atom. A complex is strict if numbers of parts matter; it is flexible if they do not. E.g., adding an oxygen atom to a water molecule would turn it into a hydrogen peroxide molecule; the molecule would cease to be a water molecule. On the other hand, a portion of water would remain a portion of water if more H_2O molecules were added, and a flock of sheep would remain a flock of sheep if more sheeps joined. Both collectives and compounds are given by the summation of non-overlapping material components. We introduce the relation **has-grain** for relating collectives with their elements and **has-component** to connect compounds with their components. As exemplified in the toy example, the region covered by a component may include immaterial entities, but whether what is contained by this space is ignored or not will depend on our definition, just as we can decide whether the dice d or only the space h belongs to the box in Fig. 8.2. As pointed out in Jansen and Schulz (2011), the classification depends on the specificity of the sortal distinctions: If the right and the left kidney are considered as two different sorts, then a pair of kidneys would be a strict compound instead of a strict collective. The cardinality criterion (i.e. strict vs. flexible in number) allows the inference that, if we take away one component of a numerically strictly defined entity, it becomes an entity of a slightly different sort, often a defective entity. For instance, if a pawn has been taken away from a complete set of chess pieces, the remainder will no longer be of the type "complete chess set". Similar examples would be the extraction of a tooth, a fingernail, the tonsils or an eye lens.

There are also compounds of collectives, such as mixtures or solutions. A mixture of water and ethanol is a compound of exactly two components, i.e. the water fraction and the ethanol fraction. Either fraction is a collection of H_2O or C_2H_5OH molecules, respectively. In biology, more complex compounds of collectives are tissues and body substances. Now compounds are generally expected to have their components clearly delimited from each other, whereas in biological systems, especially in larger anatomical structures, the exact boundaries of an object have to be drawn by fiat. The kidneys, for example, are connected to other body structures by tubular structures (vein, artery, ureter) which do not display any discontinuity that would demarcate the boundary of the kidney. In principle, it would be possible to extend the application of the term "compound" to objects that are structured by fiat divisions, such as the brain that is anatomically divided into the right and the left hemisphere. However, as this is in tension with the intuitive idea of composition, we prefer to introduce the relation **has-fiat-division** for these cases.

All of the three newly introduced relations are specialisations of inclusion, therefore:

$$\textbf{has-grain}(I, i, t) \rightarrow \textbf{includes}(I, i, t) \tag{8.10}$$

$$\textbf{has-component}(I, i, t, p) \rightarrow \textbf{includes}(I, i, t) \tag{8.11}$$

$$\textbf{has-fiat-division}(I, i, t, p) \rightarrow \textbf{includes}(I, i, t) \tag{8.12}$$

As there are many possible fiat divisions and often several ways to decompose a compound, we add the partition scheme p as a fourth argument of the relations **has-component** and **has-fiat-division**. For instance, we can consider all atoms of a small organic molecule as its components, but also its carbon chain and its functional groups. In any of these two partitions, the typical characteristics of a compound are preserved: the components do not overlap and the whole is of a different type if one of its components is missing. The relation **has-grain** is, by definition, intransitive, as are **has-component** and **has-fiat-division**, once the partition scheme p is hold fixed. With the help of these three subrelations of **has-part**, we can expose some non-contentious parthood and containment cases:

- First, if a certain H_2O molecule is a grain of a certain amount of H_2O, it is obviously one of its parts. This is an instance of the general **has-grain** relationship between a collection and its grains, as introduced above. The argument that all grains are parts of the collection derives from the fact that a collection is grounded in its elements, i.e. it ontologically depends on them. Although the removal of just one grain generally does not affect the identity and sortality of the collection (mind the looming Sorites paradox!), there is nothing on which the collection depends and which constitutes the collection apart from its grains.

- The relation between my right femur and my skeleton is an example for the relation between a component and a compound. The sortality of the compound depends on the completeness of its components. This would support the claim to axiomatically state that all component-compound relations are parthood relations.
- The relation between the lower third of my oesophagus and the oesophagus is an example for the relation between a fiat part and its whole. By definition, fiat divisions are parts of their wholes. With fiat parts, the whole is, in a way, prior to the parts.
- The relation between my brain and my body is another case of the relation between an anatomic component and the human body as a compound. Here, an even stronger case can be made for parthood: My brain is now part of my body because (i) both are physically connected, (ii) my body does not function without my brain, (iii) my brain does not function without my body, and (iv) my brain's function cannot be substituted by any other object. (It can, however, be spatially separated from my body and it could also outlive my body, but given the present state of art in medicine, it will no longer be functional in this case.) Another argument would be that a brainless body would be an entity of a different sort. Whether the brain is a component of the body in the above sense depends on how it is divided into different, non-overlapping components.

8.4 Some Possible Axioms

In effect, the hard criteria we found are either categorial criteria or derive from the logical properties of specific subrelations of **has-part**. Such criteria refer, among other things, to the ontological categories of includer and includee. We will now formulate some axioms for biological parthood based on such categorial distinctions. In these axioms, we refer to the following categories as top-level types of entities using the three-letter type names given here:

- *MAT*: material object, e.g. a cell, a molecule, an organ, the intestinal mucosa.
- *CPD*: compound, i.e. it implies a well-defined number, type, and arrangement of elements. If one part is missing, the compound is incomplete. For instance: my lens is a component of my eye.
- *CON*: container. A container has a 3-D immaterial part, the possible contents of which are not part of the container. An example is the cranium, which has the cranial cavity as an immaterial part.
- *COL*: collection of grains of the same type, the exact number of which is irrelevant, e.g. the cells in an early embryo before differentiation, or the water molecules in a litre of water. Collections are typically homomereous, i.e. they have proper parts of the same type.
- *MFI*: material fiat object, i.e. an object that is divided by a fiat (arbitrary) boundary from a larger includer. An example is the lower third of the oesophagus.

- *ART*: non-biological artefact. It means that it is the output of a manufacturing process and does not engage in metabolic exchange with biological objects.
- *SEL*: self-connected object. I.e. an uninterrupted line can be drawn between any two points within the region of the object. Abutting of subunits of the object is sufficient; there is no requirement for strong chemical bonds.
- *IMT*: immaterial object. Example: three-dimensional hollow space. As illustrated in Fig. 8.2, hollow spaces necessarily have a "host". Holes are part of their hosts (Smith et al. 2013, 2005). Immaterial objects can also have fiat boundaries. Example: the lumen of the lower third of the oesophagus.
- *SPR*: spatial region, i.e. a three-dimensional portion of space.

We will now formulate some axioms for biological parthood based on such categorial distinctions.

For a start, immaterial (*IMT*) includers do not have material (*MAT*) parts. Hence, if an instance of *MAT* is included in an instance of *IMT*, it must be contained therein. If, e.g., the cavity of my stomach includes a portion of food, it contains that food. The lumen of my aorta includes a portion of blood; therefore it contains it. In general:

$$\begin{aligned}(\mathbf{includes}(I,i,t) \wedge \mathbf{instance\text{-}of}(I,IMT,t) \wedge \mathbf{instance\text{-}of}(i,MAT,t))\\ \rightarrow \mathbf{contains}(I,i,t)\end{aligned} \tag{8.13}$$

$$\mathbf{has\text{-}part}(I,i,t) \wedge \mathbf{instance\text{-}of}(i,MAT,t) \rightarrow \mathbf{instance\text{-}of}(I,MAT,t) \tag{8.14}$$

Following these patterns, we can formalize the following restrictions:

- A material object can only be part of material objects; it cannot be part of an immaterial object or a spatial region.
- An immaterial object can be part of a material or an immaterial object, but it cannot be part of a spatial region.
- A spatial region can only be part of spatial regions; it cannot be part of a material or immaterial object.

Containers are material entities and have at least one immaterial part. An aorta, for example, is such a container; it has an immaterial part, i.e. its lumen. Hence, the existence of a container implies the existence of an immaterial part:

$$\mathbf{instance\text{-}of}(I,CON,t) \rightarrow \exists i\,(\mathbf{has\text{-}part}(I,i,t) \wedge \mathbf{instance\text{-}of}(i,IMT,t)) \tag{8.15}$$

Conversely, not every material entity that has immaterial parts is a container and not every includer of a container contains what is in the container. For instance, my aorta is a body part that is a container for some portion of blood. This portion of blood is, therefore, not part of my aorta. However, the blood is part of my body, as well as the aorta and its lumen.

Further restrictions follow from the properties of the relations **has-grain**, **has-component** and **has-fiat-division**, which are specific subrelations of **has-part**. Because of this subrelation property we can state:

$$\textbf{has-grain}(I, i, t) \rightarrow \textbf{has-part}(I, i, t) \tag{8.16}$$

$$\textbf{has-component}(I, i, t, p) \rightarrow \textbf{has-part}(I, i, t) \tag{8.17}$$

$$\textbf{has-fiat-division}(I, i, t, p) \rightarrow \textbf{has-part}(I, i, t) \tag{8.18}$$

Of these, **has-grain** holds between a collective and its grains, **has-component** between a compound and its components and **has-fiat-division** holds between a thing and its fiat parts (i.e. parts existing due to human fiat). In contrast to **has-part**, each of these relations is asymmetric and none of them is reflexive or transitive.

As a consequence of formulae (8.16), (8.17) and the transitivity of **has-part**, a grain of a component of a compound is also a part of this compound. For instance, a water molecule of a water/ethanol mixture is a part of this mixture. Also, if we regard the totality of water molecules as a component of the cytoplasm of a cell, then each individual water molecule is also part of it (and of the cell if we regard the cytoplasm as a component of the cell).

Collections and compounds may comprise sub-collections and sub-compounds. Although these are not grains or components of their includer, they can be considered parts of it. For this it is, however, necessary that all grains of the sub-collection or all components of the sub-compound are also grains or components of the original includer. For example, the collection of all mitochondria in my liver is a part of the collection of mitochondria in my body.

$$\begin{aligned}(\textbf{instance-of}(I, COL, t) \wedge \textbf{instance-of}(i, COL, t)\\ \wedge\forall x(\textbf{has-grain}(i, x, t) \rightarrow \textbf{has-grain}(I, x, t)))\\ \rightarrow \textbf{has-part}(I, i, t)\end{aligned} \tag{8.19}$$

$$\begin{aligned}(\textbf{instance-of}(I, CPD, t) \wedge \textbf{instance-of}(i, CPD, t)\\ \wedge\neg\exists p(\textbf{has-component}(i, x, t, p) \rightarrow \textbf{has-component}(I, x, t, p)))\\ \rightarrow \textbf{has-part}(I, i, t)\end{aligned} \tag{8.20}$$

We can also try to formalise the functional intuitions connected with the **part-for** relation: If an entity i only realises its function when included in an instance of I_{TYPE} and if it is presently included in I, which is actually an instance of I_{TYPE}, then i is a part of I. As we said, an ion channel protein is a **part-for** a cell membrane and if such a protein is actually included in a certain cell membrane, it is a part of it. However, a cardiac pace maker is also a part-for a human body and similar arguments could be brought forward for xenografts (e.g. a baboon heart to be transplanted to a human child). As these cases are more contentious, we exclude from this criterion all artefacts and xenogeneic material, i.e. material from different species:

$$\begin{aligned}&(\textbf{part-for}(i, I_{TYPE}) \wedge \textbf{instance-of}(I, I_{TYPE}, t) \wedge \textbf{includes}(I, i, t)\\&\wedge\neg\textbf{instance-of}(i, ART, t) \wedge \neg\textbf{xenogeneic}(I, i, t)) \rightarrow \textbf{has-part}(I, i, t)\end{aligned} \tag{8.21}$$

In a similar way we can now formalise the intuitions connected with the **whole-for** relation. According to this intuition, i is part of I if i is included in I and if i is an instance of i_{TYPE} and if I needs some instance of type i_{TYPE} in order to realise its function. For instance, the cell membrane needs ion channel proteins in order to realise its function. Hence, this intuition says, the ion channel proteins are part of the membrane. We need, however, to be careful to not overgeneralise this criterion, for a car does not only rely on its motor and wheels (which are all among its parts), but also on petrol (which is not one of its parts). Similarly, the stomach cannot realise its function without food and the lungs not without air to breathe. Petrol, food and breathing air are all taken into their includer by way of containers: The petrol tank is a container for petrol as the stomach is a container for food. Hence, we should not extend the criterion to things that are included in a container that is part of the includer. However, we must not exclude all such things, for the aorta is a container for blood, but the blood contained in the aorta is, nevertheless, a part of the body.

Again, we exclude artefacts from the criterion. Otherwise, we would be forced to admit that a drug that keeps alives a biological entity is also part of that entity:

$$\begin{aligned}&(\textbf{whole-for}(I, i_{TYPE}) \wedge \textbf{instance-of}(i, i_{TYPE}, t) \wedge \textbf{includes}(I, i, t)\\&\wedge\neg\exists c(\textbf{instance-of}(c, CON, t) \wedge \textbf{part-of}(c, I, t) \wedge \textbf{includes}(c, i, t))\\&\wedge\neg\textbf{instance-of}(i, ART, t)) \rightarrow \textbf{has-part}(I, i, t)\end{aligned} \tag{8.22}$$

8.5 Discussion

8.5.1 *Relevant Inferences*

It is mostly the combination of compounds and collections which helps to clarify important issues: For instance, a single calcium channel is part of a cell membrane because the whole collection of calcium channels of which it is a grain is necessary for the functioning of the membrane (cf. (8.16)), and is, therefore, part of the membrane. The parthood condition of the single molecule then derives from the transitivity of the part-of relation. We do not assert that the portion of urine that is included in the bladder is a part of the body because neither the urine has any function in the bladder nor does the body's functioning depend on the urine. To the contrary: The urine has to be discarded regularly. This is different for the portion of blood in the body because it has clearly described functions in the body and the body could not function without it. Even a small amount of blood would be part of the body because it constitutes a fiat division of the entire blood and is therefore part of the latter.

Our axioms also solve the puzzle of the virus that is gradually digested by a cell as depicted in Fig. 8.1. As long as the virus is dismantled to viral protein molecules or fragments, these are contained but not parts. However, if these macromolecules are split into small molecules such as amino acids or nucleic acid monomers, these are no longer distinguishable from, e.g., the collection of alanine or cytosine molecules that were already included in the cell before. The collections of these molecules are functionally relevant for the cell's functioning; therefore, a given monomer becomes part of the cell as soon as it is no longer part of any of the viral molecules.

With regard to drugs, it is, therefore, also decisive for parthood whether a single drug molecule is identical to any member of any molecule collection that forms a component of a biological entity such as a cell, an organ or an organism. For example, the collection of all insulin molecules in an organism constitutes a functionally important component of the organism. In the case where a diabetic receives injections of biosynthetic human insulin, these molecules – as they are indistinguishable from the organism's own insulin molecules – become parts of the body. In contradistinction, porcine insulin molecules in a human's body would be merely located in it. A clear fiat division was introduced regarding xenogeneic material, whereas allogeneic and autogeneic materials are allowed to be parts. Examples are venous autografts, such as coronary bypasses, which fulfil an important function for the heart and are, therefore, parts of it, as well as blood transfusions.

8.5.2 *Trade Offs and Boundary Issues*

For those cases not resolved by the above criteria, several trade-offs and boundary issues can be identified. For example, the function of a tooth is restored by an inlay artefact. According to formula (8.21), it would not be part of the tooth. This is quite intuitive for clearly identifiable materials which do not naturally occur in biological systems, like gold or amalgam. However, it would be less clear for certain composites that do not just fill a hole, but result in an aesthetically nearly complete restoration; such a material could be composed, in large part, of mineral components very similar to those contained in dentin and enamel. Another case is given with allogeneic and xenogeneic material included within a biological entity. The more this material contributes to the overall function the more it should be accepted as a part. Note that we stay content here to present this situation as a graduality of reasons for parthood. Some authors, like Buddensiek (2006) and van Inwagen (1990), infer that the parthood relation itself is a gradual affair.

A "soft" criterion is physical connection, which is not used in any of the above rules. The more an object is connected to its surrounding structure, the more it would be considered a part, even if other criteria could be used to argue against it. The encapsulated bone splinter in a muscle would be less a part of the muscle than the invasive brain metastasis, which is tightly connected by blood vessels and tissue fibres to the surrounding brain tissue. The function criterion could also be reformulated in the following way: If the malfunction of an includee does not impact the function of the includer, this would be suggestive of containment. However, this criterion will not hold if levels of granularity are crossed: Even in a healthy organism there are always dysfunctional cells and molecules.

8.5.3 The Importance of the Components of a Compound

Many entailments change if the whole, e.g. a complete organism, is defined in a different way. For instance, if we look at the human body as a 'holobiont', i.e. as the sum of all genetically human parts plus the microbiome, i.e. the genetically alien intestinal bacteria, then the latter would be part of the body, even independently of their function, just by virtue of our definition. Whereas the foetus requires the mother's organism to function, this not true the other way round; therefore, formula (8.22) applies and parthood is not asserted because the foetus is allogeneic. Examples include molecules, which are compounds of atoms with a certain structure. These atoms and nothing else belongs to the molecule. Similarly, a cell might be seen as a compound of their organelles and a body consists of a given set of parts. Considering the depiction of a canonical organism, these parts, and nothing else, are the anatomic entities of a canonical body.

8.5.4 Is the Inclusion Condition Empty?

In this paper, our fundamental assumption was that parthood implies inclusion. All parts, that is, are of necessity located within the whole. Then we asked how we could possibly distinguish between includees that are parts and those that are not, i.e. mere containees. Our result was that this is a very difficult endeavour. Hence it could be asked whether it might have been a cul-de-sac from the beginning. Here is a reason that could be given for that kind of skepticism: A standard assumption of standard mereology is universalism, according to which for any two arbitrary entities that co-exist at a certain time there exists an entity, their so-called mereological sum, which contains both of them as its parts (Rea 2008). Hence, for any water molecule and any cell, say, there is a mereological sum cell + molecule and, trivially, that water molecule is part of this sum. Such a sum exists for any arbitrary molecule; any such molecule can be regarded to be part of such a sum that is roughly the same as the cell in question. But this does not answer the question whether a certain molecule is part of a certain cell. Somewhat trivially, according to universalism there is always an entity that comprises the cell and the molecule such that the molecule is part of this whole, namely the mereological sum of the cell and this molecule. In contrast, van Inwagen (1990) posits that there are only material simples and organisms; nothing else really exists. Hence neither molecules nor organs are part of the body, as neither molecules nor organs really exist. This position is far too radical to capture adequately the biomedical domain, as most commonly purported biological parts seem to vanish into ontological irrelevance.

Even if we reject van Inwagen's position and embrace mereological universalism, the basic assumption of this paper could, nevertheless, seem to be wanting. It might be true, or so it could be objected, that parthood implies spatial inclusion, but this does not rule out the parthood of anything.

If, say, the island Helgoland is a part of Germany, then Germany simply extends as far as Helgoland. If it were not, Germany would simply not extend thus far. The implication of spatial inclusion might be valid, that is, but uninformative. Nevertheless, this is only part of the story, for in many cases we have independent evidence of the spatial borders of certain biological entities: Atoms and molecules are bounded by electron shells, cells and organelles are bounded by membranes and organisms are bounded by their epidermis (their skin). However, these (three dimensionally extended) borders are not without difficulties:

- Not everything within these borders is necessarily a part of the includer in question: If we shoot alpha particles through a cluster of atoms, this particle does not become part of any of these atoms. If a child swallows a marble, the marble does not become part of the child.
- There might even be cases where different entities co-exist within the same border. Atoms in a crystal structure, for example, share their electron shell at least partially. Siamese twins live within the same epidermis.
- The borders are fuzzy. To start with, electrons do not have a crisp location. While the skin is a clear border on the macro-level, it becomes an unequal surface with many holes when seen at the cellular or sub-cellular level.

However, if borders are set this way, another question arises: Is the basic assumption of this paper really true? That is, is the inclusion criterion necessary for biological parthood? If we take the location criterion in a strict sense, it implies that anything ceases to be part of a body once it leaves the boundary of the epidermis. E.g., a portion of blood flowing through a dialysis machine would cease to be part of the body while flowing through the pipes and tubes of the machine and, eventually, upon re-entering the body, become part of it again. This might be counterintuitive. Whoever wants to discard the necessity of inclusion for parthood may nevertheless benefit from the results of this paper, for it still answers the question: In which cases, if any, can we ascertain biological parthood given that something is included in a biological entity?

8.6 Conclusions

In this paper, we have studied parthood and containment in biology. The objective was to find criteria for inferring from a given spatial inclusion relation that holds between two biological objects i and I, whether i is part of I or whether i is merely contained in I. Our approach is built on a precursor study (Schulz et al. 2005), which, however, has not proved practicable, as it is based on difficult primitives like function, integrity and genetic sameness, and on criteria that have not been sufficiently introduced so as to constitute a solid basis for empirical investigations as the authors suggest. In our paper, we have attempted to work out more refined criteria and submit them to a more rigorous scrutiny. The most notable difference is that (Schulz et al. 2005) regards continuous historic inclusion as a criterion for parthood of genetically identical material, even in case it is functionally not relevant. According to that proposal, urine and other body substances would be part of the body wherever they are located within the body, as well as misplaced pieces of body matter in uncommon body locations, e.g. as a result of traumatic changes of structure. On the other hand, only artefacts are a priori rejected as parts, whereas xenogeneic entities are allowed as long as they are functionally relevant. In contrast, we propose that there is an, admittedly, arbitrary line between allogeneic and xenogeneic materials. There is also a fuzzy distinction regarding the criterion of "entities of the same sort", which is crucial for regarding something as a collection: If "collection of insulin molecules" is regarded as a component of a body, then a non-human insulin molecule is a body part, in contrast to a refined "collection of human insulin molecules". We also admit that the functionality criterion may exhibit borderline cases.

There are unexplored waters between the islands of parthood and containment. They will remain even if we artificially enlarge the islands. There are soft criteria that may conduce that a certain scenario tends to parthood or to containment. Any delimitation that improves ontological purity may seem problematic if we want to align ontology with human language and cognition. One could argue, of course, that ontology has nothing to do with human language or cognition. However, if ontological divisions appear cognitively plausible, this will improve the usability and acceptance of ontologies and will probably also reduce the errors that may be committed. It would therefore be useful to submit the axioms proposed in this paper as a plausibility check, i.e. a user rating based on real-world examples.

Acknowledgements We very much thank Maureen Donnelly and Marie Kaiser for detailed comments on previous versions of this paper.

References

Ashburner, M., Ball, C. A., Blake, J. A., Botstein, D., Butler, H., Cherry, J. M., Davis, A. P., Dolinski, K., Dwight, S. S., Eppig, J. T., Harris, M. A., Hill, D. P., Issel-Tarver, L., Kasarskis, A., Lewis, S., Matese, J. C., Richardson, J. E., Ringwald, M., Rubin, G. M., & Sherlock, G. (2000). Gene ontology: Tool for the unification of biology. The gene ontology consortium. *Nature Genetics*, *25*(1), 25–29.

Beisswanger, E., Schulz, S., Stenzhorn, H., & Hahn, U. (2008). BioTop: An upper domain ontology for the life sciences. A description of its current structure, contents and interfaces to OBO ontologies. *Applied Ontology*, *3*(4), 205–212.

Bittner, T. (2004). Axioms for parthood and containment relations in bio-ontologies. In U. Hahn (Ed.), *Proceedings of the KR 2004 workshop on formal biomedical knowledge representation*, Whistler (Volume 102 of CEUR workshop proceedings).

Buddensiek, F. (2006). *Die Einheit des Individuums. Eine Studie zur Ontologie der Einzeldinge*. Berlin: Walter de Gruyter.

Casati, R., & Varzi, A. (1999). *Parts and places. The structure of spatial representation*. Cambridge: MIT.

Donnelly, M. (2009). Mereological vagueness and existential vagueness. *Synthese*, *168*(1), 53–79.

Donnelly, M. (2011). Mereological mereological principles to support metaphysics. *The Philosophical Quarterly*, *243*(61), 225–246.

Grenon, P., & Smith, B. (2004). Snap and span: Towards dynamic spatial ontology. *Spatial Cognition and Computation*, *4*(1), 60–103.

Hastings, J., de Matos, P., Dekker, A., Ennis, M., Harsha, B., Kale, N., Muthukrishnan, V., Owen, G., Turner, S., Williams, M., & Steinbeck, C. (2013). The ChEBI reference database and ontology for biologically relevant chemistry: Enhancements for 2013. *Nucleic Acids Research*, *41*(Database issue), D456–D463.

Horrocks, I., Rector, A., & Goble, C. (1996). A description logic based schema for the classification of medical data. In Baader, F., Buchheit, M., Jeusfeld, M. A., & Nutt, W. (Eds.). (1996) *Proceedings of the 3rd workshop on knowledge representation meets databases (KRDB'96)*, Budapest, August 13, 1996. CEUR-WS.org 1996 CEUR Workshop Proceedings. http://ceur-ws.org/Vol-4/.

International Health Terminology Standards Development Organisation (IHTSDO). (2013). SNOMED CT. http://www.ihtsdo.org/snomed-ct.

Jansen, L. (2013). Artefact kinds need not be kinds of artefacts. In C. Svennerlind, et al. (Eds.), *Johanssonian investigations. Essays in honour of Ingvar Johansson on his seventieth birthday* (pp. 317–337). Frankfurt: Ontos.

Jansen, L., & Schulz, S. (2011). Grains, components and mixtures in biomedical ontologies. *Journal of Biomedical Semantics*, *2*(Suppl. 4), S2.

Johansson, I. (2004). On the transitivity of the parthood relations. In H. Hochberg & K. Mulligan (Eds.), *Relations and predicates* (pp. 161–181). Frankfurt: Ontos.

National Library of Medicine. (2013). Medical Subject Headings (MeSH). http://www.nlm.nih.gov/mesh/. Last visited 2013-10-26.

Rea, M. C. (1995). The problem of material constitution. *The Philosophical Review*, *104*(4), 525–552.

Rea, M. C. (2008). In defense of mereological universalism. *Philosophy and Phenomenological Research*, *58*(2), 347–360.

Rosse, C., & Mejino, J. L. (2008). The foundational model of anatomy ontology. Anatomy ontologies for bioinformatics. *Computational Biology*, *6*, 59–117.

Schark, M. (2005). *Lebewesen versus Dinge*. Berlin: Walter de Gruyter.

Schulz, S., Daumke, P., Smith, B., & Hahn, U. (2005). How to distinguish parthood from location in bio-ontologies. In P. C. Friedman (Ed.), *AMIA 2005 – Proceedings of the annual symposium of the American Medical Informatics Association. Biomedical informatics: From foundations to applications to policy*, Washington, DC (pp. 669–673).

Schulz, S., & Hahn, U. (2002). A knowledge representation view on biomedical structure and function. In I. S. Kohane (Ed.), *AMIA 2002 – Proceedings of the annual symposium of the American Medical Informatics Association. Biomedical informatics: One discipline*, San Antonio, 9–13 Nov 2002 (pp. 687–691). Philadelphia: Hanley & Belfus.

Schulz, S., & Hahn, U. (2005). Part-whole representation and reasoning in formal biomedical ontologies. *Artificial Intelligence Medicine*, *34* (3), 179–200.

Schulz, S., & Hahn, U. (2007). Towards the ontological foundations of symbolic biological theories. *Artificial Intelligence in Medicine*, *39*(3), 237–250.

Schulz, S., & Johansson, I. (2007). Continua in biological systems. *The Monist*, *90*(4), 499–522.

Schulz, S., Kumar, A., & Bittner, T. (2006). Biomedical ontologies: What part-of is and isn't. *Journal of Biomedical Informatics*, *39*(3), 350–361.

Simons, P. (1987). *Parts*. Oxford: Clarendon.
Smith, B., Almeida, M., Bona, J., Brochhausen, M., Ceusters, W., Courtot, M., Dipert, R., Goldfain, A., Grenon, P., Hastings, J., Hogan, W., Jacuzzo, L., Johansson, I., Mungall, C., Natale, D., Neuhaus, F., Petosa, A., Rovetto, R., Ruttenberg, A., Ressler, M., & Schulz, S. (2013). Basic formal ontology 2.0. Draft specification and user's guide. http://purl.obolibrary.org/obo/bfo/2012-07-20/Reference. Last visited 2013-10-26.
Smith, B., Ashburner, M., Rosse, C., Bard, J., Bug, W., Ceusters, W., Goldberg, L. J., Eilbeck, K., Ireland, A., Mungall, C. J., OBI Consortium, Leontis, N., Rocca-Serra, P., Ruttenberg, A., Sansone, S. A., Scheuermann, R. H., Shah, N., Whetzel, P. L., & Lewis, S. (2007). The OBO foundry: Coordinated evolution of ontologies to support biomedical data integration. *Nature Biotechnology*, *11*, 1251–1255.
Smith, B., Ceusters, W., Klagges, B., Köhler, J., Kumar, A., Lomax, J., Mungall, C., Neuhaus, F., Rector, A., & Rosse, C. (2005). Relations in biomedical ontologies. *Genome Biology*, *6*(5), R46.
Smith, B., & Varzi, A. C. (2000). Fiat and bona fide boundaries. *Philosophy and Phenomenological Research*, *60*(2), 401–420.
van Inwagen, P. (1990). *Material beings*. Ithaca: Cornell University Press.
Varzi, A. C. (2006). A note on the transitivity of parthood. *Applied Ontology*, *1*, 141–146.
Varzi, A. C. (2014). Mereology. In E. N. Zalta (Ed.), *The Stanford encyclopedia of philosophy* (Spring 2014 ed.). http://plato.stanford.edu/archives/spr2014/entries/mereology.
Vogt, L., Grobe, P., Quast, B., & Bartolomaeus, T. (2012). Fiat or bona fide boundary – A matter of granular perspective. *PLoS ONE*, *7*(12), e48603.

Chapter 9
Developing the Mereology of Chemistry

Jean-Pierre Llored and Rom Harré

9.1 Introduction

The mereology set out by Paul Needham in Needham (2005) deals only with principles of reasoning about mass substances. A bucket of brine is part of the sea, a gold ring is part of the extended substance 'gold'. While these mereologies are interesting for the most part, chemistry has drawn on a mereology of parts and wholes where the parts are capable of independent existence when abstracted from the whole in which they have been resident, and preserve their identity when related with other such parts in constituting the whole. Along with this part-whole layout has been a simple explanatory theory-style – the behavior of chemical wholes is explained by reference to constituents and the relations between them that create wholes. Some of these relations must be invariant if the whole is to count as a chemical entity, say a molecule. There is a further aspect of this Boylean mereology – the results of certain analytical manipulations conducted on samples of the substance in question are not only products of the manipulation but also constituents of the wholes from which they have been derived. The physics of chemical states, entities and processes suggests that this mereology is overly simplistic. Electrons are not parts of atoms in the way that legs are part of horses though saddles are not; or bucket of brine parts of the sea while water spouts are not. No doubt atoms *afford* electron phenomena but it is a more hazardous step to infer that they are constituents of atoms in the form that they are manifested in the complex complementarity phenomena characteristic of analyses that are sufficiently refined

J.-P. Llored (✉)
Ecole Polytechnique, Paris, France
e-mail: llored@poly.polytechnique.fr

R. Harré
Georgetown University, Washington, DC, USA
e-mail: harre@georgetown.edu

C. Calosi and P. Graziani (eds.), *Mereology and the Sciences*, Synthese Library 371, DOI 10.1007/978-3-319-05356-1_9,

to reveal them. We have argued that it is more in keeping with the metaphysics implicit in the development of the idea of molecular orbitals to introduce the concept of 'affordance' to describe this level of analysis. Atoms afford electrons but this does not entail that electrons are constituents of atoms, rather the contrary (Harré and Llored 2013). In this paper we propose to take these insights further in investigations of the ontologies implicit in more physically sophisticated ways of presenting chemical phenomena.

9.2 From Chemical Practices to Mereology: A Methodological Choice

We want to track the developments in chemistry in relation to the presumptions of a variety of mereologies. In doing so, we will first scrutinize what chemists currently do when they synthesize, purify, and analyse a molecule or a material. We will thus point out that chemists develop different whole/parts discourses to explain and to predict the reactivity of a novel chemical body. In line with Denis Diderot (1754) and the later Wittgenstein, we agree that a careful study of what people actually do within the terrain of their activities is of importance and should be addressed by considering their specific sites and goals. In this respect, we will highlight some chemical practices – quantum chemistry, organic and materials synthesis, and chemical analysis – and the parts/whole strategies related to them in order to query how to connect *a* form of mereology with those practices. We will focus our attention on:

1. Chemical theories and representations
2. Laboratory operations, and
3. Apparatus and instruments.

We aim at moving the instruments and experiments into the foreground of our work. Our approach is not to apply a kind of mereology developed within another area of human thought – physics, philosophy, logic, and biology – to chemistry, but to identify the requisite contents that such a mereology must display in order to fit chemical practices. We will then confront those requirements with the different forms of mereologies available – Lewis, Simon, Lesniewski, Earley and Needham – in order to draw some conclusions and to propose new refinements.

9.2.1 Quantum Chemists at Work: Parts, Wholes, and the Environment

We propose a case study that we led in the laboratory DCMR at the French Ecole Polytechnique in 2010 (Llored 2012). This laboratory aims at determining mechanisms which account for molecular reactivity by means of quantum calculations and instruments such as mass spectrometers or fluorometers. We discussed with researchers and attended quantum calculations in order to understand better some steps which are not really present within scientific papers. We wanted to meet quantum chemists at work in order to talk with them about how they use wholes and parts relations in order to carry out their calculations.

Kohn–Sham density functional theory – DFT – has become one of the most popular tools in electronic-structure theory due to its excellent performance-cost ratio as compared with correlated wave function theory, WFT. Within this theory, the molecular space is divided into grids of cubes; researchers define an electronic density for each point of this space. It is a holistic approach that enables quantum chemists to calculate molecular geometry or total energy exhaustively thanks to its electronic density – 'ρ(r)' under certain conditions. The total energy is in consequence a functional of the electronic density, that is to say a function the basic variable which is the electronic density function (Kohn et al. 1996). Approximations are required because the exact electronic density cannot be reached.

This field of research aims at creating new density functionals with broader applicability to chemistry by including, for instance, non-covalent interactions. The crucial step is the *calibration* of new functionals against benchmark databases or best theoretical estimates (Goerigk and Grimme 2010). The researchers of the laboratory DCMR explain to us that a best estimate is the best theoretical calculations available at the time of the calculation. This best estimate may be, directly or not, connected to an empirical value. The whole system consists of data bases and best estimates which are stabilized at a particular time. The whole network is coherent; the results are highly interconnected and interdependent and depend on what chemists previously learn from their experiments and their chemical reactions.

Let us just take an example to figure out what is at stake in the device of basic functionals which are tools that enable chemists to carry out their calculations. The most popular density functional, 'B3LYP', has some serious shortcomings among which are its underestimation of barrier heights by an average of 4.4 kcal/mol. This underestimation is usually ascribed to the self-interaction error – unphysical interaction of an electron with itself – in local DFT (Zhao and Truhlar 2008). Moreover, B3LYP does not allow us to model transition metals satisfyingly and is totally inaccurate for interactions dominated by medium range correlation energy, such as Van der Waals attraction, aromatic-aromatic stacking, and alkane isomerization energies.

So what do they do to improve a functional and thus to solve this problem?

Truhlar and Zhao change parameters and include new ones while shaping a new mathematical functional form that takes physical phenomena and chemical results into account. In so doing, they design a new functional by trial and error. They then use databases to appraise the reliability of a new functional within a defined purpose. The whole work is pragmatic. As a matter of fact, two databases gather all the thermodynamic quantities: (1) the data base 'TC177' is a composite database consisting of 177 data for main-group thermochemistry including atomization energies, ionization potentials, electron affinities, proton affinities of conjugated polyenes, and hydrocarbon thermochemistry, among others data; (2) 'DBH76' is database of 76 diverse barrier heights concerning for instance nucleophilic substitution and hydrogen transfer. Truhlar and Zhao then *discuss the performance* of new functionals for these databases, they conclude that functionals labeled 'MO6-2X' and 'MO5-2X' are the *'best performers' for the main-group* thermochemistry and barrier heights. They propose cases study to exemplify their statement. The isomerization energy of octane involves stereoelectronic effects; none of the previous functionals gives the right sign for the isomerization energy from 2,2,3,3-tetramethylbutane to n-octane. The functional 'B3LYP' gives an error of 10 kcal/mol while 'M05-2X' *predicts* the right sign because this later allows a better description of medium-range XC energies, which are manifested here as attractive components of the non-covalent interaction of geminal methyl and methylene groups (Zhao and Truhlar 2008). On the basis of 496 data in 32 databases, they recommend different 'best functionals' designed for transition metal thermochemistry, main-group thermochemistry, kinetics, and non-covalent interactions.

To sum-up, they change parameters before comparing their results to experiments or with best theoretical estimates. They compare the performance of a functional against previous ones and they then modify it again until a satisfying stabilization is reached. The three main words to describe this activity are: (1) pragmatism, (2) context-dependence, and (3) performativity.

How do quantum chemists then choose one functional from another to carry out calculations?

The researchers from DCMR carried out a calculation in our presence. Choosing a functional of electron density depends upon: (1) the necessary accuracy, (2) the chemical system, and (3) the time of calculation. It also requires choosing a set of functions called a *basis* to achieve calculations for *each atom.* The basis changes according to the type of atoms, three main types are available depending on the situation to describe. They can also pragmatically use those three types within a linear combination according to the case. The basis also changes with other effects such as diffusion, polarization, pseudo potentials for chore electrons, and the size of functions – double, triple zeta. All depends on circumstances. There is no generality within this work.

The functional and its relative basis set define a 'level of calculation'. The calculation process requires choosing a computer program such as Gaussian type or Turbomole. An auto-coherent calculation can then start. Loops of calculations are carried out until the whole values reach a convergent minimum value. If calculations are not convergent, researchers can change the functional, the size of the grids, and convergence thresholds in order to optimize geometry or to calculate molecular energy. Each step reveals know-how, chemical culture, and pragmatic compromises.

Whatever the nature of the tools – molecular orbital, functional, and so on – may be, the calculation always uses both the *Variation Principle* to minimize the energy and the molecular structure which must have been determined by means of experiments. The variation principle is a technical device which allows researchers to calculate the lowest average energy. Beyond its technical aspect, the very justification of this principle requires one to envisage *the relation of the whole*

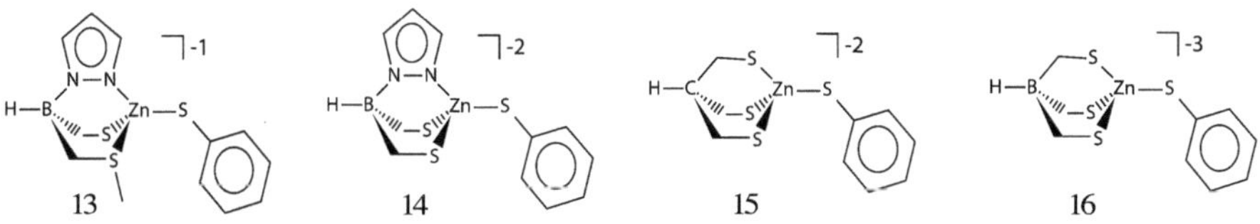

Fig. 9.1 Thiophenolate zinc complexes studied by Picot (2008, 164)

molecule with its environments – molecules, photons – in order to rationalize the whole process. This interaction with the environment may be explicit or implicit within the formalisms according to the case being studied. It depends on the situation. The interaction with the solvent may thus also modeled and sometimes taken into account. The very process of calculation thus holds together information about: (1) the whole system – geometry that chemists first define from the outset of the calculation, (2) the parts – a functional for each atom or groups of nuclei, a molecular orbital, and so on, (3) and the solvent – that is to say what is outside the whole. The calculation uses the three levels – the molecular whole, its parts, and the environment called the solvent – *at the same time* in order to: (1) minimize energy, (2) calculate an energy barriers, (3) determining a transition state, and (4) for postulating a geometry by means of different levels of calculation, respectively. One should bear this constitutive entanglement in mind before further discussion about mereology.

The model of solvent is related to the salvation free energy of each compound. This quantity is always defined as the required amount of energy necessary to transfer a molecule of gaseous solute into the solvent. The crucial step is to appraise how the solvent gets involved in a chemical reaction. Its action can be direct if some molecules of solvent take part in the chemical process or indirect if the solvent – then labeled the 'bulk medium' – only modifies reactants reactivity compared with that of the same molecules in the gas phase.

A particular level of calculation – functional and its basis – is only designed to reach a particular aim – final energy, geometry optimization, barrier height, and so on – and excludes the others. One cannot study all the molecular characteristics by means of a single level of calculation. When quantum chemists study the energy barrier of a chemical reaction, they have to choose a functional, its basis, and best estimates for each level – atoms, molecule, and the solvent. They thus tailor a useful tool to understand and predict a defined characteristic and not the other. 'Complementary' levels of calculation are thus necessary to achieve a global molecular description. Let us just quote a typical part of quantum paper to grasp the situation related to the following Figs. 9.1–9.3:

> Calculations were performed with Gaussian 03. Geometry optimizations were conducted using the B3LYP method at the 6-31 G(d,p) level for the B, N, C, O, S, H atoms. The CRENBL relativistic effective core potential and the associated valence basis set were employed to model the iodine atom. This basis set is referred to as BS1. Each stationary point has been characterized with frequency analysis and shows the correct number of negative eigenvalues (o for local minimum and one for a transition state). Energies were

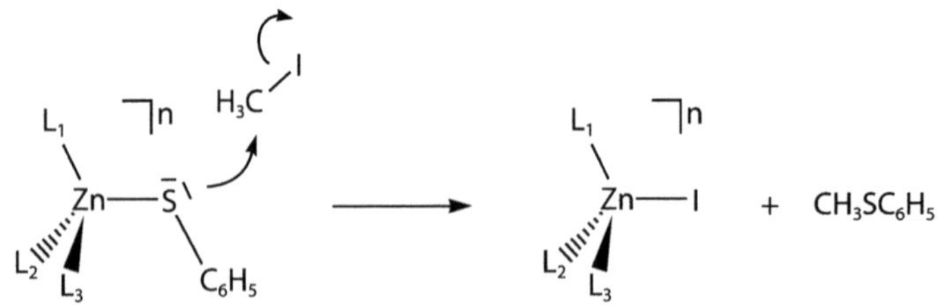

Fig. 9.2 Chemical transformation studied by Picot. The iodine replaces the substituent which contains the sulfur (Picot 2008, 166)

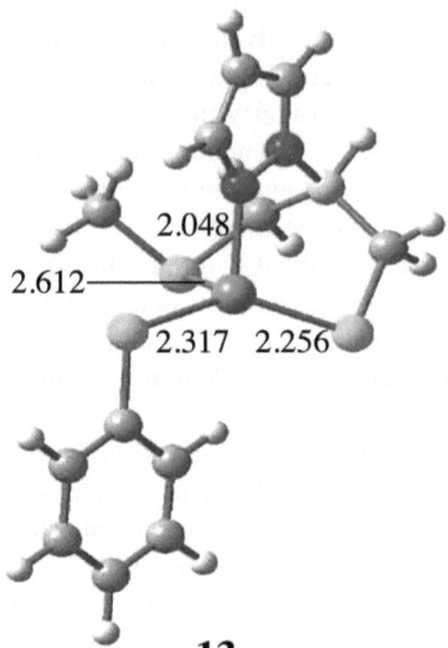

13

Fig. 9.3 Optimized structure of the previous complex n° 13 at the B3LYP/BS1 level. Bonds are expressed in Angstrom. (Picot 2008, 166)

> calculated for the stationary points at the B3LYP level using an extended basis labeled BS2. It consists in the 6-311+G(2d, 2p) for B, N, C, O, S, H, the extended Wachters basis [15s11p6d2f/10s7p4d2f] for Zn and the Aug-cc-pVTZ-PP basis set and pseudo-potential for I. We have demonstrated previously that this level of calculation gives reliable geometries and relative energies on zinc complexes (Picot 2008, ch. 5).

This typical calculation clearly illustrates what explaining a structure or a mechanism, and what predicting a transformation amount to. Quantum chemists use a lot of interrelated tools within a large and sophisticated network which holds together mathematical functions and devices, empirical outcomes, computer engineering, quantum and classical physics, and chemical knowledge and know-how. They tailor highly specific methods to account for chemical transformation. To do so, they use analogy between families of similar compounds. This work is utterly relational. Not only does it assemble multifarious tools within a coherent and performative practice of articulation but it also entangles the molecule, its parts, and its environment within the minimization of the total energy. Their method is relative to a family of chemical compounds and depends on the entanglement of inter-calibrated tools. Their method is a way and not *the* way to account for a characteristic of the whole from its parts and its environment. In this respect, a quantum method is a practice of articulation which negotiates an explanation of the whole/parts relation (Stengers 2003). Isabelle Stengers asserts:

> As soon as the question of emergence is at stake, the whole and its parts must thus co-define themselves, and mutually negotiate what an explanation of the one from the others means[1] (Stengers 2003, 207).

In quantum chemistry, the mereology used by chemists entangles the whole, its parts, and, sometimes, its environment. It is not a classical transitive mereology in so far as the whole interacts with its environments. It is not merely a holistic description within which the whole is necessary to define the parts or the kinds of parts involved. It is not merely a classical mereology that only needs the parts to define a whole univocally. Quantum chemical practices need the whole, the parts, and the other entities at the same time, that is to say, a molecular structure, atoms or nuclei and electrons, and the solvent, respectively. Quantum methods are neither purely holistic nor purely reductionist. They always negotiate the articulation of different levels of description within a network which assembles chemistry, physics, computers, and mathematics. *Chemists have contrived specific methods within which the whole and its parts are constitutively codefined.*

Quantum chemists then classify compounds by means of chemical reactions, say, by means of relations. This work fits what chemists have continued to do in other areas of chemistry since the very beginning of this science. Relations remain the cornerstone of chemical operations. The principles of alchemists, the *table de rapports* of Etienne-François Geoffroy's during the eighteenth century, the elements of Mendeleev, the nanomachines of our present days, are as many examples of the networks of interdependencies produced by chemists in order to act upon and to explain matter. The whole needs other wholes to be defined, not just intrinsic parts. In this respect, philosophers need a relational mereology not a classic one which cuts off the whole from its environments.

Parts may differ. They can be atoms, electrons, nuclei, groups of atoms according to the group theory approach, and so on. Methods of calculation can differ as well. They can be based on molecular orbital, functional, or even consist of a hybrid of the two. The whole can differ depending on the project involved. It can be either a protein or an active site within it. It can also be a local complex with includes some molecules of solvent or a metal ion into the structure. It is also possible to model different parts with classical models while using quantum methods to

[1]Our translation of the French sentence: Dès qu'il est question d'émergence, le tout et les parties doivent donc s'entre-définir, négocier entre eux ce que signifie une explication de l'un par les autres.

study others depending on the goal they wish to reach. The synchronic use of incompatible models or methods depends on the scale of description involved. Is there any 'ground' or 'foundational level' in this mereological scheme? Pragmatism prevents quantum chemists from attributing a unique basic ground from which the higher level chemical bodies derive or emerge. The pragmatic notion of levels of calculation replaces that of levels of reality and of description.

In short, quantum chemists entangle the whole, its parts, and sometimes, explicitly or not, its environment within the calculations. Quantum calculations deal with types of patterns as possible values of contextual observables and not of monadic properties. Electrons, nuclei, atoms, groups of equivalent nuclei according to the NMR scale – an instrumental technique – or according to the group theory – a cognitive tool, basins or attractors from the theory of electronic density standpoint, are not intrinsic but are constituted by the modes of access involved. In this respect, they are relational and not monadic.

Let us just develop further how quantum chemists can study a molecule-whole using another quantum method, say, the molecular orbital approach. Mulliken proposed an approach based on molecular spectroscopy. For example, rejecting the interpretation of the concept of valence as an intrinsic property of the atom, Mulliken opposed the use of 'energy state' deduced from molecular spectra on the basis of an *electronic configuration*, i.e., of a distribution of the molecular electrons in different orbits. According to Mulliken, each *orbit* is delocalized over all the nuclei and can contribute, depending on each specific case, a stabilizing or destabilizing energy contribution to the total energy of the molecule (Llored 2010). For Mulliken, the atom did not exist in a molecule. There ensued the key semantic shift from the concept of molecular orbit to that of molecular *orbital*. Mulliken wrote: 'By an atomic orbital is meant an orbital corresponding to the motion of an electron in the field of a single nucleus plus other electrons, while a molecular orbital corresponds to the motion of an electron in the field of two or more nuclei plus other electrons' (Mulliken 1932). This notion has been further refined to the textbook definition of 'orbital' as 'the region in which an electron is likely to be found' – whatever that is supposed to mean. The notion of probability for an electron to belong to a particular volume replaces the notion of atoms as intrinsic parts. If the criterion of identity for an atom or the ionic residue of such an atom is the composition of the electron shells then these criteria could not be satisfied by the components of a complex molecule. The relevant nuclei form a doublet which, speaking in the accent of Mulliken, are a unit without parts, using the molecular orbital theory of electrons as the criterion for an individual part (Harré and Llored 2011). A molecule does not have atoms or ions or even the nuclei of ions as its parts. It does have nuclei duplets however, identified as molecular parts *with respect* to molecular orbitals. Furthermore, new energy levels emerge within the molecular whole that didn't exist in the previous atoms. Mulliken proposes the concept of electronic promotion to construe molecular electronic configurations fitting with empirical data. A molecular electronic configuration is all but a 'molecular part' within this heuristic and holistic approach but a tool for prediction (Llored 2010).

But we can go further and deeper if we consider the use of group theory based on the logic of *sets and subsets* in such a molecular approach. Mulliken develops *the fragment method* in 1933, two fragments can interact provided they have the same kind of *symmetry* and that their energy gap, measured by spectroscopy, is not too high. For the ethylene molecule 'C_2H_4' Mulliken considers two 'fragments CH_2' and determines suitable molecular orbital by using the irreducible representations of ethylene. He can thus propose a representation of molecular orbital of ethylene by increasing order of energy as well as its correlation diagram thanks to those of the two 'fragments'. In so doing, he grasps all the characteristics of molecular orbital diagram of the ethylene molecule (Mulliken 1932). The possibility of an experimental support was all the more important as the nature of the initial fragments can change depending on each specific case. To model the molecule 'C_2H_2', Mulliken could just as easily have considered a fragment 'C_2' and another 'H_4' of adapted symmetries. The relation between the whole 'C_2H_2' and its 'fragments' is of secondary interest provided that the energy diagram of the 'molecular whole' is in agreement with the experiment.

This molecule can also be described using the different atoms taken separately (Pauling 1928, 1931). The two descriptions are tantamount to the same thing as van Vleck and Coulson have demonstrated it. It can also be described in terms of electron density around nuclei (method DFT). In this respect, the whole underdetermines the parts to use Quine's terminology.

Depending on the approach, a molecule can be described as a whole (Mulliken 1932), or as an aggregate of atoms (Pauling), or by means of electronic density within a space divided into grids of a particular type and size (Kohn et al. 1996). In brief, the catalogue of parts involved may change while the concept of the whole remains the same. Moreover, the process of calculation is based on the Variation Principle which is a device for minimizing energy. This calculation entangles the parts, the whole, and the environment at the same time. As ever in quantum chemistry, we need a piece of information about the whole to define the parts. Here the information is provided by a mathematical tool which is the mode of access. Mereology is thus reversed within this context. Parts and wholes enter into a new kind of relation, that is to say that they are *constitutively* co-dependent *vis-à-vis* a particular mode of access. We should nevertheless bear in mind that the apparent asymmetry from parts to whole is broken in so far as the environment plays also a key role in the chemists' calculation at stake.

Parts are not intrinsic components of the whole from which they are derived in the sense that they depend on the mode of access - cognitive and instrumental. Parts are not monadic, pre-existing, once and for all, inside a chemical 'geometral'. They are embedded into an engaged chemical practice of investigation which explores the world. Mereology needs to include modes of access dependence to envisage how a whole/parts relation is used. In doing so, mereologists should not leave chemists' pragmatism aside. In this respect, mereology seems relational, perhaps dispositional, but surely not ontological in the strong sense of this word, that is to say the shaping principle of all that is. A pragmatic account of the parts/whole dependence on the modes of access seems to be of primary importance to investigate chemical processes of transformation.

Analytic and synthetic reasonings are intertwined. It is interesting to notice that chemistry and the kind of mereology related to it may help philosophers to redefine the frontier they drawn between analytic discourse strategy and that of synthetic discourse and reasoning in so far as chemists are neither purely holistic nor purely analytic. In this respect, 'chemistry is very much a mixed science' to use Frederic L. Holmes and Trevor H. Levere's turn of phrase (Holmes and Levere 2000, introduction, XV). Theoretical chemical approaches are neither ontic nor purely epistemic. They do not express exclusively the structure of reality out there, or the form of our own knowledge, but their *active interface*. We have previously illustrated this point when focusing on the pragmatism of chemists and the role of trial and error within the open-ended device of new functionals of density. Chemical mereology deals with this interaction between a whole, its parts, and the environment within chemists' work. Moreover, the whole is known relative to its relation to other wholes within families of chemical compounds. Its acidity for instance is defined in comparison with other compounds. Its structure is relative to the 'solvent' can be studied by means of analytic methods such as NMR or radiocristallography of X-ray. Let us see how wholes and parts are studied thanks to chemical instruments.

9.2.2 Chemists at Work: Process, Instruments, and Time

Let us first point out both the role of the proportions of the chemical reagents involved and that of the experimental device. When chemists want to synthesize a solid sample of $CaCO_3$ they need to saturate a solution. They can achieve this precipitation by various means. The most common are solvent evaporation, or chemical reactions which produce insoluble species. Starting from a different number of nuclei, particles will grow to attain different final sizes and morphologies; thus by adding a reactive chemical at once or in a continuous way, the final materials may appear completely different. By adding a small amount of fine material to be precipitated – seeds, the apparently chaotic nucleation step can be better controlled. For example, adding calcite seeds allows the precipitation of pure calcite, whereas a mixture of calcite and vaterite with a larger particle size distribution and various morphologies are obtained without seeds. The whole $CaCO_3$ depends on the process and on the time factor. It can be a mixture of calcite and vaterite obtained by precipitation – case a below, or a pure obtained by precipitation using a seed source – case b below. The structure and the size of the final whole are different (Llored et al. 2013).

We know the initial 'parts' of the mixture, say, the reagents, and depending on the process, the new whole may be different. Chemists also know how to separate initial parts from the new whole by means of operations. The structure of the new whole $CaCO_3$ is totally different. The environment, the device, and the time factor have shaped it differently. Those two compounds share the same formula $CaCO_3$ but are different wholes. The composition is the same but the relatedness has been influenced by the environment. Once again, the whole, its parts, and the environment are intertwined within a process. The mode operation cannot be eliminated. It determines the whole and its correlative parts. We could give other examples extracted from other chemical areas – in particular in chemical kinetics or chemical engineering – to illustrate the role of devices and time factor. We have even given an example regarding dissipative structure elsewhere (Harré and Llored 2011). This example is sufficient to call for a mereology which includes the process and the operative framework of chemistry into the investigation of chemical bodies in terms of wholes and parts.

This is not how the story ends because the instrumentation used by chemists has thoroughly evolved. New microscopes and kinds of spectroscopy, Auger for instance, allow chemists to identify heterogeneity where they formally consider matter to be homogeneous. For example, the study of the surface of an alloy of aluminium by electron microscopy enables chemists to identify heterogeneity located at an intergranular joint. The correlations between different instruments – optical microscopy and scanning probe microscopy to quote but a few – and the sample figure out that 'parts' can diffuse from one place to another and from one into another. Some 'zones' are impoverished in precipitate. Instrumentation interacts with matter at smaller and smaller scales and affords new parts among seemingly homogeneous parts. Instruments afford parts from parts inside a whole. Parts evolve, move, and transform themselves. The whole/parts relation is dynamic and non-transitive. In this respect, a transitive and aggregative mereology based on mass is of no avail because the same parts, the same ingredients with additive mass, may compose different precipitates, that is to say generate different kinds of parts depending on the experimental devices. In brief, they depend on the mode of access which does not simply revealing the parts but which, on the contrary, constitutes them, in Kant's sense. Different modes of access afford complementary information about parts.

A chemical 'segregation' is about the difference of local composition, and the migrations related to it. An interface – a free surface, an internal surface, an intergranular joint, or interphase joints – can display differences in chemical composition *vis-à-vis* the massive phases they delineate. These differences result from transformations called interfacial 'segregation'. Parts are multifarious and act upon others within interfaces. The study of a surface by means of Auger Spectrometry allows one to 'contextualize' parts by their extraction energy which is relative to the nature and the topology of the particular site in which a particular part is located. It thus depends on the mode of access and the local environment. A chemical whole, its parts, and the environment are afforded by finer grained instruments which act upon them within new scales – within space and time. Microconstituents of a chemical body are explanatory, but not intrinsic. They result from the interaction with the instruments which affords them.

Chemists need complementary affordances to describe the whole/parts relation better: parts are relative to modes of access, but matter is inexhaustible. Should we write 'matters' instead of 'matter'? The same chemical body can act differently depending on the context. The same composition does not even lead to the same whole/parts analysis. The mass-aggregative mereology does not meet the practices of chemistry and materials science. Parts entangle the mode of access, the individual, and its environment. Parts are local affordances constituted by the instrument. Parts are afforded by the world/apparatus complex (Harré 2013) and we should avoid mereology fallacies that arise if these basic features of science are overlooked (Harré and Llored 2013).

We need to contrast the concept of 'affordance' with that of 'product' and both with that of 'constituent'. Affordances are certainly products of the interaction of equipment and the world, but in many cases they are not constituents of that which affords them, neither as properties such as 'colour' nor as entities such as 'parts', nor as processes such as 'walking' (Harré and Llored 2013). We have warmed philosophers against two kinds of mereological fallacies as soon as the wholes/parts dichotomy depends on the operative framework. In doing so, we point out that applying to a part of an entity or chunk or mass a predicate that gets its meaning from its use for ascribing an attribute to the whole from which the part comes is a the first mereological fallacy to avoid. A holistic predicate is not necessarily a part predicate, the Wittgensteinian notion of 'use' is crucial because all those predications are context-sensitive. When we discover that sodium vapour – or sodium salts in a flame – affords two adjacent yellow spectrum lines with the help of a spectrometer how do we justify the conclusion that this phenomenon reveals a feature of the electron *components* of sodium atoms? Our response is that electrons are products only of the indissoluble unit – world-apparatus. Neither can be detached from the electron affording complex (Harré and Llored 2013). We also point out a second mereological fallacy which consists in inferring that substantive products of an analytical procedure are parts of the substance on which the procedure was performed. They are affordances!

9.2.3 Chemical Analysis and the Relativity of Afforded Parts

Let us envisage just the examples of NMR and column chromatography regarding the following chemical transformation (Fig. 9.4):

The first step after a chemical synthesis remains the purification of the compound that is its separation from initial reagents, the solvents, and secondary products. Chemists use two different phases. A solid phase called the stationary phase, here silica in the column, and a mobile phase, here a liquid that is cooled and then percolated through the solid. The mobile phase is often a mixture of solvents with a global particular polarity. This mobile phase dissolves our initial mixture and allows it to go through the silica column by means of gravity. The different chemical bodies can interact with the silica differently depending on their own polarity. They are thus separated in so far as they don't have the same velocity. The column and the mobile phase thus *afford* the different parts of the whole mixture (Fig. 9.5).

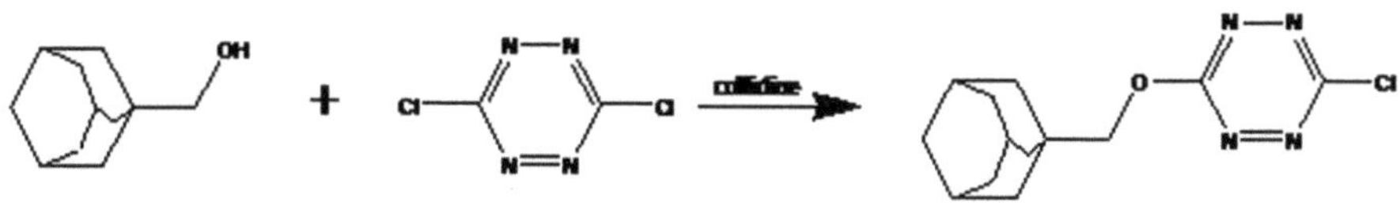

Fig. 9.4 Chemical reaction under study in our example

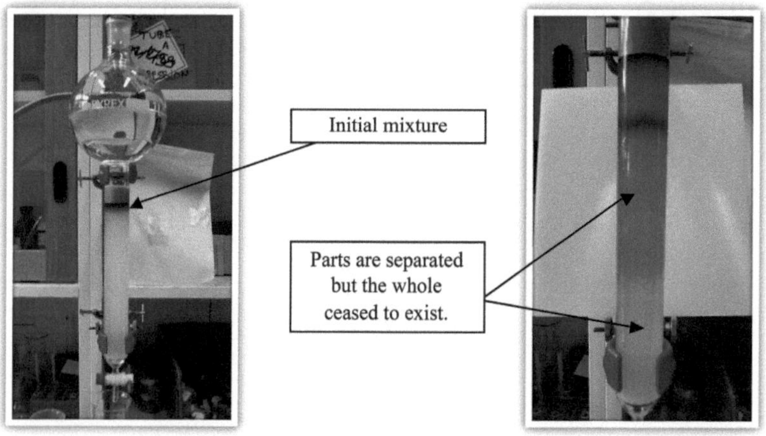

Fig. 9.5 Chromatographic separation carried out by our student Benoît Marcillaud at the Ecole Normale Supérieure de Cachan (France). Reproduction authorized by the professor Pierre Audebert who supervised this work within the PPSM laboratory

Some parts/wholes analyses are sometimes destructive, others not. This example fits the classical mereology better because chemists separate additive parts. There is a crucial difference with Needham's perspective nevertheless, that is to say the role of *affordance*. The separation is not that of a mere aggregate which results from an addition of its parts. It is the result of an interaction with a column, a liquid, and depends on chemical polarities. When chemists perform purification by means of chromatography, they develop a parts/wholes discourse that is to say a contextualized mereology which takes the constitutive relation with the instrument into account. It is not a resultant parts/whole dependence from nowhere, but the result of a chemical operation. Parts/wholes relations are defined through chemical operations and not intrinsically.

Once the purification occurred, chemists can analyze their product by means of NMR – nuclear magnetic resonance. The 'spectrum' that results from this technology displays the radio frequency response of the different atoms ^1H inside the molecule (Harré and Llored 2013). Different parts emerge from the spectrum. Each signal represents a set of atoms ^1H which are chemically equivalent according the ^1H NMR magnetic scale. This method affords different sets of atoms of hydrogen. Every signal depends on the locality, the environment of each particular

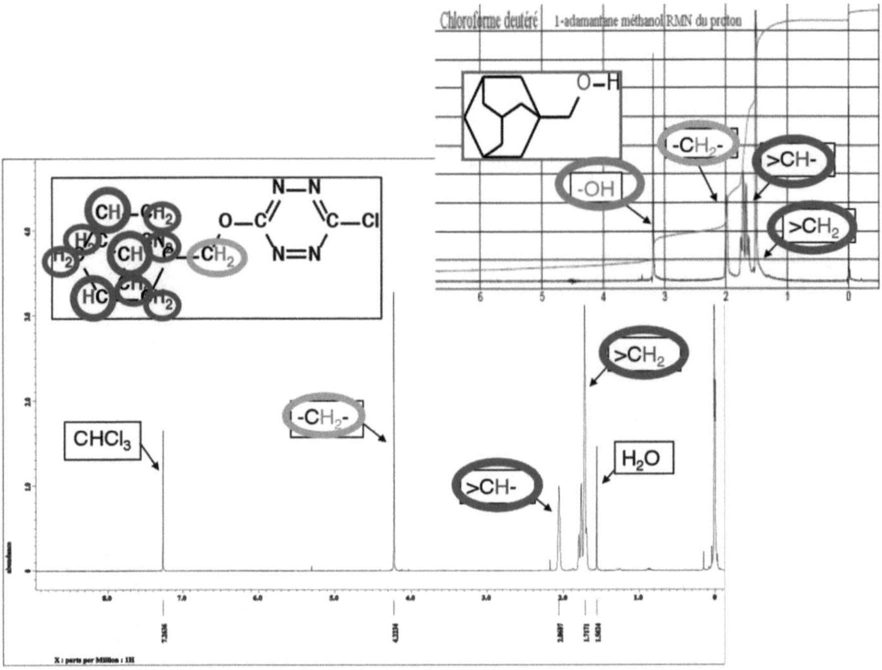

Fig. 9.6 ^{1}H NMR analysis carried out by our student Benoît Marcillaud at the Ecole Normale Supérieure de Cachan. Reproduction authorized by the professor Pierre Audebert who supervised this work within the PPSM laboratory

nucleus of proton. The set of different signals enables chemist to identify a chemical body and thus to assess its degree of purity (Fig. 9.6).

But chemists can also afford a complementary description of equivalent parts that is to say of equivalent sets of nuclei of carbon using NMR ^{13}C. The sets of different atoms of carbon absorb a precise radio frequency depending on their locality in the whole. Whole/parts structure can thus be afforded by the conjoint use of a magnetic field and a wave. The chemical shift on the axis of abscises provides information about the local environment relative to a reference, the tetramethylsilane (TMS). The parts of a structure can be afforded relative to a mode of access, an environment, and a chemical standard. Parts are constituted by the interaction between the world and our apparatus, they are not intrinsic. We can only know interactions, and not the reality isolated from us. It is an old philosophical debate.

Chemists are also able to combine different NMR (^{1}H, ^{13}C, ^{19}F, ^{31}P, etc.) in order to correlate the different parts between them. In brief, they associate each kind of carbon with hydrogen nuclei to identify the global structure and the connectivity inside the whole. They then compare their structure with that derived from other methods and previous data bases. They can use what we call two or three dimensional NMR. This is really the current style of investigation chemists

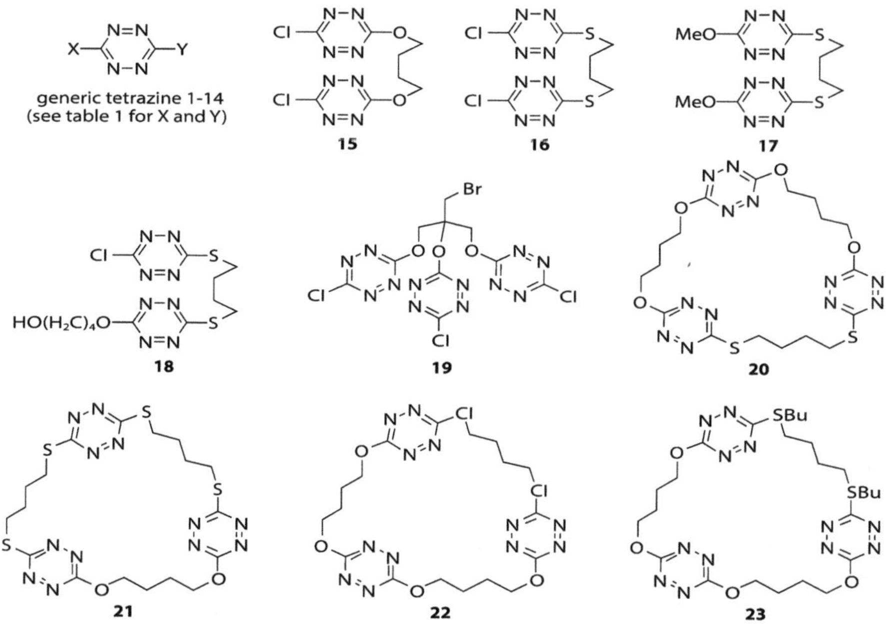

Fig. 9.7 Different molecules used as sensors (Yong-Hua et al. 2009)

use in order to relate parts and wholes! They always connect affordances in order to determine relations between parts and wholes. They are always trying to determine the structure with precision. In doing so, they articulate heterogeneous information constituted by different modes of access. Let us envisage a last example before entering into philosophical developments.

9.2.4 The Structure/Reactivity Discourse Within a Typical Chemical Investigation

The following family of compound derives from the s-tetrazine. They are used as sensors in biochemistry (Yong-Hua et al. 2009) (Fig. 9.7).

Most of those molecules display fluorescence. Acting globally (or coarsely, at a large scale) yields consequences that can be detected by experiments bearing on local levels. A substituent is changed by means of a chemical transformation – please see below, and new chemical fluorescence is measured by means of fluorometer. Two modalities of *action* exerted on a process make possible the study of correlation between properties – chemists compare the fluorescence of different compounds with the standard molecule in order to analyze the effect of a substituent-change on the global fluorescence of the whole. If one intervenes in the structure of

Sustituent X	Sustituent Y	UV-VIS absorbance			Fluorescence $\lambda^{émis}$ (nm)
		λ_1^{abs} (nm)	λ_2^{abs} (nm)	λ_3^{abs} (nm)	
OCH_3	OCH_3	524	345	275	575
OCH_3	$O(CH_2)_4OH$	526	347	No absorbance	572
OCH_3	SCH_3	528	394	No absorbance	No fluorescence

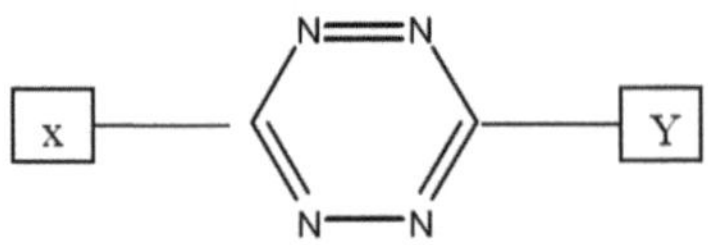

Fig. 9.8 Effects on parts-substituents on molecular absorbance or fluorescence

the whole, some effects of this action can then be detected by a mode of access specifically aimed at the fluorescence study. When chemists act upon the molecule by means of UV-VIS radiation using a wavelength λ_1, the whole replies by emitting a radiation of fluorescence the wavelength of which is λ_2 (superior to λ_1) (Yong-Hua et al. 2009, 6125) (Fig. 9.8).

Depending on parts –X and Y, wholes display global behaviors which are classified into a correlation web of relations *vis-à-vis* a standard molecule. Chemists thus propose the following parts/wholes/reactivity correlation (Fig. 9.9 below)

Chemists at work use modalities of action to correlate whole structure, parts, and chemical reactivity. To do so, they investigate relational webs of interdependencies and classify compound by reference to a molecular standard. Mereology is relative to modalities of action, that is to say, to affordances. That is not how the story ends. Those whole can trap ions specifically, it thus displays a novel *causal power* (Fig. 9.10).

The fluorescence of the new whole is different from the previous structures without the ion. The quantification of this difference allows quantifying a pollutant, the ion Pb^{2+} for example in a liquid sample. A new *part* generates a new *whole* with different causal powers and different behaviors depending on the instrumentation. Therefore, saying that some intervention at a higher level *downwardly causes* alterations detected at a lower level (or, conversely that some intervention at a lower level *upwardly causes* alterations detected at a higher level) is an accurate expression of a dual mode of *operational definition* of the levels. This analysis of the modalities of action calls for an interventionist notion of causation that widens the space of reflection proposed by Harré and Madden while paving the way for both a relational form of emergence, and a non-transitive form of mereology that fits chemists' work.

Let us now query current mereologies by using our chemical investigation.

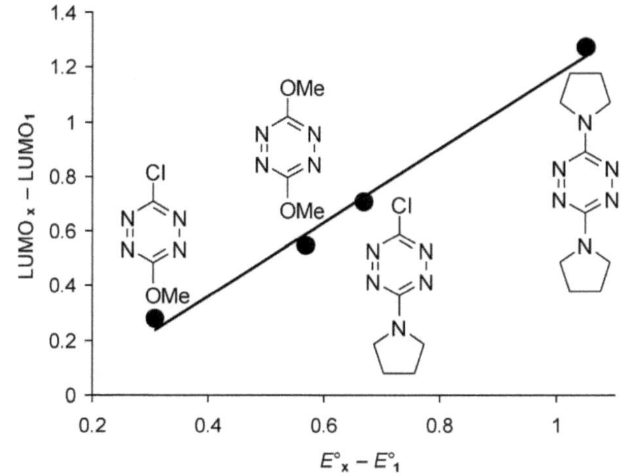

Fig. 9.9 The energy difference of the highest unoccupied molecular orbital is correlated to the difference of electrochemical standard potential. The reference (number 1) is the s-dichlorotetrazine (X=Y= Cl= Chloride). The determination of LUMO used the Variation Principle which entangles the whole, its parts and its environment in a calculation. (Yong-Hua et al. 2009, 2019) (Reproduction authorized by professor Pierre Audebert)

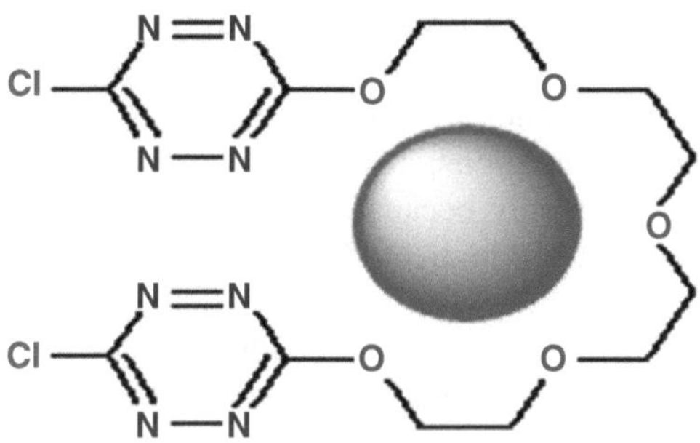

Fig. 9.10 Ion trapped within the molecule

9.3 Connecting Mereological Work with Chemical Practices: A Discussion

9.3.1 The S-Mereology and the C-Mereology

Let us briefly recall what a classical mereology demands regarding wholes, parts, and their relations. We have asserted elsewhere that classical mereology – the C-mereology according to our own taxonomy – rests on two basic principles (Harré and Llored 2011). First, the Principle of Unique Composition according to which there is a unique being, the sum or 'fusion' of a certain collection of beings, of which every such being is a part and which has no parts other than such a part. As a matter of fact, some chemists such as Linus Pauling consider molecule to be a unique collection of just these chemical atoms, and only these chemical atoms. We note that the composition of such a collection does not serve to uniquely identify a molecule as a being of certain *kind* – the properties of molecules include structures as well as components. In practice sums can be uniform or disparate if parts belong to the same type or not. Axiom MA3 in Simons (1987). Second, the Principle of Mereological Transitivity which states that if B is a part of A and C is a part of B, then C is a part of A. Simons (1987), Axiom MA2. If we include *function* among the attributes that define how a being is a constituent of another being, that is how it is a part, then clearly a tooth is a part of a gear wheel in a different way from the way a gear wheel is part of a gear box, and transitivity of that part-whole relation fails. Some philosophers (for example Oppenheim and Rescher (1995)) face this problem by proposing functional mereological principles.

Even though constituents lose their *actualised* functional attributes when removed from the whole of which they have been parts, they do not cease to exist. Nor do they lose the core attributes that enabled them to count as parts of the relevant whole. In the light of our knowledge of how a component fits into a whole, we may want to hold that *potential* functionality survives some ways of decomposing the original whole. Chemists can decompose HCl into Cl_2 and H_2 by means of a specific reaction even if Cl_2 and H_2 cease to exit as such when assembling under the whole! Mulliken shows, for instance, how the atom of helium He disappears during the synthesis of the molecule HeH. He also explains how HeH can afford He in certain conditions, just as well as an electrolysis of molten $NaCl$ affords sodium atoms in plenty because $NaCl$ affords the sodium nuclei that are essential to the formation of sodium, atom by atom. There is no sodium in salt. But salt *affords* sodium.

Oppenheim and Rescher suggest three conditions on wholes: (1) a whole must possess an attribute that is peculiar to it as a whole, (2) the parts of a whole must stand in some special relationship to one another, and (3) a whole must have a structure. Electrons are part of chloride which is part of HCl. But the gas Cl_2 has relational properties that a plasma of electrons does not display *ceteris paribus*. HCl *reacts* differently – for instance owing to its acidity – than Cl_2 whenever in contact with other chemical reagents *ceteris paribus*. Transitivity faces hardships as soon as emergent *relational properties* are at stake.

Lewis proposes to develop a mereology about sets, subsets, and supersets. The fusion of all cats is composed of all the cats there are, and nothing else (Lewis 1991, 1). According to him: (1) one set is a part of another if and only if the first is a subset of the second, (2) no set has any part that is not a set, (3) reality divides exhaustively into individuals and sets, (4) no sets is part of an individual, and (5) any fusion of individuals is an individual (Lewis 1991, 7). Since the member of a member of a set is not in general a member of that set, membership is not the same relation as part to whole. We referred this system as the S-Mereology (Harré and Llored 2011). It follows from Principles 1 and 2 that fusions of individuals are not sets the membership of which is extensionally equivalent to the individuals that constitute the fusion. Using the concept of a fusion, Lewis refines the simple Lesniewskian scheme with alternative axioms for a mereology of sets and subsets (Lewis 1991, 74). The reasoning is based on three foundations: (1) transitivity – if x is a part of some part of y, then x is a part of y, (2) unrestricted composition – whenever there are some things, then there exists a fusion of those things, and (3) the uniqueness of composition – it never happens that the same things have two different fusions.

Lewis introduces an ontologically and mereologically significant concept of the 'singleton', the single membered set. Here we have a genuine alternative ontology – are the atomic constituents of molecules and polyatomic ions single member subsets that are subsets of molecular sets?

Which version of mereology should we use to express the structures of molecules and molecular ions – is a molecule a thing of which its parts are also things, a structured collective – or is a molecule a set of which its constituents are subsets? The former would require the classical mereology of Lesniewski, the C-mereology in its functional form. The latter would require one to make sense of the mereologised set theory of Lewis, the S-mereology, in chemical contexts. In that mereology there no provision for distinguishing parts functionally with respect to their roles in the wholes of which they are parts. Is the grammar of chemistry classical mereology or mereologised set theory? (Harré and Llored 2011).

9.3.2 The Case for S-Mereology

The case for adopting set theory as the Mereology for chemistry begins with the predictions by Odling and Mendeleev of the properties of elements yet to be discovered. At the time of their proposals only the intensions of the set of atoms of eka-iodine was available in chemical discourse. The set had a null extension for the users of the grammar appropriate to the situation as it then stood, since the set had no members, and conceivably might never had any. Obviously there cannot be a fusion or a sum of which there are no parts. To talk of eka-iodine in the grammar of classical Mereology made no sense. It does seem to make sense in a discourse in which the parts of sets are subsets. Are hydrogen and oxygen atoms (ions) subsets of the water molecule set? Each water molecule would be a subset of the superset, the stuff water. However, what is the intension of the set of which two sets, pair of hydrogen atoms and a singleton oxygen atom are the subsets? Well, it is the properties of whatever it takes to be a subset of the set of water molecules that is the water stuff. The hydrogen atoms (ions) are members of the set of all hydrogen atoms, while the oxygen atom (ion) is a member of the set of all oxygen atoms. Does this have any advantage over the classical mereological grammar? (Harré and Llored 2011).

Maybe it can be useful to use a mereology of set when group symmetry and chemical analysis are at stake. We point out that sets of different equivalent fragments with appropriate symmetry and energy can be used to obtain molecular orbitals. Those sets of fragments provide the description of the whole molecule without being material parts as such. We also notice that the NMR affords sets of nuclei which are *equivalent* or similar as regards this technique – ^{1}H, ^{13}C, ^{19}F, ^{31}P, and so on. Subsets also appeared from spectra depending on the wavelength of the spectrometer. A finer structure then arises which enables chemists to scrutinize the coupling between different kinds of nuclei-sets. This work is relative to the spectrometer resolution and is of importance to find out the relatedness within the whole from spectra. The major difference from Lewis's point of view is that sets are afforded and are not identified solely by useful abstraction from description. As we also see, chemists always classify compounds by means of chemical operation and reactions. Their systems are webs of interrelations. Chemists thus define sets of chemicals the characterization of which depends on similar relations. Lewis's mereology should be thus adapted to fit those sets of similar compounds co-defined by means of relations. This mereology could thus fit the requirement to call for the environment when defining a whole and its sets. It could finally allow philosophers to connect chemical whole/parts discourses with the practices from which they originate. As a matter of fact, chemists use sets of operations which allow us to synthesise, to purify, and to characterize a set of comparable wholes by means of chemical reactions or similarity in correlations. It is also the case when they formulate new polymers by means of sets of available *similar* molecules. The repetition of a member of a set, say a monomer, allows defining new compounds formally depending on the number of monomers included into a particular polymer. They can also use different sets of polymers to describe copolymers and so on. The set mereology could be interesting as regards chemical discourse within which the repetition of a basic molecule enables chemists to develop analogies. The chemical notation – $(M)_a - (N)_b$ – where M and N are monomers and a and b entire numbers fits a logic of sets where M and N can be replaced by similar monomers which are at hand and which enables chemists to achieve their goal. The chemical notation -$HX + H_2O \rightarrow H_3O^+ + X^-$ – follows a strategy of sets where X can be replaced by *Cl* or *F* for example. Lewis's mereology is interesting to write transformation by means of alternative and economic formula.

9.3.3 The Case for the C-Mereology

The first argument for the C-mereology depends on the possibility of a whole having emergent properties as a result of some structural invariants. A set only accidentally has structural properties because as a conceptual object, membership is fixed by an entity satisfying the intension of the set. A whole has structural properties because is a material entity, with real relations between its parts. Sets are held together by similarity relations,[2] not by real relations between the parts of wholes such as

[2] A mere collection is of no interest to the philosophy of chemistry.

material connectivity (the parts of a chair) or causality, the parts of a molecule. A set can have only similar members, while a whole can have dissimilar parts. A set is a logical object while a whole is a material object.

The second argument for C-mereology depends on the criteria for set membership that is the intensionality component of the set concept. If H^+ and O^{2-} are subsets of the water molecule set what is their common property that makes them members of this set? It can only be that they are constituents of a water molecule. Hence the S-mereology treatment of chemical unity in multiplicity depends on a C-mereological understanding of the relation between atoms (ions) and the molecules of which they are parts. The S-mereology also depends on what van Brakel calls the manifest image of what water is according our current experience of it (Van Brakel 2000). In this respect, both mereologies are complementary. In the grammar of classical Mereology, the three atoms are the parts of a water molecule which is their (disparate) fusion or sum. The water in the sea is the mereological fusion of certain varieties of hydrogen-oxygen conglomerates as parts. But it is not the sum of these conglomerates, which is a being of much greater dimensions being all the water there is. A bucket of brine as a part of the sea is a fusion of the 'water' ionic conglomerates which are its parts. As Earley has argued the same does not apply to the Na^+ and Cl^- ions in the sea (Earley 2005). Here we need to supplement C-mereology with *dispositional* concepts as illustrated in the simple case of the parts of the chair. The concept of the whole, the chair, cannot be eliminated from the criteria for ascribing dispositional properties to the chair parts. This is indeed a reflection of a general condition for practically oriented cognition. We also need to replace dispositions by affordances according to our previous work from chemical practices.

In classical mereology the Principle of Unique Composition runs up against the conclusion that the parts of chemical wholes like molecules and atoms are affordances not themselves concrete entities. However, those same atoms which Mulliken's approach transforms into affordances are the parts of elements modelled as fusions, that is obey the Unique Composition Principle. It seems to us that Transitivity of the Part-Whole Relation as defined in C-mereology does hold because electron affordances are used to construct heuristic models by means of imagined parts of atoms, while so long as we confine ourselves to such models and refrain from drawing metaphysical conclusions, we find it useful to say that atom affordances are parts of molecules, electron affordances are parts of atoms and so of molecules. That is the conclusion to be drawn from Mulliken's demonstration of the power of the molecular orbital set (Harré and Llored 2011).

As we argued above classical mereology can be extended to include rules for the use of a whole – part relation for contexts in which the parts are functionally distinct relative to the whole of which they form parts. In the absence of the concept of the whole, the shapes of the parts, for example, are mereologically irrelevant to their mereological status. They have no role as parts. We are forced to conclude therefore that a new set of mereological rules is required for the logic of chemical discourses. It is neither wholly a C- nor wholly an S-mereology. The new mereology requires a revision of the basic principles that are definitive of each of the mereologies set out above.

Chemistry also makes use of 'ephemeral individuals' as parts of wholes. For instance, the swiftly composing and decomposing hydrogen-oxygen structures of which real water is really composed are ephemeral individuals. Water is made up of these beings. As such they are constituents of a certain whole. Here is a mereological set-up for which neither C-mereology nor S-mereology seems well adapted as discourse grammars because they do not integrate the conceptual time-dependence between process of transformation and dissipative structure. As we have noticed it, the chemical device is also at stake. Parts and sets are relational. They are entangled with the whole and its environments and depend on the mode of access – cognitive or instrumental.

And that is why again, the topological chemical quantum turn is of the utmost importance for philosophical enquiries such as this. That is why, too, philosophers need to go back to laboratories of research to grasp what scientists are really doing with their new models and apparatus.

9.4 Concluding Remarks

We have argued and illustrated with important examples of contemporary chemistry that to take electrons to be constituents of atoms is to commit the product-process fallacy, or, in contemporary terminology, a mereological fallacy, that of presuming that products of an analytical procedure exercised on matter at the level of atoms are constituents of those atoms. Electrons are not entities in the mereological sense – they yield incompatible phenomena – tracks and interference patterns, though the ontological relation between these phenomena is complex. The knowledge engendering relation is the two way comparison between analogues (Harré 2004). Probes are directed to natural objects, substances, and processes. The results of probing can be either constituents of that which is probed (presumed to exist independently of the probe) or affordances that is states, processes and so on the character of which is not independent of the nature of the probe. To presume that all affordances are constituents is the source of a metaphysical fallacy. However, *pseudo-constituents* play an indispensable role in chemical thinking as the content of powerful and heuristic working models.

The standard mereology for chemical compounds involves the presumption that just as molecules are ultimate constituents or parts of material things, so atom-cores are parts or constituents of molecules. However, we have argued elsewhere that atoms do not have entity-like constituents – it is a mistake to treat electrons as constituents of anything (Harré and Llored 2013). Nevertheless, entity-like beings form the basis for the standard *model* that continues to reappear in textbooks – nucleus and planetary electrons (Sukumar 2013). Drawing on Wittgenstein's terminology (Wittgenstein 1979), we could say that such representations are *local hinges* without which chemists' knowledge vanishes. Chemists work with matter,

matter *stands fast* for them, it is not the result of an inference but a basic belief that belongs to the *bedrock* of the chemical world before any chemical knowledge comes up to the fore. We have developed the way chemists establish correlations between wholes, sets or parts, and environments. We hope the future mereology will include those perspectives into the philosophical debate by starting from chemical practices. More than ever, different ways of doing philosophy, analytical or not, should be articulated in order to grasp what chemist really do. To do so, mereology should integrate time and modality to deal with chemical transformation and the role of chemical devices. Chemical relations are not merely formal; sums and incomplete fragments lack chemical work. The thesis that objects with the same parts are identical, say, mereological extensionality, cannot work within this context. Chemical mereology should include contingency which enters into functional wholes by means of the device-dependence and the relation with other chemical bodies. We call for a relational and dynamic form of mereology.

References

Diderot, D. (1754). *Pensées sur l'interprétation de la nature*. Paris: Pissot.

Earley, J. (2005). Why there is no salt in the sea. *Foundations of Chemistry, 7*, 85–102.

Goerigk, L., & Grimme, S. (2010). A general database for main group thermochemistry, kinetics and non covalent interactions. Assessment of common and reparametrized meta-gga density functionals. *Journal of Chemical Theory and Computation, 6*, 107–126.

Harré, R. (2004). Modeling: Gateway to the unknown. In R. Rothbart (Ed.), *Studies in multidisciplinarity* (Vol. 1). Amsterdam: Elsivier.

Harré, R. (2013). Affordances and hinges: New tools for the philosophy of chemistry. In J.-P. Llored (Ed.), *The philosophy of chemistry: Practices, methodologies, and concepts* (pp. 580–595). New Castle: Cambridge Scholars Publishing.

Harré, R., & Llored, J.-P. (2011). Mereologies as the grammars of chemical discourses. *Foundations of Chemistry, 13*(1), 63–76.

Harré, R., & Llored, J.-P. (2013). Molecules and mereology. *Foundations of chemistry, 15*(2), 127–144.

Holmes, F. D., & Levere, T. H. (2000). *Intruments and experiments in the history of chemistry*. Cambridge/London: MIT.

Kohn, W., Becke, A. D., & Parr, R. G. (1996). Density functional theory of electronic structure. *Journal of Physical Chemistry, 100*, 12974–12980.

Lewis, D. (1991). *Parts of classes*. Oxford: Blackwell.

Llored, J.-P. (2010). Mereology and quantum chemistry: The approximation of molecular orbital. *Foundations of Chemistry, 12*, 203–221.

Llored, J.-P. (2012). Emergence and quantum chemistry. *Foundations of Chemistry, 14*(3), 245–274.

Llored, J.-P., et al. (2013). Chemical interfaces. In J.-P. Llored (Ed.), *The Philosophy of chemistry: Practices, methodologies, and concepts* (pp. 172–201). New Castle: Cambridge Scholars Publishing.

Mulliken, R. S. (1932). Electronic structures of polyatomic molecules and valence iii. Quantum theory of the double bond. *Physical Review, 41*, 754.

Needham, P. (2005). Mixtures and modality. *Foundations of Chemistry, 7*, 103–118.

Oppenheim, P., & Rescher, N. (1995). Logical analysis of gestalt concepts. *British Journal for the Philosophy of Science, 6*, 89–106.

Pauling, L. (1928). The shared-electron chemical bond. *Proceedings of the national academy of sciences, 14* , 359–362.
Pauling, L. (1931). The nature of the chemical bond. Application of results obtained from the quantum mechanics and from a theory of paramagnetic susceptibility to the structure of molecules. *Journal of American Chemical Society, 53*, 1367–1400.
Picot, D. (2008). *Modélisation de la réaction d'alkylation du motif zinc-thiolate*. PhD thesis, Ecole Polytechnique, France.
Simons, P. (1987). *Parts*. Oxford: Clarendon Press.
Stengers, I. (2003). La vie et l'artifice: Visages de l'émergence. In *Cosmopolitiques II*. Paris: Éditions La découverte.
Sukumar, N. (2013). The atom in a molecule as a mereological construct in chemistry. *Foundations of Chemistry, 15*(3), 303–309.
Van Brakel, J. (2000). *Philosophy of chemistry*. Leuven: Leuven University Press.
Wittgenstein, L. (1979). *On certainty*. Oxford: Blackwell.
Yong-Hua, G., Miomandre, F., Meallet-Renault, R., Badre, S., Galmiche, L., Tang, J., Audebert, P., & Clavier, G. (2009). Synthesis and physical chemistry of s-tetrazines: Which ones are fluorescent and why? *Chemical European Journal of Organic Chemistry, 35*, 6121–6128.
Zhao, Y., & Truhlar, D. G. (2008). Density functionals with broad applicability in chemistry. *Accounts of Chemical Research, 41*(2), 157–167.

Part IV
Computer Sciences and Engineering

Introduction to Part IV: Mereology, Computer Science and Engineering

It has been one of the greatest philosophical problems since the very birth of philosophy that of organizing vast and diverse collections of data into systematic, comprehensive and rigorous frameworks. One of the main results of this organizational attempt has been the development of several regional ontologies on the one hand and the search for a formal ontological core on the other. Whereas in the past these areas of research have been mostly confined within the philosophical sphere, the recent advancement of computer science has inherited most of the classical problems and has produced new ones. Examples of the significant role played by ontological considerations within computer sciences are knowledge engineering, database design, information modeling, knowledge management and organization, spatial qualitative reasoning, biological information system. And the list could go on, as Guarino (2010) points out. These research areas come with their own problems, methods and structures. It is an important and rather general problem (i) provide a rigorous vocabulary to investigate those problems and methods, and (ii) develop regional ontologies to account and systematize those architectural structures. It is clear that mereology could be a valuable instrument to accomplish both tasks. The papers in this section implement exactly such strategy.

In the first one (*Mereology in Engineering and Computer Science*) Polkowski provides a very comprehensive account of the potential applications of mereology to computer science. They include areas as diverse as Spatial Reasoning, Behavioral Robotics, Knowledge Engineering and many others. One of the greatest challenge of these applications is the difficulty of finding when and whether something is part of something else. This calls for a profound innovation both from the technical and the philosophical point of view, i.e. the introduction of some measure of overlapping in our mereological theory. This is accomplished via the introduction of the new primitive notion of *parthood to a degree* that is essential in developing a new mereological theory called *rough mereology*. This can be seen as an example of a very general and deep philosophical problem, that of indeterminacy and vagueness which was somehow already present in the last work of the previous section.

The second paper (*Discrete Mereotopology*) by Galton provides an overview and a development of atomistic mereotopology. Mereotopology has been studied particularly within the Qualitative Spatial Reasoning community. It has been maintained however within that community that entities are infinitely divisible. Galton shows that interpreting mereotopological predicates over discrete domains renders impossible to define the mereological notion of parthood in terms of the topological relation of connection. He thus develops discrete meretopology taking both of them as primitives. He is able to show that familiar definitions and axioms take a much simpler form within this framework and also establishes important theorems such as the *extensionality principle* for discrete mereotopology, i.e. *having the same atomic parts* is both *sufficient* and *necessary* for *identity*. After considering a particular model of the theory, namely that of adjacent spaces, and introducing notions of distance and size for such spaces, he is able to show important applications of discrete mereotopology in mathematical morphology and digital imaging.

In the final paper (*A Role for Mereology in Domain Science and Engineering*) Bjørner explores the relation between mereology and the relatively new domain science. A domain can be roughly understood as a universe of discourse, or an area of activities for which some form of computing and communication is desirable. The very general idea behind the work is that before developing the software that does the computing one has to understand the structure of the domain in question, and most of the relevant structure is simply mereological structure. He focuses on concrete examples of specific domains such as financial service industry, health care and security IT systems to mention a few. He provides a model-oriented specification of the different mereological structures of those domains and he is able to show that to each mereology it corresponds a program of cooperating sequential process. Finally, he conjectures that the converse holds as well.

To get a detailed perspective on rough mereology, its applications and its relation to fuzzy-set-theory there is probably nothing better than Polkowski (2011). An excellent survey of many topics included in this section is contained in Aiello et al. (2007). For computer science and software engineering we refer to the monumental Bjørner (2006). To get a more philosophical perspective on many of these issues the reader can start from Floridi (2003).

References

Aiello, M., Pratt-Hartmann, E., & Van Benthem, J. (Eds.). (2007). *Handbook of spatial logics*. Dordrecht: Springer.

Bjørner, D. (2006). *Software engineering* (3 Vols.). Berlin/New York: Springer.

Floridi, L. (Ed.). (2003). *The Blackwell guide to the philosophy of computing and information*. Wiley-Blackwell: Oxford.

Guarino, N. (2010). *Formal ontology and information systems, proceedings of FOIS'98*, Trento (pp. 3–15). Amsterdam: IOS Press.

Polkowski, L. (2011). *Approximate reasoning by parts. An introduction to rough mereology*. Berlin: Springer.

Chapter 10
Mereology in Engineering and Computer Science

Lech Polkowski

PART I. FOUNDATIONS

10.1 Introduction

This Chapter is intended as a survey of some basic applications of mereological schemes of reasoning in computer science and engineering. In accordance with the specificity of surveys, it does collect in one text various topical applications, not necessarily related by means of a logical order of things and this calls for a unifying thread stemming from a general discussion of mereological view on things and their relations. In language of mereology, we can state that one thing is a part of the other, that a thing is a fusion of some other things, its parts, or, more generally that some things approximate parts of another thing to a specified degree.

Particular cases are: spatial reasoning where things are figures in space, semantics of spatial locutions addressing spatial relationships of mutual positions like 'in', 'out', 'on', 'under', etc., planning and navigation by intelligent agents (e.g., mobile autonomous robots) in which environment is presented as a collection of polygons which should be bypassed in order to reach a goal, granulation of knowledge represented in data tables along with applications to synthesis of decision or classification rules, logics for reasoning about knowledge, problems of design and assembling of artifacts from parts, problems of representation of spatial features of things like dents, holes, joints. All these problems require adequate notions of parts and relations induced with their help.

L. Polkowski (✉)
Polish-Japanese Institute of Information Technology, Warszawa, Poland

Dept. Mathematics and CS, University of Warmia and Mazury, Olsztyn, Poland
e-mail: polkow@pjwstk.edu.pl

C. Calosi and P. Graziani (eds.), *Mereology and the Sciences*, Synthese Library 371, 217
DOI 10.1007/978-3-319-05356-1_10,

One can justly say that we aim at knowledge expressed in language of underlying mereology theory, and, knowledge we understand here in the sense of I. M. Bocheński (1954), i.e., as a collection of true statements about *things*, their collections, i.e., *concepts*, and about *relations* among concepts.

A discussion of relations among concepts and things and concepts goes back to Aristotle (1989) who singled out four basic types of relations among concepts, i.e., containment or not, intersection or not, and established twenty four valid figures of reasoning called *syllogisms*, see Łukasiewicz (1939, 1957) for a modern logical rendering of the Aristotle system including an axiomatization and Słupecki (1949–1950) for a proof of completeness of the system. In Aristotle's system, knowledge was rendered as a collection of syllogisms, i.e., relations solely among concepts.

Georg Cantor introduced to this discussion things as *elements* and *sets* as collections of things. Naïve set theory of Cantor (from 1873 on), in which one assumed that to each property of things there existed the collection (set) of things which satisfied the property, gave way to *antinomies*, i.e., statements arising from valid deduction rules and seemingly valid assumptions, but contradicting some other assumptions, like the Cantor antinomy (the existence of the set of all sets), the Burali–Forti antinomy (the existence of the set of all ordinal numbers), and most notably the Russell antinomy, see Frege (1903); this antinomy, stated in elementary terms of the theory, defined the set X of things x having the property $x \notin x$, concluding that $X \in X \Leftrightarrow X \notin X$. In the aftermath of this antinomy, set theory became an axiomatized system due in particular to Zermelo (the Zermelo–Fraenkel system) (1908) and Gödel (the Gödel–Bernays system) (1951); the independence of some statements like continuum hypothesis, Souslin hypothesis, etc., see Balcar and Štěpánek (1986) caused emergence of many systems of set theory.

Another way of resolving the problem of the Russell antinomy was pointed by Stanisław Leśniewski (1916), see a translation in Leśniewski (1982), who developed a theory of Mereology in which the primitive notion was that of a part. This notion appeared in Aristotle and Leśniewski defined it as a relation between *individuals* which, in turn, are defined in His Ontology Leśniewski (1927, 1930), see expositions in Słupecki (1955), Lejewski (1958), or, Iwanuś (1973). Ontology is founded on the predicate ϵ to be read "is" (in Greek, *ei* "you are", cf., Plutarch 1936) which is required to satisfy the Ontology Axiom AO, formulated by Leśniewski as early as 1920, see Słupecki (1955)

$$\text{AO } x\epsilon\, y \Leftrightarrow \exists z.(z\epsilon\, x) \wedge \forall z.(z\epsilon\, x \Rightarrow z\epsilon\, y) \wedge \forall z, w.(z\epsilon\, x \wedge w\epsilon\, x \Rightarrow z\epsilon\, w).$$

This axiom determines the meaning of the copula ϵ in the way adopted by Leśniewski: in spite of the copula occurring on either side of the equivalence, its meaning can be revealed by requiring the equivalence to be true as adopted by Leśniewski in His Protothetics, see Słupecki (1955), Miéville (1984). AO means that the thing x is an individual called by a singular name responding to the singular or collective (possibly) name of y. In mereology, the predicate *part* is applied to individual things.

The meaning of AO is that x responds to the name of y if and only if x is non–void ($\exists z.(z\epsilon\ x)$), all x is y, and, x is an individual ($\forall\ z, w.(z\epsilon\ x \wedge w\epsilon\ x \Rightarrow z\epsilon\ w)$). In particular, the relation $x\epsilon x$ does characterize individuals.

The relation of a part being non–reflexive and transitive, the union of the part relation and the identity relation, called the *ingredient* relation *ingr* does satisfy the relation $ingr(x, x)$ for every thing x which implies that each thing is its element (see (10.3) in Sect. 10.2.1); this fact eliminates the Russell antinomy, cf., a formal analysis in Sobociński (1950).

Let us call a *complex thing* a thing which has other things as parts; a question poses itself, in the Henri Poincare sense, of mutual spatial relations of these parts: some may be over, some under other parts, some may be in, some out, some before, some after, some parts may have interiors, some not, etc.

The relation of mereology to spatial analysis was recognized early, e.g., in Tarski's axiomatization of geometry of solids (Tarski 1929). Spatial considerations led Alfred North Whitehead (1916, 1919, 1920, 1929) in His attempt at capturing some notion of a 'continuity of events' to the notion of a relation of being connected, rendered in full generality in Clarke's Connection Calculus (Clarke 1981), and explicating the intuitive notion of an 'external contact' (see below). By considering spatial relations among parts, mereology comes in touch with topology, a mathematical discipline investigating spatial properties of things; hence, attempts at introducing into mereological universa of *mereotopology*, i.e., of a topological structure induced by part structure. Mereotopology in turn may be applied in semantics of utterances describing spatial relations, see, Aurnague and Vieu (1995). The reader will find an extensive discussion of spatial aspects of mereology in Casati and Varzi (1999).

Synthesis of a complex thing from its parts involves two stages, viz., *design* as well as *assembling*; either of these stages requires a specification of the order in which parts enter the process, their mutual orientation and connections, timing of particular operations etc. This belongs in the domain of engineering.

Orientation in space and movement planning are one of principal intelligent behaviors of humans; artificial intelligence and machine learning which are studied within computer science devote much attention to those problems and one of principal applications of methods elaborated in those studies is the field of intelligent (behavioral) robotics. Robots which may move freely in space allotted to them, endowed with sensors, are examples of intelligent agents, capable of performing autonomous tasks. This requires *planning* of paths and subsequent *navigation* in the environment. As environment is modeled as a collections of continua (i.e., compact connected sets) spatial analysis is a principal component in the process of planner synthesis.

Other intelligent behavior is the ability to solve problems, classify things to categories. In studies of this problem, a few paradigms were defined like *fuzzy set theory* of Zadeh (1965), *rough set theory* of Pawlak (1991) along with cognitive paradigms like *neural networks*. In particular, fuzzy set theory is built on the notion of a *membership to a degree*; transferring this idea to mereology brought forth *rough mereology* whose primitive notion is a relation of a part to a degree (Polkowski and

Skowron 1994), also see Polkowski (2011). Rough mereology allows for a more precise description of spatial relations among parts and its applications in classifier synthesis and planning for intelligent robots are described in this Chapter.

In the process of synthesis of a classifier/decision algorithm, one is forced to cope with noise immanent in data; one means of reducing noise is *granulation of data/knowledge* consisting of forming *granules of knowledge* which consist of things similar one to another to a satisfactory degree; rough mereology delivers effective tools for granulation of knowledge and permits to obtain effective classifiers at reduced complexity.

In our opinion, the topics outlined above, which belong in the so–called cognitive technologies and knowledge engineering, are among the most important ones in computer science and engineering, and, in consequence, we discuss them in more detail in the following sections.

The unifying thread which can be defined as the existence of part or part to a degree relations on concepts as the basic starting component in reasoning is strongly evident in those applications.

Preliminary to applications, we want to acquaint the reader with necessary background in theoretical foundations of mereology, rough mereology and mereotopology as bases for further developments.

10.2 Mereology

In this Chapter, we propose to introduce the reader to the notions and methods of mereology. We begin with the mereology in the sense of Leśniewski, based on the notion of a part, and then we present the version of mereology based on the notion of being connected, Connection Calculus proposed in Clarke (1981) on the lines of Whitehead (1916, 1919, 1920, 1929), de Laguna (1922), and Leonard and Goodman (1940). Connection relation captures the notion of being in contact, e.g., having parts in common, or intersecting, and within Connection Calculus one redefines a part relation which leads to a richer theory with a topological flavor. Mereotopology along with necessary background on topology are presented in a further section.

The reader may be aware of the existence of a vast literature on philosophical and ontological aspects of mereology which cannot be mentioned nor discussed here, and, we advise them to consult Simons (2003) and Casati and Varzi (1999) for discussions of those aspects.

10.2.1 Mereology of Leśniewski

Mereology due to Leśniewski arose from attempts at reconciling antinomies of naïve set theory, see Leśniewski (1916, 1927, 1982), Srzednicki et al. (1992), and Sobociński (1950, 1954–1955). Leśniewski (1916) was the first presentation of the foundations of his theory as well as the first formally complete exposition of mereology.

10.2.1.1 On the Notion of Part

The primitive notion of mereology in this formalism is the notion of a *part*. Given some category of things, a relation of a part is a binary relation π which is required to be

M1 *Irreflexive*: *For each thing* x, *it is not true that* $\pi(x,x)$.
M2 *Transitive*: *For each triple* x,y,z *of things, if* $\pi(x,y)$ *and* $\pi(y,z)$, *then* $\pi(x,z)$.

Remark. In the original scheme of Leśniewski, the relation of parts is applied to *individual things* as defined in Ontology of Leśniewski, see Leśniewski (1930), Iwanuś (1973), Słupecki (1955) (see Introduction for the Ontology Axiom AO).

The relation of *part* induces the relation of an *ingredient* (the term is due to T. Kotarbiński), *ingr*, defined as

$$ingr(x,y) \Leftrightarrow \pi(x,y) \vee x = y \tag{10.1}$$

The relation of ingredient is a partial order on things, i.e.,

1. $ingr(x,x)$.
2. $ingr(x,y) \wedge ingr(y,x) \Rightarrow (x = y)$.
3. $ingr(x,y) \wedge ingr(y,z) \Rightarrow ingr(x,z)$.

We formulate the third axiom with a help from the notion of an ingredient.

M3 (*Inference*) For each pair of things x,y, if the property
$I(x,y)$: *For each* t, *if* $ingr(t,x)$, *then there exist* w,z *such that* $ingr(w,t)$, $ingr(w,z)$, $ingr(z,y)$ *hold*,
is satisfied, then $ingr(x,y)$.

The predicate of *overlap*, *Ov* in symbols, is defined by means of

$$Ov(x,y) \Leftrightarrow \exists z.ingr(z,x) \wedge ingr(z,y) \tag{10.2}$$

Using the overlap predicate, one can write $I(x,y)$ down in the form
$I_{Ov}(x,y)$: *For each* t *if* $ingr(t,x)$, *then there exists* z *such that* $ingr(z,y)$ *and* $Ov(t,z)$

10.2.1.2 On the Notion of a Class

The notion of a *mereological class* follows; for a non–vacuous property Φ of things, the *class of* Φ, denoted $Cls\Phi$ is defined by the conditions

C1 *If* $\Phi(x)$, *then* $ingr(x, Cls\Phi)$.
C2 *If* $ingr(x, Cls\Phi)$, *then there exists* z *such that* $\Phi(z)$ *and* $I_{Ov}(x,z)$.

In plain language, the class of Φ collects in an individual thing all things satisfying the property Φ. The existence of classes is guaranteed by an axiom.

M4 *For each non–vacuous property Φ there exists a class $Cls\Phi$.*

The uniqueness of the class follows by M3. M3 implies also that, for the non–vacuous property Φ, if for each thing z such that $\Phi(z)$ it holds that $ingr(z, x)$, then $ingr(Cls\Phi, x)$.

The notion of an overlap allows for a succinct characterization of a class: for each non–vacuous property Φ and each thing x, it happens that $ingr(x, Cls\Phi)$ if and only if for each ingredient w of x, there exists a thing z such that $Ov(w, z)$ and $\Phi(z)$.

Remark. Uniqueness of the class along with its existence is an axiom in the Leśniewski (1916) scheme, from which M3 is derived. Similarly, it is an axiom in the Tarski (1929, 1935, 1937) scheme.

Please consider two examples.

1. The strict inclusion $\subset$ on sets is a part relation. The corresponding ingredient relation is the inclusion $\subseteq$. The overlap relation is the non–empty intersection. For a non–vacuous family F of sets, the class $ClsF$ is the union $\bigcup F$.
2. For reals in the interval $[0, 1]$, the strict order $<$ is a part relation and the corresponding ingredient relation is the weak order $\leq$. Any two reals overlap; for a set $F \subseteq [0, 1]$, the class of F is $supF$.

10.2.1.3 Notions of Element, Subset

The notion of an element is defined as follows

$$el(x, y) \Leftrightarrow \exists \Phi . y = Cls\Phi \wedge \Phi(x) \tag{10.3}$$

In plain words, $el(x, y)$ means that y is a class of some property and x responds to that property. To establish some properties of the notion of an element, we begin with the property $INGR(x) = \{y : ingr(y, x)\}$, for which the identity $x = ClsINGR(x)$ holds by M3. Hence, $el(x, y)$ is equivalent to $ingr(x, y)$. Thus, each thing x is its own element. This is one of means of expressing the impossibility of the Russell paradox within the mereology, cf., Leśniewski (1916), Thms. XXVI, XXVII, see also Sobociński (1950).

We observe the extensionality of overlap: *For each pair x, y of things, $x = y$ if and only if for each thing z, the equivalence $Ov(z, x) \Leftrightarrow Ov(z, y)$ holds.* Indeed, assume the equivalence $Ov(z, x) \Leftrightarrow Ov(z, y)$ to hold for each z. If $ingr(t, x)$ then $Ov(t, x)$ and $Ov(t, y)$ hence by axiom M3 $ingr(t, y)$ and with $t = x$ we get $ingr(x, y)$. By symmetry, $ingr(y, x)$, hence $x = y$.

The notion of a subset follows

$$sub(x, y) \Leftrightarrow \forall z . [ingr(z, x) \Rightarrow ingr(z, y)] \tag{10.4}$$

It is manifest that for each pair x, y of things, $sub(x, y)$ holds if and only if $el(x, y)$ holds if and only if $ingr(x, y)$ holds.

For the property $Ind(x) \Leftrightarrow ingr(x, x)$, one calls the class *ClsInd*, *the universe*, in symbols V.

10.2.1.4 The Universe of Things, Things Exterior, Complementation

It follows that (1) The universe is unique. (2) $ingr(x, V)$ holds for each thing x. (3) For each non–vacuous property Φ, it is true that $ingr(Cls\Phi, V)$.

The notion of an *exterior* thing x to a thing y, $extr(x, y)$, is the following

$$extr(x, y) \Leftrightarrow \neg Ov(x, y) \tag{10.5}$$

In plain words, x is exterior to y when no thing is an *ingr*edient both to x and y.

Clearly, the operator of exterior has properties (1) No thing is exterior to itself. (2) $extr(x, y)$ implies $extr(y, x)$. (3) If for a non–vacuous property Φ, an thing x is exterior to every thing z such that $\Phi(z)$ holds, then $extr(x, Cls\Phi)$.

The notion of a complement to a thing, with respect to another thing, is rendered as a ternary predicate $comp(x, y, z)$, cf., Leśniewski (1916), par. 14, Def. IX, to be read:'x is the complement to y with respect to z', and it is defined by means of the following requirements (1) $x = ClsEXTR(y, z)$. (2) $ingr(y, z)$, where $EXTR(y, z)$ is the property which holds for an thing t if and only if $ingr(t, z)$ and $extr(t, y)$ hold.

This definition implies that the notion of a complement is valid only when there exists an ingredient of z exterior to y. Following are basic properties of complement (1) If $comp(x, y, z)$, then $extr(x, y)$ and $\pi(x, z)$. (2) If $comp(x, y, z)$, then $comp(y, x, z)$.

We let for a thing x, $-x = ClsEXTR(x, V)$. It follows that (1) $-(-x) = x$ *for each thing* x. (2) $-V$ *does not exist.*

We conclude this paragraph with two properties of classes useful in the following

$$\text{If } \Phi \Rightarrow \Psi \text{ then } ingr(Cls\Phi, Cls\Psi) \tag{10.6}$$

and a corollary

$$\text{If } \Phi \Leftrightarrow \Psi \text{ then } Cls\Phi = Cls\Psi \tag{10.7}$$

10.2.2 Mereology Based on Connection

In Whitehead (1916, 1919, 1920), a proposition of the notion of 'x *extends over* y', appeared, dual to that of a part. Th. de Laguna (1922) published a variant of the Whitehead scheme, which led Whitehead (1929) to another version of his approach, based on the notion of 'x *is extensionally connected to* y'. Connection Calculus based on the notion of a 'connection' was proposed in Clarke (1981), which we outline here.

10.2.2.1 On the Connection Predicate

The *predicate of connection* C is subject to basic requirements

CN1 $C(x, x)$ *for each thing* x.
CN2 *If* $C(x, y)$, *then* $C(y, x)$ *for each pair* x, y *of things.*

It follows that connection is reflexive and symmetric. This theory is sometimes called *Ground Topology* T, cf., Casati and Varzi (1999). The additional *extensionality* requirement

CN3 *If* $\forall z.[C(z, x) \Leftrightarrow C(z, y)]$, *then* $x = y$.

produces the *Extensional Ground Topology* ET., see, op. cit.

Let us observe that the predicate C can be realized by taking $C = Ov$; clearly, CN1–CN3 are all satisfied with Ov. We call this model of connection mereology, the *Overlap model*, denoted *OVM*. Also, letting $C(x, y)$ if and only if $x \cap y \neq \emptyset$, defines a connection relation on non-empty sets.

In the universe endowed with C, satisfying CN1, CN2, one defines the notion of an ingredient $ingr_C$ by letting

$$IC\ ingr_C(x, y) \Leftrightarrow \forall z.[C(z, x) \Rightarrow C(z, y)] \tag{10.8}$$

Then, the following properties of $ingr_C$ hold

1. $ingr_C(x, x)$.
2. $ingr_C(x, y) \wedge ingr_C(y, z) \Rightarrow ingr_C(x, z)$.
3. In presence of CN3, $ingr_C(x, y) \wedge ingr_C(y, x) \Rightarrow x = y$.
4. $ingr_C(x, y) \Leftrightarrow \forall z.[ingr(z, x) \Rightarrow ingr(z, y)]$.
5. $ingr_C(x, y) \wedge C(z, x) \Rightarrow C(z, y)$.
6. $ingr_C(x, y) \Rightarrow C(x, y)$.

10.2.2.2 Introducing Notions of a Part, an Ingredient, Overlapping Things and Things Exterior

The notion of a *C–part* π_C can be introduced as

$$PC\ \pi_C(x, y) \Leftrightarrow ingr_C(x, y) \wedge x \neq y \tag{10.9}$$

The predicate of *C–overlapping*, $Ov_C(x, y)$ is defined by means of

$$OC\ Ov_C(x, y) \Leftrightarrow \exists z.[ingr_C(z, x) \wedge ingr_C(z, y)] \tag{10.10}$$

Basic properties of C–overlapping follow.

1. $Ov_C(x, x)$.
2. $Ov_C(x, y) \Leftrightarrow Ov_C(y, x)$.
3. $Ov_C(x, y) \Rightarrow C(x, y)$.
4. $ingr_C(x, y) \wedge Ov_C(z, x) \Rightarrow Ov_C(z, y)$.
5. $ingr_C(x, y) \Rightarrow Ov_C(x, y)$.

The notion of an *C–exterior* things, $extr_C$ is defined by means of

$$EC\ extr_C(x, y) \Leftrightarrow \neg Ov_C(x, y) \tag{10.11}$$

10.2.2.3 Notions Derived from C

A new notion is *C–external connectedness*, *EC*, defined as follows

$$EC\ EC(x, y) \Leftrightarrow C(x, y) \wedge extr(x, y) \tag{10.12}$$

It is easy to see that in the model *OVM*, *EC* is a vacuous notion. Clearly, by definition (10.12),

1. $\neg EC(x, x)$.
2. $EC(x, y) \Leftrightarrow EC(y, x)$.
3. $C(x, y) \Leftrightarrow EC(x, y) \vee Ov_C(x, y)$.
4. $Ov_C(x, y) \Leftrightarrow C(x, y) \wedge \neg EC(x, y)$.
5. $\neg EC(x, y) \Leftrightarrow [Ov_C(x, y) \Leftrightarrow C(x, y)]$: This is a logical rendering of our remark that in *OVM*, no pair of things is in EC, hence, $\neg EC(x, y) = TRUE$ for each pair of things.
6. $\neg \exists z.EC(z, x) \Rightarrow \{ingr_C(x, y) \Leftrightarrow [\forall w.Ov_C(w, x) \Rightarrow Ov_C(w, y)]\}$.

A comment in the way of proof. The implication

$$ingr_C(x, y) \Rightarrow [\forall w.Ov_C(w, x) \Rightarrow Ov_C(w, y)]$$

is always true. Thus, it remains to assume that (i) $\neg \exists z.EC(z, x)$ and to prove that

$$(*)\ [\forall w.Ov_C(w, x) \Rightarrow Ov_C(w, y)] \Rightarrow ingr_C(x, y)$$

(i) can be written down as (ii) $\forall z.\neg C(z, x) \vee Ov_C(z, x)$. To prove that $ingr_C(x, y)$ it should be verified that (iii) $\forall z.(C(z, x) \Rightarrow C(z, y))$.

Consider an arbitrary thing z'; either $\neg C(z', x)$ in which case implication in (iii) is satisfied with z', or, $Ov_C(z', x)$, hence, $Ov_C(z', y)$ by the assumed premise in (*), which implies that $C(z', y)$. The implication (iii) is proved and this concludes the proof.

The richer structure of connection based calculus allows for some notions of a topological nature; the first is the notion of a *tangential ingredient*, $Tingr_C(x, y)$, defined by means of

$$TI\ Tingr_C(x, y) \Leftrightarrow ingr_C(x, y) \wedge \exists z.EC(z, x) \wedge EC(z, y) \qquad (10.13)$$

Basic properties of tangential parts follow by (10.13)

1. $\exists z.EC(z, x) \Rightarrow Tingr_C(x, x)$.
2. $\neg\exists z.EC(z, y) \Rightarrow \neg existsx.Tingr_C(x, y)$.
3. $Tingr_C(z, x) \wedge ingr_C(z, y) \wedge ingr_C(y, x) \Rightarrow Tingr_C(y, x)$.

For Property 3, some argument may be in order; consider w such that $EC(w, x)$, $EC(w, z)$ existing by $Tingr_C(z, x)$. hence, $C(w, y)$. As $\neg Ov_C(w, x)$, it follows that $\neg Ov_C(w, y)$, hence, $EC(w, y)$, and $Tingr_C(y, x)$.

These properties witness the fact that if there is some thing externally connected to x, then x is its tangential ingredient. This fact shows that the notion of a tangential ingredient falls short of the idea of a boundary. Dually, in absence of things externally connected to y, no ingredient of y can be a tangential ingredient.

A thing y is a *non–tangential ingredient* of a thing x, $NTingr_C(y, x)$, in case it is an ingredient but not any tangential ingredient of x,

$$NTI\ NTingr_C(y, x) \Leftrightarrow \neg Tingr_C(y, x) \wedge ingr_C(y, x) \qquad (10.14)$$

Basic properties of the operator NTI are

1. $NTingr_C(y, x) \Rightarrow \forall z.\neg EC(z, y) \vee \neg EC(z, x)$.
2. $\neg\exists z.EC(z, x) \Rightarrow NTingr_C(x, x)$.

In absence of externally connected things, each thing is a non–tangential ingredient of itself, hence, in the model *OVM* each object is its own non–tangential ingredient and it has no tangential ingredients. To produce models in which *EC*, $NTingr_C$, $Tingr_C$ will be exhibited, we resort to topology, see Sect. 10.4.3.

Further properties of the predicate $NTingr_C$ are

1. $NTingr_C(y, x) \wedge C(z, y) \Rightarrow C(z, x)$.
2. $NTingr_C(y, x) \wedge Ov_C(z, y) \Rightarrow Ov_C(z, x)$.
3. $NTingr_C(y, x) \wedge C(z, y) \Rightarrow Ov_C(z, x)$.
4. $ingr_C(y, x) \wedge NTingr_C(x, z) \Rightarrow NTingr_C(y, z)$.
5. $ingr_C(y, z) \wedge NTingr_C(x, y) \Rightarrow NTingr_C(x, z)$.
6. $NTingr_C(y, z) \wedge NTingr_C(z, x) \Rightarrow NTingr_C(y, x)$.

For Property 3, from already known $\forall z.\neg EC(z, y) \vee \neg EC(z, x)$, it follows

$$(i)\ \forall w.\neg C(w, x) \vee Ov_C(w, x) \vee \neg C(w, y) \vee Ov_C(w, y)$$

As $C(z, y)$, one obtains $C(z, x)$. Thus, by (i), $Ov_C(z, y) \vee Ov_C(z, x)$ and $Ov_C(z, x)$.

For Property 4, assume $ingr_C(y,x), ingr_C(x,z)$ and hence, $ingr_C(y,z)$ (otherwise there is nothing to prove), consider $\neg NTingr_C(y,z)$, i.e., for some w: $EC(w,z), EC(w,y)$. Thus, $C(w,z)$, $\neg Ov_C(w,z)$, $C(w,y)$, $\neg Ov_C(w,y)$. Then, $C(w,x)$ and $\neg Ov_C(w,x)$, hence, $EC(w,x)$ and $\neg NTingr_C(x,z)$, a contradiction. Similarly, one justifies Properties 5 and 6.

10.3 Rough Mereology

A scheme of mereology, introduced into a collection of things, sets an exact hierarchy of things of which some are (exact) parts of others; to ascertain whether a thing is an exact part of some other thing is in practical cases often difficult if possible at all, e.g., a robot sensing the environment by means of a camera or a laser range sensor, cannot exactly perceive obstacles or navigation beacons. Such evaluation can be done approximately only and one can discuss such situations up to a degree of certainty only. Thus, one departs from the exact reasoning scheme given by decomposition into parts to a scheme which approximates the exact scheme but does not observe it exactly.

Such a scheme, albeit its conclusions are expressed in an approximate language, can be more reliable, as its users are aware of uncertainty of its statements and can take appropriate measures to fend off possible consequences.

Imagine two robots using the language of connection mereology for describing mutual relations; when endowed with touch sensors, they can ascertain the moment when they are connected; when a robot has as a goal to enter a certain area, it can ascertain that it connected to the area or overlapped with it, or it is a part of the area, and it has no means to describe its position more precisely.

Introducing some measures of overlapping, in other words, the extent to which one thing is a part to the other, would allow for a more precise description of relative position, and would add an expressional power to the language of mereology. Rough mereology answers these demands by introducing the notion of a *part to a degree* with the degree expressed as a real number in the interval [0, 1]. Any notion of a part by necessity relates to the general idea of *containment*, and thus the notion of a part to a degree is related to the idea of *partial containment* and it should preserve the essential intuitive postulates about the latter.

The predicate of a part to a degree stems ideologically from and has as one of motivations the predicate of an element to a degree introduced by L. A. Zadeh as a basis for fuzzy set theory (Zadeh 1965); in this sense, rough mereology is to mereology as the fuzzy set theory is to the naive set theory. To the rough set theory, owes rough mereology the interest in concepts as things of analysis.

The primitive notion of rough mereology is the notion of a *rough inclusion* which is a ternary predicate $\mu(x, y, r)$ where x, y are *things* and $r \in [0, 1]$, read '*the thing x is a part to degree at least of r to the thing y*'. Any rough inclusion is associated with a mereological scheme based on the notion of a part by postulating that $\mu(x, y, 1)$ is equivalent to $ingr(x, y)$, where the ingredient relation is defined by the adopted mereological scheme. Other postulates about rough inclusions stem from intuitions about the nature of partial containment; these intuitions can be manifold, a fortiori, postulates about rough inclusions may vary. In our scheme for rough mereology, we begin with some basic postulates which would provide a most general framework. When needed, other postulates, narrowing the variety of possible models, can be introduced.

10.3.1 Rough Inclusions

We have already stated that a rough inclusion is a ternary predicate $\mu(x, y, r)$. We assume that a collection of things is given, on which a part relation π is introduced with the associated ingredient relation *ingr*. We thus apply inference schemes of mereology due to Leśniewski, presented above.

Predicates $\mu(x, y, r)$ were introduced in Polkowski and Skowron (1994, 1997); they satisfy the following postulates, relative to a given part relation π and the induced by π relation *ingr* of an ingredient, on a set of things

RINC1 $\mu(x, y, 1) \Leftrightarrow ingr(x, y)$.

This postulate asserts that parts to degree of 1 are ingredients.

RINC2 $\mu(x, y, 1) \Rightarrow \forall z[\mu(z, x, r) \Rightarrow \mu(z, y, r)]$.

This postulate does express a feature of partial containment that a 'bigger' thing contains a given thing 'more' than a 'smaller' thing. It can be called a *monotonicity condition* for rough inclusions.

RINC3 $\mu(x, y, r) \wedge s < r \Rightarrow \mu(x, y, s)$.

This postulate specifies the meaning of the phrase 'a part to a degree at least of r'. From postulates RINC1–RINC3, and known properties of ingredients some consequences follow

1. $\mu(x, x, 1)$.
2. $\mu(x, y, 1) \wedge \mu(y, z, 1) \Rightarrow \mu(x, z, 1)$.
3. $\mu(x, y, 1) \wedge \mu(y, x, 1) \Leftrightarrow x = y$.
4. $x \neq y \Rightarrow \neg\mu(x, y, 1) \vee \neg\mu(y, x, 1)$.
5. $\forall z \forall r[\mu(z, x, r) \Leftrightarrow \mu(z, y, r)] \Rightarrow x = y$.

Property 5 may be regarded as an *extensionality postulate* in rough mereology.

By a *model* for rough mereology, we mean a quadruple

$$M = (V_M, \pi_M, ingr_M, \mu_M)$$

where V_M is a set with a part relation $\pi_M \subseteq V_M \times V_M$, the associated ingredient relation $ingr_M \subseteq V_M \times V_M$, and a relation $\mu_M \subseteq V_M \times V_M \times [0, 1]$ which satisfies RINC1–RINC3.

We now describe some models for rough mereology which at the same time give us methods by which we can define rough inclusions, see Polkowski (2002, 2003, 2004a,b, 2005a, 2007, 2008, 2009a), a detailed discussion may be found in Polkowski (2011).

10.3.1.1 Rough Inclusions from t–norms

We resort to *continuous t–norms* which are continuous functions $T : [0, 1]^2 \rightarrow [0, 1]$ which are (1) symmetric. (2) associative. (3) increasing in each coordinate. (4) satisfying boundary conditions $T(x, 0) = 0, T(x, 1) = x$, cf., Polkowski (2011), Chs. 4 and 6, Hájek (1998), Ch. 2. Classical examples of continuous t–norms are

1. $L(x, y) = max\{0, x + y - 1\}$ (the *Łukasiewicz's t–norm*).
2. $P(x, y) = x \cdot y$ *(the product t–norm)*.
3. $M(x, y) = min\{x, y\}$ *(the minimum t–norm)*.

The *residual implication* $\Rightarrow_T$ induced by a continuous t–norm T is defined as

$$x \Rightarrow_T y = max\{z : T(x, z) \leq y\} \tag{10.15}$$

One proves that $\mu_T(x, y, r) \Leftrightarrow x \Rightarrow_T y \geq r$ is a rough inclusion; particular cases are

1. $\mu_L(x, y, r) \Leftrightarrow min\{1, 1 - x + y \geq r\}$ (the *Łukasiewicz implication*).
2. $\mu_P(x, y, r) \Leftrightarrow \frac{y}{x} \geq r$ when $x > 0$, $\mu_P(x, y, 1)$ when $x = 0$ (the *Goguen implication*).
3. $\mu_M(x, y, r) \Leftrightarrow y \geq r$ when $x > 0$, $\mu_M(x, y, 1)$ when $x = 0$ (the *Gödel implication*).

A particular case of continuous t–norms are *Archimedean t–norms* which satisfy the inequality $T(x, x) < x$ for each $x \in (0, 1)$. It is well–known, see Ling (1965), that each archimedean t–norm T admits a representation

$$T(x, y) = g_T(f_T(x) + f_T(y)) \tag{10.16}$$

where the function $f_T : [0, 1] \rightarrow R$ is continuous decreasing with $f_T(1) = 0$, and $g_T : R \rightarrow [0, 1]$ is the *pseudo–inverse* to f_T, i.e., $g \circ f = id$.

It is known, cf., e.g., Hájek (1998), that up to an isomorphism there are two Archimedean t–norms: L and P. Their representations are

$$f_L(x) = 1 - x; \; g_L(y) = 1 - y \tag{10.17}$$

and

$$f_P(x) = exp(-x); \; g_P(y) = -ln\ y \tag{10.18}$$

For an Archimedean t–norm T, we define the rough inclusion μ^T on the interval [0, 1] by means of

$$(ari)\ \mu^T(x, y, r) \Leftrightarrow g_T(|x - y|) \geq r \tag{10.19}$$

equivalently,

$$\mu^T(x, y, r) \Leftrightarrow |x - y| \leq f_T(r) \tag{10.20}$$

It follows from (10.20), that μ^T satisfies conditions RINC1–RINC3 with *ingr* as identity =.

To give a hint of proof: for RINC1: $\mu^T(x, y, 1)$ if and only if $|x - y| \leq f_T(1) = 0$, hence, if and only if $x = y$. This implies RINC2. In case $s < r$, and $|x - y| \leq f_T(r)$, one has $f_T(r) \leq f_T(s)$ and $|x - y| \leq f_T(s)$.

Specific recipes are

$$\mu^L(x, y, r) \Leftrightarrow |x - y| \leq 1 - r \tag{10.21}$$

and

$$\mu^P(x, y, r) \Leftrightarrow |x - y| \leq -ln\ r \tag{10.22}$$

Both residual and archimedean rough inclusions satisfy the *transitivity condition*

$$\text{(Trans) } if\ \mu(x, y, r)\ and\ \mu(y, z, s), then\ \mu(x, z, T(r, s)).$$

In the way of a proof, assume, e.g., $\mu^T(x, y, r)$ and $\mu^T(y, z, s)$, i.e., $|x - y| \leq f_T(r)$ and $|y - z| \leq f_T(s)$. Hence, $|x - z| \leq |x - y| + |y - z| \leq f_T(r) + f_T(s)$, hence, $g_T(|x - z|) \geq g_T(f_T(r) + f_T(s)) = T(r, s)$, i.e., $\mu^T(x, z, T(r, s))$. Other cases go on same lines. Let us observe that rough inclusions of the form (ari) are also *symmetric*.

10.3.1.2 Rough Inclusions in Information Systems (Data Tables)

An important domain where rough inclusions will play a dominant role in our analysis of reasoning by means of parts is the realm of *information systems* of Pawlak (1991), cf., Polkowski (2011), Ch. 6. We will define information rough inclusions denoted with a generic symbol μ^I.

We recall that an *information system* (a *data table*) is represented as a pair (U, A) where U is a finite set of things and A is a finite set of *attributes*; each attribute $a : U \to V$ maps the set U into the *value set* V. For an attribute a and a thing v, $a(v)$ is the value of a on v.

For things u, v the *discernibility set* $DIS(u, v)$ is defined as

$$DIS(u, v) = \{a \in A : a(u) \neq a(v)\} \tag{10.23}$$

For an (ari) μ_T, we define a rough inclusion μ_T^I by means of

$$(airi)\ \mu_T^I(u, v, r) \Leftrightarrow g_T(\frac{|DIS(u, v)|}{|A|}) \geq r \tag{10.24}$$

Then, μ_T^I is a rough inclusion with the associated ingredient relation of identity and the part relation empty.

For the Łukasiewicz t–norm, the *airi* μ_L^I is given by means of the formula

$$\mu_L^I(u, v, r) \Leftrightarrow 1 - \frac{|DIS(u, v)|}{|A|} \geq r \tag{10.25}$$

We introduce the set $IND(u, v) = A \setminus DIS(u, v)$. With its help, we obtain a new form of (10.25)

$$\mu_L^I(u, v, r) \Leftrightarrow \frac{|IND(u, v)|}{|A|} \geq r \tag{10.26}$$

The formula (10.26) witnesses that the reasoning based on the rough inclusion μ_L^I is the probabilistic one which goes back to Łukasiewicz (1970). Each (airi)–type rough inclusion μ_T^I satisfies the transitivity condition (Trans) and is symmetric.

10.3.1.3 Rough Inclusions on Sets and Measurable Sets

Formula (10.26) can be abstracted to set and geometric domains. For finite sets A, B,

$$\mu^S(A, B, r) \Leftrightarrow \frac{|A \cap B|}{|A|} \geq r \tag{10.27}$$

where $|X|$ denotes the cardinality of X, defines a rough inclusion μ^S. For bounded measurable sets X, Y in an Euclidean space E^n,

$$\mu^G(A, B, r) \Leftrightarrow \frac{||A \cap B||}{||A||} \geq r \tag{10.28}$$

where $||A||$ denotes the area (the Lebesgue measure) of the region A, defines a rough inclusion μ^G. Both μ^S, μ^G are symmetric but not transitive.

Other rough inclusions and their weaker variants will be defined in later chapters.

10.4 Mereotopology and Mereogeometry

Both mereology and topology address problems of mutual relations among things like 'being external', 'being inside' etc., hence, as the language of topology is well established, it is desirable to trace topological constructs in mereological universa. First, we would like to introduce the reader to rudiments of topology necessary in order to follow our exposition.

10.4.1 A Topological Background

We begin with the notion of a *topological space* which is a pair (X, τ) where X is a set and τ a family of subsets of X; sets in the family τ are called *open sets* provided the following are satisfied (1) τ is closed on finite intersections. (2) unions of sub–families of τ belong in τ. Examples are provided, e.g., by metric spaces; given a metric ρ on a set X, *open* balls are defined as sets of the form $B(x, r) = \{y \in X : \rho(x, y) < r\}$ for $x \in X$ and $r > 0$. Open sets are defined in this case as unions of families of open balls.

Closed sets are complements to open sets; a set $C \subseteq X$ is closed if and only if the set $X \setminus C$ is open. Clearly, intersections of arbitrary families of closed sets are closed and finite unions of closed sets are closed.

10.4.1.1 Approximations: Interior and Closure of a Set

In a given topological space, (X, τ), open as well as closed sets may be called *definable* as the membership problem for them is decidable; other sets can be *approximated* only by open, respectively, closed sets. To this end, topology offers operators of *interior*, *Int*, respectively, of *closure*, *Cl*. The operator *Int* provides the approximation from below, whereas *Cl* yields the approximation from above.

For a subset $A \subseteq X$, the interior of A is $IntA = \bigcup\{V \in \tau : V \subseteq A\}$; the dual operator of closure can be defined as $ClA = X \setminus Int(X \setminus A)$. The well–known properties of those operators follow (1) $Int\emptyset = \emptyset$ $(ClX = X)$. (2) $IntA \subseteq A$ $(A \subseteq ClA)$. (3) $IntIntA = A$ $(ClClA = ClA)$. (4) $(A \subseteq B) \Rightarrow (IntA \subseteq IntB)$ $(A \subseteq B \Rightarrow ClA \subseteq ClB)$, where formulas in parentheses give properties of the closure operator, dual to those of the interior operator. Properties 1–4 may be taken as axioms for interior, respectively, closure, operators. Each set $A \subseteq X$ is sandwiched between *IntA* and *ClA*.

10.4.1.2 Boundaries

The difference $ClA \setminus IntA$, denoted *BdA*, is the *boundary* of A. Clearly, $BdA = ClA \cap Cl(X \setminus A)$. In case $BdA = \emptyset$, the set A is *closed–open (clopen)*, A is *meager* in case $IntBdA = \emptyset$, and A is *nowhere–dense* in case $IntBdClA = \emptyset$. Let us observe that the boundary operator requires for its definition either interior and complement or closure and complement operators, i.e., it is of open–and–closed character.

We close this paragraph with an essential property of *Int* and *Cl* operators

$$V \cap ClA \subseteq Cl(V \cap A) \tag{10.29}$$

for each set A and each open set V in a topological space (X, τ). For the proof, it suffices to observe that given $x \in V \cap ClA$ and an arbitrary open $W \ni x$, one has $(W \cap V) \cap A \neq \emptyset$, i.e., $W \cap (V \cap A) \neq \emptyset$, hence, $x \in Cl(V \cap A)$.

Topological spaces are classified also with respect to their *separation properties*; T_0 (or, Kolmogorov) property consists in $Clx \neq Cly$ when points $x \neq y$; T_1 means $Clx = x$, each x; T_2 (Hausdorff) means that each pair of distinct points can be separated by disjoint open sets; T_3 (regularity) means that each pair x, F, F closed and $x \notin F$, can be separated by disjoint open sets(regularity); a topological space (X, τ) is *regular* if and only if for each pair $x \in V$, where V open, there exists an open W such that $x \in W \subseteq ClW \subseteq V$.

Of interest to us are particular sub–categories of open or closed sets. We introduce *regular closed* sets which allow for non–trivial examples of C–mereological operators of overlap, interior and closure.

10.4.2 Regular Open and Regular Closed Sets

A set A in a topological space (X, τ) is *regular open* if it is of the form $IntClB$ for some set B; then, by property (iii) of operators *Int*, *Cl*, $IntClA = IntClB$, i.e., $A = IntClA$. Hence, regular open sets are characterized by the identity $A = IntClA$. Dually, a set C is *regular closed* if it satisfies the identity $C = ClIntC$. It follows that A is regular open (resp. regular closed) if and only if the set $X \setminus A$ is regular closed (resp. regular open). As the set $IntBd(ClV \setminus V)$ with an open V is empty, each regular open or regular closed set has a nowhere–dense boundary.

Examples of regular closed sets are closed disks in the 2D–space or closed balls in the 3D–space and the open complements are examples of regular open sets; more generally, regular closed are compact convex regions in Euclidean spaces and their interiors are specimens of regular open sets.

A very basic property of regular sets is that they form complete Boolean algebras; regular open sets form the Boolean algebra, denoted $RO(X)$. In order to justify this claim, we let

$$A^{\perp} = X \setminus ClA$$

The set A is regular open if and only if $A = A^{\perp\perp}$. Indeed, $A^{\perp\perp} = X \setminus Cl(X \setminus ClA) = IntClA$. Properties of the operation $A^{\perp}$ are (in proofs, one uses (10.29)

1. If $A \subseteq B$, then $B^{\perp} \subseteq A^{\perp}$.
2. If A is an open set, then $A \subseteq A^{\perp\perp}$.
3. If A is an open set, then $A^{\perp} = A^{\perp\perp\perp}$, hence, $A^{\perp\perp} = A^{\perp\perp\perp\perp}$.
4. If A, B are open sets, then $(A \cap B)^{\perp\perp} = A^{\perp\perp} \cap B^{\perp\perp}$.
5. $(A \cup B)^{\perp} = A^{\perp} \cap B^{\perp}$.
6. If A is an open set, then $(A \cup A^{\perp})^{\perp\perp} = X$.

Now, we define in the family $RO(X)$ of regular open sets operations $\wedge, \vee, '$

1. $A \vee B = (A \cup B)^{\perp\perp} = IntCl(A \cup B)$.
2. $A \wedge B = A \cap B$.
3. $A' = A^{\perp} = X \setminus ClA$.

and constants $\mathbf{0} = \emptyset, \mathbf{1} = X$.

All operations listed above give regular open sets by properties of $(.)^{\perp}$. It remains to check that axioms of a Boolean algebra are satisfied. Commutativity laws $A \vee B = B \vee A, A \wedge B = B \wedge A$ are satisfied evidently. The laws $A \vee \mathbf{0} = A, A \wedge \mathbf{1} = A$ are also manifest. We have $A \wedge A' = A \cap A^{\perp} = A \setminus ClA = \emptyset = \mathbf{0}$ as well as $A \vee A' = (A \cup A^{\perp\perp})^{\perp\perp} = X = \mathbf{1}$. The distributive laws $A \vee (B \wedge C) = (A \vee B) \wedge (A \vee C)$ as well as $A \vee (B \wedge C) = (A \vee C) \wedge (A \vee C)$ hold by Property 5.

A particular sub–algebra of $RO(X)$ is the algebra $CO(X)$ of *clopen sets* in X. In case of $CO(X)$ boolean operations $\vee, \wedge, '$ specialize to usual set–theoretic operations $\cup, \cap, \setminus$ i.e. $CO(X)$ is a field of sets.

The basic distinction between $RO(X)$ and $CO(X)$ is the fact that $RO(X)$ is a *complete Boolean algebra* for any X whereas $CO(X)$ needs not be such.

Let us observe that the boolean ordering relation $\leq$ is in this case the inclusion $\subseteq$.

Consider $\mathscr{A} \subseteq RO(X)$. Let $s(\mathscr{A}) = (\bigcup \mathscr{A})^{\perp\perp}$; we check that $s(\mathscr{A})$ is the supremum of $\mathscr{A}$.

Indeed, for $A \in \mathscr{A}$, we have $A \in \bigcup \mathscr{A}$ hence $A = A^{\perp\perp} \subseteq (\bigcup \mathscr{A})^{\perp\perp}$ i.e. $A \leq s(\mathscr{A})$. It follows that $s(\mathscr{A})$ is an upper bound for $\mathscr{A}$.

Now, assume that $B \in RO(X)$ is an upper bound for $\mathscr{A}$, i.e., $A \subseteq B$ for each $A \in \mathscr{A}$. Hence $\bigcup(\mathscr{A}) \subseteq B$ and thus $(\bigcup \mathscr{A})^{\perp\perp} \subseteq B^{\perp\perp} = B$ i.e. $s(\mathscr{A}) \leq B$ proving that $s(\mathscr{A})$ is the supremum of $\mathscr{A}$. Finally, by duality it follows that $i(\mathscr{A}) = (\bigcap \mathscr{A})^{\perp\perp}$ is the infimum of $\mathscr{A}$.

By duality applied to the family $RC(X)$ of regular closed sets in X, we obtain a dual proposal that $RC(X)$ is a complete boolean algebra under operations $\wedge, \vee, '$ defined as follows

1. $A \vee B = A \cup B$.
2. $A \wedge B = ClInt(A \cap B)$.
3. $A' = X \setminus IntA$.

and with constants $\mathbf{0} = \emptyset, \mathbf{1} = X$.

It follows that $RO(X), RC(X)$ are *mereological categories* as they are closed on formation of classes; contrariwise, closed sets do not form any mereological category; this is in part responsible for difficulties with boundaries in mereology.

A complete axiomatization of mereotopology interpreted in regular open sets is given in Asher and Vieu (1995).

10.4.3 An Application: The Model ROM for Connection

We define in the space $RO(X)$ of regular open sets in a regular space X the connection C by demanding that

$$C(A, B) \Leftrightarrow ClA \cap ClB \neq \emptyset.$$

For simplicity sake, we assume that the regular space X is connected, so no set in it is clopen, equivalently, the boundary of each set is non–empty.

10.4.3.1 Ingredient in ROM

First, we investigate what $ingr_C$ means in ROM. By definition IC in (10.8), for $A, B \in RO(X)$,

$$ingr_C(A, B) \Leftrightarrow \forall Z \in RO(X).ClZ \cap ClA \neq \emptyset \Rightarrow ClZ \cap ClB \neq \emptyset.$$

This excludes the case when $A \setminus ClB \neq \emptyset$ as then we could find a $Z \in RO(X)$ with

$$Z \cap A \neq \emptyset = ClZ \cap ClB$$

(as our space X is regular). It remains that $A \subseteq ClB$, hence, $A \subseteq IntClB = B$. It follows finally that in model ROM, $ingr_C(A, B) \Leftrightarrow A \subseteq B$.

10.4.3.2 Overlap in ROM

Now, we can interpret overlapping in ROM. For $A, B \in RO(X)$, $Ov_C(A, B)$ means that there exists $Z \in RO(X)$ such that $Z \subseteq A$ and $Z \subseteq B$ hence$Z \subseteq A \cap B$, hence

$$A \cap B \neq \emptyset.$$

This condition is also sufficient by regularity of X. We obtain that in ROM,

$$Ov_C(A, B) \Leftrightarrow A \cap B \neq \emptyset.$$

10.4.3.3 External Connectedness in ROM

The status of *EC* in *ROM* is

$$EC(A, B) \Leftrightarrow ClA \cap ClB \neq \emptyset \wedge A \cap B = \emptyset.$$

This means that closed sets *ClA*, *ClB* do intersect only at their boundary points.

10.4.3.4 Tangential Ingredient in ROM

We can address the notion of a tangential ingredient: $Tingr_C(A, B)$ means the existence of $Z \in RO(X)$ such that

$$ClZ \cap ClA \neq \emptyset \neq ClZ \cap ClB$$

and

$$Z \cap A = \emptyset = Z \cap B$$

along with $A \subseteq B$.

In case

$$ClA \cap (ClB \setminus B) \neq \emptyset$$

letting $Z = X \setminus ClB$ we have

$$ClZ = Cl(X \setminus ClB)$$

and

$$BdZ = ClZ \setminus Z = Cl(X \setminus ClB) \setminus (X \setminus ClB)$$

which in turn is equal to

$$Cl(X \setminus ClB) \cap ClB = Cl(X \setminus B) \cap ClB = BdB.$$

Hence, $ClB \setminus B \subseteq ClZ$, and $ClZ \cap ClA \neq \emptyset$; a fortiori, $ClB \cap ClZ \neq \emptyset$. As $Z \cap B = \emptyset$, a fortiori $Z \cap A = \emptyset$ follows.

We know, then, that

$$ClA \cap (ClB \setminus B) \neq \emptyset \Rightarrow Tingr_C(A, B)$$

Was to the contrary, $ClA \subseteq B$, from $Z \cap ClA \neq \emptyset$ it would follow that $Z \cap B \neq \emptyset$, negating $EC(A, B)$.

It follows finally that in ROM, $Tingr_C(A, B)$ if and only if $A \subseteq B$ and $ClA \cap (ClB \setminus B) \neq \emptyset$, i.e.,

$$Tingr_C(A, B) \Leftrightarrow A \subseteq B \wedge ClA \cap BdB \neq \emptyset.$$

From this analysis we obtain also that $NTingr_C(A, B)$ if and only if $ClA \subseteq IntB$.

10.4.4 Mereotopology in Part Mereology

We assume now a Leśniewski–style universe with part and ingredient relations and derived notions. Topological structures which arise in this context can be induced from overlap relations.

As in topology, interior as well as closure operators act on unions and intersections of sets, we recall here two *fusion operators* due to Tarski (1935), cf., Clay (1974). These operators are the sum $x + y$ and the product $x \cdot y$ defined by means of

$$ingr(z, x + y) \Leftrightarrow ingr(z, x) \vee ingr(z, y), \tag{10.30}$$

and,

$$ingr(z, x \cdot y) \Leftrightarrow ingr(z, x) \wedge ingr(z, y) \tag{10.31}$$

10.4.4.1 On Closures

As the first approximation to topology, let us define for each thing x, its *closure* $c(x)$ by means of

$$c(x) = ClsOv(x) \tag{10.32}$$

where the property $Ov(x)$ is defined by $Ov(x)(y) \Leftrightarrow Ov(x, y)$, i.e., we build the closure $c(x)$ as the class of things which overlap with x.

The closure operator $c(.)$ has the following properties

Cl1 $ingr(x, c(x))$.
Cl2 If $ingr(x, y)$, then $ingr(c(x), c(y))$.
Cl3 $ingr(c(x \cdot y), c(x) \cdot c(y))$.
Cl4 $c(x + y) = c(x) + c(y)$.

In way of proof, we observe that Cl1 and Cl2 follow from definition of the overlap relation and the class definition. For Cl3, if $ingr(t, c(x \cdot y))$, then there is z such that $Ov(t, z)$ and $Ov(z, x \cdot y)$ thus for some w one has $Ov(z, w)$ and $ingr(w, x)$, $ingr(w, y)$ which imply that $ingr(t, c(x))$, $ingr(t, c(y))$ and finally $ingr(t, c(x) \cdot c(y))$. By M3,

$ingr(c(x \cdot y), c(x) \cdot c(y))$. For Cl4, it suffices to observe that $Ov(z, x + y) \Leftrightarrow Ov(z, x) \vee Ov(z, y)$.

Another possibility for a topology is in iteration of the operator c, viz, we let

$$Ov^{n+1}(x, y) \Leftrightarrow \exists z.Ov(x, z) \wedge Ov^{n}(z, y); Ov^{1}(x, y) \Leftrightarrow Ov(x, y) \quad (10.33)$$

and we define

$$OVLP(x)(y) \Leftrightarrow \exists n.Ov^{n}(x, y) \quad (10.34)$$

The closure $Cl(x)$ is defined as the class of the property $OVLP(x)$, i.e.,

$$Cl(x) = ClsOVLP(x) \quad (10.35)$$

The operator $Cl(x)$ has the following properties

CL1 $Cl(Cl(x)) = Cl(x)$.
CL2 $ingr(x, Cl(x))$.
CL3 $ingr(x, y)$ implies $ingr(Cl(x), Cl(y))$.
CL4 $Cl(x + y) = Cl(x) + Cl(y)$.

Please observe that CL2, CL3 follow straightforwardly from definitions. For CL1, observe that $ingr(t, Cl(x))$ if and only if $OVLP(x)(t)$. Thus, $ingr(t, Cl(Cl(x)))$ if and only if $OVLP(Cl(x))(t)$ if and only if $OVLP(x)(t)$ if and only if $ingr(t, Cl(x))$.

For CL4, assume first that $ingr(t, Cl(x + y))$ hence $OVLP(t, x + y)$ and thus $OVLP(t, x) \vee OVLP(t, y)$, i.e., $ingr(t, Cl(x)) \vee ingr(t, Cl(y))$ and thus $ingr(t, Cl(x) + Cl(y))$. Assume now that $ingr(t, Cl(x) + Cl(y))$, i.e., $OVLP(t, x) \vee OVLP(t, y)$, so there exists m such that $Ov^{m}(t, x) \vee Ov^{m}(t, y)$, i.e., $Ov^{m}(t, x + y)$, hence, $ingr(t, Cl(x + y))$.

It follows that the operator Cl is a genuine closure operator; its properties are weak, as it in fact delineates components of things with respect to the overlap property: it is not even a T_0–closure operator.

10.4.4.2 On Boundaries

A definition of a *boundary* can be attempted on the lines of topological boundary concept. For a thing x, let a property $\Upsilon(x)$ be defined as follows

$$\Upsilon(x)(t) \Leftrightarrow ingr(t, x) \wedge \forall z.[Ov(z, x) \wedge Ov(z, -x) \Rightarrow Ov(z, t)] \quad (10.36)$$

We may define the *boundary of* x, $Fr(x)$, by letting

$$Fr(x) = Cls\Upsilon(x) \quad (10.37)$$

Properties of $Fr(x)$ following directly from definitions above are

1. $ingr(Fr(x), x)$.
2. $\forall z. Ov(z, x) \wedge Ov(z, -x) \Rightarrow Ov(z, Fr(x))$.

The above notion of a boundary has a topological flavor though by definition the boundary of the thing must be its ingredient contrary to topological reality; however, the notion of a boundary has a much wider scope. It can also support the idea of a *separator* between two things within a third, which does encompass either, like a river flowing through a town separates parts on opposite banks. To implement this idea, for things x, y, such that $extr(x, y)$, we define the property

$$\Omega(x, y)(t) \Leftrightarrow extr(t, x) \wedge extr(t, y) \tag{10.38}$$

and we let

$$Bd(x, y) = Cls\Omega(x, y) \tag{10.39}$$

Then the boundary operation Bd has properties

1. $Bd(x, y) = Bd(y, x)$.
2. $Bd(x + y, z) = Bd(x, z) \cdot Bd(y, z)$.

Property 1 is obvious. Property 2 follows from the equivalence $extr(x + y, z) \Leftrightarrow extr(x, z) \wedge extr(y, z)$. A relative variant can be defined; assuming that $ingr(x, z)$, $ingr(y, z)$ and $extr(x, y)$, a *boundary relative to z between x and y*, $Bd_z(x, y)$, is the class of things t such that $ingr(t, z)$, $extr(t, x)$, $extr(t, y)$ provided this property is non–vacuous.

10.4.5 Connection Mereotopology

Topological operators are constructed in connection mereology under same caveat as quasi–Boolean operators: absence of the null thing causes the need for reservations concerning existence of some things necessary for topological constructions. We will make this reservations not trying to add new axioms which would guarantee existence of some auxiliary things. We follow Clarke (1981) in this exposition.

10.4.5.1 On the Notion of C–interior

The *C–interior* $Int_C(x)$ of a thing x is defined as the class of non–tangential ingredients of x.

We define the property $NTP(x)$

$$NTP(x)(z) \Leftrightarrow NTP(z, x) \tag{10.40}$$

The interior $Int_C(x)$ is defined by means of

$$INT_C \quad Int_C(x) = ClsNTP(x) \tag{10.41}$$

hence, properties follow

1. $C(z, Int_C(x)) \Leftrightarrow \exists w.NTingr_C(w, x) \wedge C(z, w)$ by the class definition.
2. $\neg\exists z.EC(z, x) \Rightarrow (Int_C(x) = x)$. In particular, in the model *OVM*, $Int_C(x) = x$ for each thing x.
3. $ingr_C(Int_C(x), x)$ as $C(z, Int_C(x)) \Rightarrow C(z, x)$.
4. $C(z, Int_C(x)) \Rightarrow Ov_C(z, x)$.
5. $EC(z, x) \Rightarrow \neg C(z, Int_C(x))$.
6. $ingr_C(z, Int_C(x)) \Leftrightarrow NTingr_C(z, x)$.
7. $ingr_C(z, x) \Rightarrow ingr_C(Int_C(z), Int_C(x))$.
8. $Int_C(x) = x \Leftrightarrow C(z, x) \Rightarrow Ov_C(z, x)$.
9. $Int_C(x) = x \Leftrightarrow NTingr_C(x, x)$.

An *open* thing is x such that $Int_C(x) = x$.

Under additional axiomatic postulate that the boolean product of any two open sets is open, see Clarke (1981), A2.1, one can prove that $Int_C(x \cdot y) = Int_C(x) \cdot Int_C(y)$.

10.4.5.2 On the Notion of C–Closure

The notion of a topological *closure* $Cl_C(x)$ of x, can be introduced by means of the standard duality

$$Cl_C \quad Cl_C(x) = -Int_C(-x) \tag{10.42}$$

By properties of the interior and by duality (10.42), one obtains dual properties of closure

1. $ingr_C(x, Cl_C(x))$.
2. $Cl_C(Cl_C(x)) = Cl_C(x)$.
3. $ingr_C(x, y) \Rightarrow ingr_C(Cl_C(x), Cl_C(y))$.
4. $Int_C(x \cdot y) = Int_C(x) \cdot Int_C(y) \Leftrightarrow Cl_C(x + y) = Cl_C(x) + Cl_C(y)$.
5. $C(z, Cl_C(x)) \Leftrightarrow \exists w.NTingr_C(w, -x) \wedge C(z, w)$.

10.4.5.3 C–Boundaries and a Barry Smith's Proposal for Mereotopology

The notion of a *boundary* can be introduced along standard topological lines

$$Bd_C \quad Bd_C(x) = -(Int_C(x) + Int_C(-x)) \tag{10.43}$$

We collect basic properties of the boundary

1. Under Property 4, $Bd_C(x) = Cl_C(x) \cdot -Int_C(x)$, i.e., it can be expressed as the difference between the closure and the interior of the thing.
2. $Bd_C(x) = Bd_C(-x)$.
3. $ingr_C(Bd_C(x), Cl_C(x))$.

An interesting current in mereotopology is that by Smith (1996). It is situated in a universe endowed with part π and ingredient *ingr* relations, no matter in what way introduced. It departs from the above schemes for mereotopology by introducing the notion of an *interior part*, $\iota\pi$, which is supposed to satisfy the requirements

1. $\iota\pi(x, y) \Rightarrow \pi(x, y)$.
2. $\iota\pi(x, y) \wedge \pi(y, z) \Rightarrow \iota\pi(x, z)$.
3. $\pi(x, y) \wedge \iota\pi(y, z) \Rightarrow \iota\pi(x, z)$.
4. $\iota\pi(x, y) \wedge \iota\pi(y, z) \Rightarrow \iota\pi(x, y \cdot z)$.
5. For a non–void property (collection) F, if $F(x) \Rightarrow \iota\pi(x, y)$ then $\iota\pi(Cls F, y)$.
6. There exists y such that $\iota\pi(x, y)$ for each x.
7. $\iota\pi(x, y) \Rightarrow \iota\pi(x, Cls\{t : \iota\pi(t, y)\})$.

The interior of x can be defined as $Intx = Cls\{t : \iota\pi(t, x)\}$, and, one can declare the thing x as *open* when $x = Intx$.

From these postulates one derives in the standard way the following properties

1. $\iota\pi(V, V)$.
2. $\iota\pi(x, V)$.
3. $\iota\pi(x, y) \Leftrightarrow \pi(x, Cls\{t : \iota\pi(t, y)\})$.
4. $\iota(Cls\{t : \iota\pi(t, y)\}, y)$.

An approach to the notion of a boundary follows in two steps. First, the relation $\dagger(x, y)$, 'x *crosses* y' is defined as

$$\dagger(x, y) \Leftrightarrow Ov(x, y) \wedge Ov(x, -y) \tag{10.44}$$

Observe that no x can cross the universe V. The second notion of *straddling* involves open sets; one says that $Str(x, y)$, x *straddles* y, when

$$\iota\pi(x, z) \Rightarrow \dagger(z, y) \tag{10.45}$$

for each z.

It follows that

1. $\pi(x, y) \Rightarrow \iota\pi(x, y) \vee Str(x, y)$.
2. $\iota\pi(x, x) \vee Str(x, x)$.

The notion of *to be boundary* , $B(x, y)$ is derived by means of

$$B(x, y) \Leftrightarrow \pi(z, x) \Rightarrow Str(z, y) \tag{10.46}$$

for each z.

The boundary Bdx of x is then defined as

$$Bdx = Cls\{y : B(y, x)\} \tag{10.47}$$

Closure of x, $cl(x)$ is defined as the union $x + Bdx$. It then satisfies the postulates for topological closure operator,

1. $\pi(x, cl(x))$.
2. $cl(cl(x)) = cl(x)$.
3. $cl(x + y) = cl(x) + cl(y)$.

10.4.6 Rough Mereotopology

We analyze now topological structures in rough mereological framework. We consider the case of transitive and symmetric rough inclusions here, for more general discussion, cf., Polkowski (2011), Ch. 6. Here belong rough inclusions of types (ari), (airi). We use a generic symbol μ to denote either of these forms. μ is transitive with some t–norm T

$$\mu(x, y, r), \mu(y, z, s) \Rightarrow \mu(x, z, T(r, s)) \tag{10.48}$$

and symmetric

$$\mu(x, y, r) \Leftrightarrow \mu(y, x, r) \tag{10.49}$$

10.4.6.1 The Notion of an Open Set

For each thing x, we define $O_r(x)$ as the class of property $M(x, r)$, where

$$M(r, x)(y) \Leftrightarrow \mu(y, x, r), \tag{10.50}$$

and we let

$$O_r(x) = ClsM(x, r) \tag{10.51}$$

Hence: *ingr*$(z, O_r(x))$ *if and only if* $\mu(z, x, r)$. Indeed, *ingr*$(z, O_r(x))$ if and only if there exists t such that $Ov(z, t)$ and $\mu(t, x, r)$, hence, there exists w such that *ingr*(w, z), *ingr*(w, t), hence $w = z = t$, and finally $\mu(z, x, r)$.

We regard the thing $O_r(x)$ as an analogue of the notion of the 'closed ball about x of the radius r'. To define the analogue of an open ball, we consider the property

$$M_r^+(x)(y) \Leftrightarrow \exists q > r.\mu_{s,t}(y, x, q) \tag{10.52}$$

The class of the property $M_r^+(x)$ will serve as the open ball analogue

$$Int(O_r(x)) = ClsM_r^+(x) \tag{10.53}$$

Then: *ingr*$(z, Int(O_r(x)))$ *if and only if* $\exists q > r.\mu_{s,t}(z, x, q)$. We follow the lines of the preceding proof. It is true that

$$ingr(z, Int(O_r(x)))$$

if and only if there exists t such that $Ov(z,t)$ and there exists $q > r$ for which $\mu_T(t, x, q)$ holds, hence, there exists w such that $ingr(w, z)$, $ingr(w, t)$, which implies that $w = z = t$, and finally $\mu_T(z, x, q)$.

It follows

1. $ingr(Int(O_r(x)), O_r(x))$.
2. If $s < r$, then $ingr(O_r(x), O_s(x))$, $ingr(Int(O_r(x)), Int(O_s(x)))$.

Consider z with $ingr(z, Int(O_r(x)))$. $\mu_T(z, x, s)$ holds with some $s > r$. We can choose $\alpha \in [0, 1]$ with the property that $T(\alpha, s) > r$. For any thing w with $ingr(w, O_\alpha(z))$, we can find an thing u such that $\mu_T(u, z, \alpha)$ and $Ov(w, u)$. For a thing t such that $ingr(t, u)$ and $ingr(t, w)$, we have $\mu_T(t, w, 1)$, $\mu_T(t, u, 1)$, hence, $\mu_T(t, x, T(\alpha, s))$, i.e, $ingr(t, Int(O_r(x)))$. As $t = w$, we find that $ingr(w, Int(O_r(x)))$. We have verified that

(P) *for each* z *with* $ingr(z, Int(O_r(x)))$, *there exists* $\alpha \in [0, 1]$ *such that*

$$ingr(O_\alpha(z), Int(O_r(x))).$$

For any thing z, when $ingr(z, Int(O_r(x)))$ and $ingr(z, O_s(y))$, one finds $\alpha, \beta \in [0, 1]$ such that $ingr(O_\alpha(z), Int(O_r(x)))$ and $ingr(O_{beta}(z), Int(O_s(y)))$, hence, $ingr(O_q(z), Int(O_r(x)))$, $ingr(O_q(z), Int(O_s(y)))$, for $q = max\{\alpha, \beta\}$.

We can sum up the last few facts: *the collection* $\{Int(O_r(x)) : x$ a thing, $r \in [0, 1]\}$ *is an open basis for a topology on the collection of things.*

A thing x is *open*, $Open(x)$ in symbols, in case it is a class of some property of things of the form $Int(O_r(x))$. Hence,

1. If Φ is any non–vacuous property of things of the form $Open(x)$, then $Open(Cls\Phi)$.
2. If $Ov(Open(x), Open(y))$, then $Open(Open(x) \cdot Open(y))$.

10.4.6.2 On Closures and Interiors

We define closures of things, and to this end, we introduce a property $\Gamma(x)$ for each thing x

$$\Gamma(x)(y) \Leftrightarrow \forall s < 1.Ov(O_s(y), x) \tag{10.54}$$

Closures of things are defined by means of

$$Cl(x) = Cls\Gamma(x) \tag{10.55}$$

Then one verifies that

1. $ingr(z, Cl(x))$ if and only if $Ov(O_r(z), x)$ for every $r < 1$.
2. $ingr(z, Cl(O_w(x)))$ if and only if $Ov(O_r(z), O_w(x))$ for every $r < 1$.
3. $ingr(z, Cl(O_w(x)))$ if and only if $ingr(z, O_w(x))$.
4. $Cl(O_w(x)) = O_w(x)$.

We define $Int(x)$, the *interior of x* as

$$ingr(z, Int(x)) \Leftrightarrow \exists w.[Ov(z, w) \wedge \exists r < 1.ingr(O_r(w), x)] \tag{10.56}$$

A standard reasoning shows: $ingr(z, Int(x))$ if and only if there exists $r < 1$ such that $ingr(O_r(z), x)$.

10.4.6.3 On Boundaries

We can now address the problem of a *boundary* of any thing of the form $O_r(x)$. We define the boundary $Bd(O_r(x))$ as

$$Bd(O_r(x)) = O_r(x) \cdot -Int(O_r(x)) \tag{10.57}$$

We have a characterization of boundary ingredients: $ingr(z, Bd(O_r(x)))$ if and only if

$$\mu(z, x, r) \wedge \neg\exists q > r.\mu_{s,t}(z, x, q).$$

Hence,

$$ingr(Bd(O_r(x)), O_r(x)).$$

10.4.7 Mereogeometry

This section introduces mereogeometry modeled on classical axiomatization of geometry by Tarski (1959). It will serve us in the sequel in building tools for defining and navigating formations of intelligent agents (robots).

Elementary geometry was defined by Alfred Tarski in His Warsaw University lectures in the years 1926–1927 as a part of Euclidean geometry which can be described by means of the 1st order logic.

There are two main aspects in formalization of geometry: one is metric aspect dealing with the distance underlying the space of points which carries geometry and the other is affine aspect taking into account the linear structure.

In Tarski axiomatization, Tarski (1959), the metric aspect is expressed as a relation of *equidistance* (congruence) and the affine aspect is expressed by means of the *betweenness* relation. The only logical predicate required is the identity $=$. Equidistance relation denoted $Eq(x, y, u, z)$ (or, as a congruence: $xy \equiv uz$) means that the distance from x to y is equal to the distance from u to z (pairs x, y and u, z are equidistant).

Betweenness relation is denoted $B(x, y, z)$, (x is between y and z). Van Benthem (1983) took up the subject proposing a version of betweenness predicate based on the nearness predicate and suited, hypothetically, for Euclidean spaces.

We are interested in introducing into the mereological world defined by μ of a geometry in whose terms it will be possible to express spatial relations among things. We first introduce a notion of a distance κ, induced by a rough inclusion μ

$$\kappa(X, Y) = min\{max\, r, max\, s : \mu(X, Y, r), \mu(Y, X, s)\} \tag{10.58}$$

Observe that the mereological distance differs essentially from the standard distance: the closer are things, the greater is the value of κ: $\kappa(X, Y) = 1$ means $X = Y$ whereas $\kappa(X, Y) = 0$ means that X, Y are either externally connected or disjoint, no matter what is the Euclidean distance between them.

10.4.7.1 On the Notion of Betweenness in Tarski and Van Benthem Sense

The notion of *betweenness in the Tarski sense* $B(Z, X, Y)$ in terms of κ is

$$B(Z, X, Y) \Leftrightarrow \text{for each region W}, \kappa(Z, W) \in [\kappa(X, W), \kappa(Y, W)] \tag{10.59}$$

Here, $[a, b]$ means the non–oriented interval with endpoints a, b.

We use κ to define in our context the relation N of *nearness* proposed in Van Benthem (1983)

$$N(X, U, V) \Leftrightarrow \kappa(X, U) > \kappa(V, U) \tag{10.60}$$

Here, $N(X, U, V)$ means that X *is closer to* U *than* V *is to* U.

Then, N does satisfy all axioms for nearness in Van Benthem (1983)

1. NB1 $N(Z, U, V)$ and $N(V, U, W)$ imply $N(Z, U, W)$ (transitivity).
2. NB2 $N(Z, U, V)$ and $N(U, V, Z)$ imply $N(U, Z, V)$ (triangle inequality).
3. NB3 $N(Z, U, Z)$ is false (irreflexivity).
4. NB4 $Z = U$ or $N(Z, Z, U)$ (selfishness).
5. NB5 $N(Z, U, V)$ implies $N(Z, U, W)$ or $N(W, U, V)$ (connectedness).

We provide a sketch of proof.

For NB1, assumptions are $\kappa(Z,U) > \kappa(V,U)$ and $\kappa(V,U) > \kappa(W,U)$; it follows that $\kappa(Z,U) > \kappa(W,U)$ i.e. the conclusion $N(Z,U,W)$ follows.
For NB2, assumptions $\kappa(Z,U) > \kappa(V,U)$, $\kappa(V,U) > \kappa(Z,V)$ imply $\kappa(Z,U) > \kappa(Z,V)$, i.e., $N(U,Z,V)$.
For NB3, it cannot be true that $\kappa(Z,U) > \kappa(Z,U)$.
For NB4, $Z \neq U$ implies in our world that $\kappa(Z,Z) = 1 > \kappa(Z,U) \neq 1$.
For NB5, assuming that neither $N(Z,U,W)$ nor $N(W,U,V)$, we have $\kappa(Z,U) \leq \kappa(W,U)$ and $\kappa(W,U) \leq \kappa(V,U)$ hence $\kappa(Z,U) \leq \kappa(V,U)$, i.e., $N(Z,U,V)$ does not hold.

We introduce a *betweenness* relation in the sense of Van Benthem T_B modeled on betweenness proposed in Van Benthem (1983)

$$T_B(Z,U,V) \Leftrightarrow [\text{for each } W\ (Z = W) \text{ or } N(Z,U,W) \text{ or } N(Z,V,W)] \quad (10.61)$$

10.4.7.2 Example: The Case of Betweenness for Robots in 2D Space

The principal example bearing, e.g., on our approach to robot control deals with rectangles in 2D space regularly positioned, i.e., having edges parallel to coordinate axes. We model robots (which are represented in the plane as discs of the same radii in 2D space) by means of their safety regions about robots; those regions are modeled as squares circumscribed on robots. One of advantages of this representation is that safety regions can be always implemented as regularly positioned rectangles.

Given two robots a, b as discs of the same radii, and their safety regions as circumscribed regularly positioned rectangles A, B, we search for a proper choice of a region X containing A, and B with the property that a robot C contained in X can be said to be between A and B. In this search we avail ourselves with the notion of betweenness relation T_B.

Taking the rough inclusion μ^G defined in (10.28), for two disjoint rectangles A, B, we define the *extent*, *ext*(A,B) of A and B as the smallest rectangle containing the union $A \cup B$. Then we have the claim, obviously true by definition of T_B: *given two disjoint rectangles C, D, the only thing between C and D in the sense of the predicate T_B is the* extent *ext*(C,D) *of* C, D.

For a proof, as linear stretching or contracting along an axis does not change the area relations, it is sufficient to consider two unit squares A, B of which A has (0,0) as one of vertices whereas B has (a,b) with $a, b > 1$ as the lower left vertex (both squares are regularly positioned). Then the distance κ between the extent *ext*(A,B) and either of A, B is $\frac{1}{(a+)(b+1)}$.

For a rectangle $R : [0,x] \times [0,y]$ with $x \in (a, a+1), y \in (b, b+1)$, we have that

$$\kappa(R,A) = \frac{(x-a)(y-b)}{xy} = \kappa(R,B) \quad (10.62)$$

For $\phi(x, y) = \frac{(x-a)(y-b)}{xy}$, we find that

$$\frac{\partial \phi}{\partial x} = \frac{a}{x^2} \cdot (1 - \frac{b}{y}) > 0 \tag{10.63}$$

and, similarly, $\frac{\partial \phi}{\partial y} > 0$, i.e., ϕ is increasing in x, y reaching the maximum when R becomes the extent of A, B.

An analogous reasoning takes care of the case when R has some (c,d) with $c, d > 0$ as the lower left vertex.

Further usage of the betweenness predicate is suggested by the Tarski (1959) axiom of *B,Eq–upper dimension*, which implies collinearity of x, y, z. Thus, a line segment may be defined via the auxiliary notion of a pattern; we introduce this notion as a relation $Pt(u, v, z)$ which is true if and only if $T_B(z, u, v)$ or $T_B(u, z, v)$ or $T_B(v, u, z)$.

We will say that a finite sequence $u_1, u_2, \ldots, u_n$ of things *belong in a line segment* whenever $Pt(u_i, u_{i+1}, u_{i+2})$ for $i = 1, \ldots, n-2$; formally, we introduce the functor $Line$ of finite arity defined by means of

$$Line(u_1, u_2, \ldots, u_n) \textit{ if and only if } Pt(u_i, u_{i+1}, u_{i+2}) \textit{ for } i < n-1.$$

For instance, any two disjoint rectangles A, B and their extent $ext(A, B)$ form a line segment.

PART II. APPLICATIONS

10.5 Mereology in Engineering: Artifacts, Design and Assembling

Mereology plays a fundamental role in problems of design and assembling as basic ingredients in those processes are parts of complex things. The process of synthesis involves sequencing of operations of fusion of parts into more complex parts until the final product – artifact. We propose a scheme for assembling and a parallel scheme for design; the difference is in the fact that design operates on *abstracta*, i.e. categories of things whereas assembling deals with *concreta*, i.e., with real things. The interplay between abstracta and concreta will be described as a result of our analysis.

10.5.1 On the Notion of an Artifact

The term *artifact* means, etymologically, *a thing made by art*, which covers a wide specter of things from man–made things of everyday usage to abstract pieces of mathematical proofs, software modules or sonnets, or concertos. All those distinct things are unified in a scheme dependent on some common ingredients in their making, cf., e.g., a concise discussion in SEP (2012). We cannot include here a discussion of vast literature on ontological, philosophical and technological aspects of this notion, see, e.g., Baker (2004), Hilpinen (1995), Margolis and Laurence (2007), we point only to a thorough analysis of ontological aspects of artifacts in Borgo and Vieu (2009) in which authors propose also a scheme defining artifacts. It follows from discussion by many authors that important in analysis of artifacts are such aspects as: authorship, intended functionality, parthood relations. Analysis of artifacts is closely tied to design and assembly, cf., Boothroyd (2005) and Boothroyd et al. (2002) as well as Salustri (2002), Kim et al. (2008) and Seibt (2009). A discussion of mereology with respect to its role in domain science and engineering and computer science can be found in Björner and Eir (2010) and in Chapter by Björner in this Volume.

We attempt at a definition of an artifact as a thing obtained over a collection of things as a most complex thing in the sense of not being a part of any thing in the collection; to aspects of authorship (operator) and functionality, we add a temporal aspect, which allows for well–foundedness of the universe of parts, and seems to be a natural aspect of the assembling or design process. We regard processes leading to artifacts as *fusion processes* in which a by–product is obtained from a finite number of substrats. Though processes, e.g., of assembling a bike from its parts or a chemical reaction leading to a product obtained from a mixture of substances are very distinct to the observer, yet the formal description is identical for the two; it does require a category of *operators* P, a category of *functionalities* F, a *linear time* T with the *time origin* 0. The domain of things is a category *Things(P, F, π)* of things endowed with a part relation π. The assignment operator S acts as a partial mapping on the Cartesian product $P \times F \times Things(P, F, \pi)$ with values in the category *Tree* of rooted trees.

The act of assembling is expressed by means of a predicate

$$Art(p(u), < v_1(u), \cdots, v_k(u) >, u, f(u), t(u), T(u)),$$

which reads: *an operator* $p(u)$ *assembles at time* $t(u)$ *a thing* u *with functionality* $f(u)$ *according to the assembling scheme* $T(u)$ *organized by* $p(u)$ *which is a tree with the root u, from things* $v_1(u), \cdots, v_k(u)$ *which are leaves of* $T(u)$. *The thing* $v_i(u)$ enters in the position i *the assembling process for* u.

The predicate *ART* is subject to the following requirements.

ART1. If $Art(p(u), < v_1(u), \cdots, v_k(u) >, u, f(u), t(u), T(u))$ and for any i in $\{1, \ldots, k\}$, it holds that

$$Art(p(v_i(u)), < v_{i_1}(v_i(u)), \cdots, v_{i_k}(v_i(u)) >, v_i(u), f(v_i(u)), t(v_i(u)), T(v_i(u))),$$

then $t(v_i(u)) < t(u)$, $f(u) \subseteq f(v_i(u))$, $p(v_i(u)) \subseteq p(u)$, and $T(v_i(u))$ attached to $T(u)$ at the leaf $v_i(u)$ yields a tree, called an *unfolding of T(u) via the assembling tree for* $v_i(u)$.

The meaning of ART1 is that for each substrate v entering the assembly process for u, v is assembled at time earlier than time for u, functionality of u is lesser than that of v, the operator for u has a greater operating scope than that of v, and the assembly tree for u can be expanded at the leaf v by the assembly tree for v.

ART2. $Art(p(u), < v_1(u), \cdots, v_k(u) >, u, f(u), t(u), T(u)) \Rightarrow \pi(v_i(u), u)$ for each $v_i(u)$.

Meaning that each thing can be assembled only from its parts.

We introduce an auxiliary predicate $App(v, i(v), u, t(u))$ meaning: v enters in the position i the design process for u at time $t(u)$.

ART3. $\pi(v, u) \Rightarrow \exists w_1(v, u),$

$$\cdots, w_k(v, u), t(w_2(v, u)), \cdots, t(w_k(v, u)), i(w_1(v, u)), \cdots, i(w_{k(v,u)-1}(v, u))$$

such that $v = w_1(v, u), t(w_2(v, u)) < \cdots < t(w_k(v, u), w_k(v, u)) = u,$

$$App(w_j(v, u)), i(w_j(v, u)), w_{j+1}(v, u), t(w_{j+1}(v, u))$$

for $j = 1, 2, k(v, u) - 1$.

This means that each thing which is a part of the other thing will enter the assembly tree of the thing.

ART4. Each thing used in assembling of some other thing can be used in only one such thing in only one position at only one time.

This requirement will be referred to as the *uniqueness requirement*.

ART5. Values $t(u)$ belong in the set $T = \{0, 1, 2, \cdots\}$ of time moments.

Corollary 1. *By ART1, ART2, ART5: The universe of assembly things is well–founded, i.e., there is no infinite sequence* $\{x_i : i = 1, 2, \ldots\}$ *of things with* $\pi(x_{i+1}, x_i)$ *for each i.*

From this Corollary, it follows that our notion of identity of artifacts (EA) is equivalent to extensionality notions (EP), (EC), (UC) discussed in Varzi (2008).

For a tree $T(u)$, the *ART–unfolding* of $T(u)$ is the tree $T(u, 1)$ in which leaves $v_1(u), v_2(u), \cdots, v_k(u)$ are expanded by attaching those trees $T(v_1(u))$, $\cdots, T(v_k(u))$ which are distinct from their roots. For a tree $T(u)$, the *maximal ART–unfolding* $T(u, max)$ is the tree obtained from $T(u)$ by repeating the operation of ART–unfolding until no further ART–unfolding is possible.

Corollary 2. *Each leaf of the tree* $T(u, max)$ *is an atom.*

We now define an artifact: an *artifact over the category Things(P, F, π)* of assembly things is a thing u such that $\pi(u, v)$ holds for no thing v in *Things(P, F, π)*. Thus artifacts are 'final things' in a sense.

We define the notion of *identity for artifacts*:

(Extensionality of artifacts EA) *artifacts a, b are identical if and only if trees* $Tree(a, max)$, $Tree(b, max)$ *are isomorphic and have identical things at corresponding under the isomorphism nodes.*

10.5.2 Design Artifacts

We regard the process of design as analogous to the assembly process; the only difference between the two which we introduce is that in design, the designer works with not the things but with classes of equivalent things. Thus, to begin with, we introduce an equivalence relation on things. To this end, we let

$$u \sim v \text{ if and only if } [\pi(u,t) \text{ if and only if } \pi(v,t)] \text{ for each } thing\ t \tag{10.64}$$

and

$$Cat(u) = Cat(v) \text{ if and only if } u \sim v \tag{10.65}$$

Things in the same category Cat are 'universally replaceable'. It is manifest that the part relation π can be factored through categories, to the relation Π of part on categories,

$$\Pi(Cat(u), Cat(v)) \text{ if and only if } \pi(u, v) \tag{10.66}$$

In our formalism, design will imitate assembling with things replaced with categories of things and the part relation π replaced with the factorization Π. We need only to repeat the procedure with necessary replacements. We use the designer set D, the functionality set F, and the time set T as above.

The act of design is expressed by means of a predicate,

$$Des(d, < Cat_1, \cdots, Cat_k >, Cat, f(Cat), t(Cat), T(Cat))$$

which reads: a designer d designs at time t a category of things *Cat* with functionality $f(Cat)$ according to the design scheme $T(Cat)$ organized by d which is a tree with the root *Cat*, from categories $Cat_1, \cdots, Cat_k$ which are leaves of $T(Cat)$. The category Cat_i *enters in the position i* the design process for *Cat*.

The predicate *Des* is subject to the following requirements.

DES1. If $Des(d, < Cat(v_1(u)), \cdots, Cat(v_k(u)) >, Cat(u), f(u), t(u), T(u))$ and for any i in $\{1, \cdots, k\}$, it holds that

$$Des(p(Cat(v_i(u))), < Cat(v_{i_1}(v_i(u))), \cdots, Cat(v_{i_k}(v_i(u))) >,$$

$$Cat(v_i(u)), f(v_i(u)), t(v_i(u)), T(v_i(u))),$$

then $t(v_i(u)) < t(u)$, $f(u) \subseteq f(v_i(u))$, $p(v_i(u)) \subseteq p(u)$, and $T(v_i(u))$ attached to $T(u)$ at the leaf $Cat(v_i(u))$ yields a tree, called the *unfolding of T(u) via the design tree for* $Cat(v_i(u))$.

DES2.

$$Des(d, < Cat(v_1(u)), \cdots, Cat(v_k(u)) >, Cat(u), f(u), t(u), T(u)) \Rightarrow$$

$$\Pi(Cat(v_i(u)), Cat(u))$$

for each $v_i(u)$.
Meaning that each thing can be designed only from its parts.
We introduce an auxiliary predicate $App(v, i(v), u, t(u))$ meaning: $Cat(v)$ enters in the position i the design process for $Cat(u)$ at time $t(u)$.

DES3. $\Pi(Cat(v), Cat(u)) \Rightarrow \exists Cat(w_1(v,u)), \cdots, Cat(w_k(v,u))$, and,

$$t(w_2(v,u)), \cdots, t(w_k(v,u)), i(w_1(v,u)), \cdots, i(w_{k(v,u)-1}(v,u))$$

such that $v = w_1(v,u), t(w_2(v,u)) < \cdots < t(w_k(v,u), w_k(v,u)) = u$,

$$App(w_j(v,u)), i(w_j(v,u)), w_{j+1}(v,u), t(w_{j+1}(v,u))$$

for $j = 1, 2, \cdots, k(v,u) - 1$.
This means that for each thing which is a part of the other thing the category of the former will enter the design tree for the category of the latter.
For ART4, we may not have the counterpart in terms of DES: clearly, things of the same category may be used in many positions and at many design stages of some other category. We may only repeat our assumption about timing.

DES4. Values $t(u)$ belong in the set $T = \{0, 1, 2, \cdots\}$ of time moments.

Corollary 1. *The universe of categories is well–founded.*

We define a *design artifact* as a category $Cat(u)$ such that $\Pi(Cat(u), Cat(v))$ is true for no v.

We are approaching the notion of identity for design artifacts. To begin with, for a design artifact a, denote by the symbol $art(a)$ the artifact obtained by filling in the design tree for a all positions $Cat(v)$ with things v for some choices of v. We state the identity condition for design artifacts.

(Extensionality for design artifacts ED) *design artifacts a, b are identical if and only if there exist artifacts $art(a), art(b)$ which are identical.*

From the principle of identity for artifacts, a corollary follows.

Corollary 2. *If design artifacts a, b are identical then a, b have isomorphic design trees and categories at corresponding nodes are identical.*

Corollary 3. *If design artifacts a, b have isomorphic design trees and categories at corresponding nodes are identical, then a, b are identical.*

Indeed, consider two design artifacts a, b which satisfy the condition in the corollary. There is at least one category *Cat*(v) in the same position in design trees of a and b. Choose a thing x in *Cat*(v) and let $a(x), b(x)$ be artifacts assembled according to a, b, respectively. Having a thing in common, $a(x), b(x)$ are identical hence a, b are identical.

10.5.3 Action of Things on Design Abstracta

The interplay between concreta and abstracta in design and assembly can be exhibited by action of things on design artifacts. We define a partial mapping ι on the product $Things(P, F, \pi) \times Design\_Artifacts$ into *Artifacts*: for a thing v and a design artifact a, we define the value $\iota(v, a)$ as NIL in case category *Cat*(v) is not any node in the design tree for a, and, the unique artifact $a(v)$ in the contrary case. The inverse $\iota^{-1}(\iota(v, a))$ is the set $\{(u, b) : b \in Design\_Artifacts, Cat(u)\ a\ node\ in\ b\}$; thus, abstracta are equivalent in this sense to collections of concreta.

10.6 Mereology in Spatial Reasoning

Spatial orientation of a thing depends on the real world in which things are immersed, hence, to, e.g., discern among sides of a thing, one needs additional knowledge and structures. An example of this approach is found, e.g., in Aurnague et al. (1997), where it is proposed to exploit in determining orientation, e.g., the direction of gravity ('haut–grav', 'bas–grav') or peculiar features of things (like the neck of a bottle) suggesting direction, and usage of geometric predicates like equidistance in definitions of, e.g., orthogonal directions.

10.6.1 Properties of Artifacts: Mereological Theory of Shape and Orientation

It is manifest that mereology is amorphous in the sense that decomposition of a thing into parts does not depend of orientation, isometric transformations etc. Hence, to exhibit in things additional features like shape, side, one needs *augmented mereology*.

Particular features of shape like existence of 'dents' or 'holes' in a thing resulting from removal of other things can be accounted for within mereology.

We define the predicate $hole(x, y)$ reading *a thing x constitutes a hole in a thing y* as follows,

$$hole(x, y) \Leftrightarrow \exists z.NTP(x, z) \wedge comp(y, x, z) \tag{10.67}$$

i.e., x is a non–tangential thing in z and y *comp*lements x in z.

The predicate $dent(x, y)$, reading *a thing x constitutes a dent in a thing y* is defined as

$$dent(x, y) \Leftrightarrow \exists z.TP(x, z) \wedge comp(y, x, z) \tag{10.68}$$

i.e., x is a tangential thing in z and y complements x in z. The notion of a dent may be useful in characterizing things that 'fit into a thing': the predicate $fits\_into(x, y)$ may be defined as

$$fits\_into(x, y) \Leftrightarrow \exists z.dent(z, y) \wedge ingr(x, z) \tag{10.69}$$

i.e., x is an ingredient of a thing which is a dent in y. A particular case of fitting is 'filling' i.e., a complete fitting of a dent. We offer a predicate *fills(x, y)*

$$fills(x, y) \Leftrightarrow \exists z.dent(z, y) \wedge z = x \cdot y \tag{10.70}$$

i.e., dent–making z is the product of x and y. Following this, the notion of a *join* can be defined as

$$joins(x, y, z) \Leftrightarrow \exists w.w = x + y + z \wedge fills(x, y) \wedge fills(x, z) \tag{10.71}$$

i.e., x joins y and z when there is a thing $x + y + z$ and x fills both y and z.

This predicate can be inductively raised to

$$join(n)(x_1, x_2, \ldots, x_n; y_1, y_2, \ldots, y_n, y_{n+1})$$

via

$$join(1)(x_1; y_1, y_2) \Leftrightarrow join(x_1, y_1, y_2)$$

and

$$join(k + 1)(x_1, x_2, \ldots, x_{k+1}; y_1, y_2, \ldots, y_{k+1}, y_{k+2}) \Leftrightarrow$$

$$join(x_{k+1}, join(k)(x_1, x_2, \ldots, x_k; y_1, y_2, \ldots, y_{k+1}), y_{k+2})$$

in which we express sequentially a possibly parallel processing.

In case x joins y and z, possibility of assembling arises which may be expressed by means of modal operator $\Diamond$ of 'possibility', with an extended operator Asmbl to the form *Asmbl*$(x, i, y, j, \ldots w, p, f, t)$ meaning that w can be assembled from x in position i, y in position j,... by an operator p with functionality f at time t,

$$join(x, y, z) \Rightarrow \Diamond\exists w, p, f, t, i, j, k.Asmbl(x, i, y, j, z, k; w, p, f, t) \qquad (10.72)$$

Assuming our mereology is augmented with environment endowed with directions N, S, E, W, we may represent these directions by means of mobile agents endowed with laser or infrared beams of specified width; at the moment when the beam range reaches the thing x, it marks on its boundary a region which we denote as *top* in case of N, *bottom* in case of S, *left-side* in case of W, and *right-side* in case of E. Thus we have $top(x), bottom(x), left - side(x), right - side(x)$ as areas of the boundary of x; these are not parts of x. To express relations among sides of things we need a distinct language; for the sake of this example let us adopt the language of set theory regarding sides as sets.

Then we may say that the thing y

1. Is *on* the thing x in case *bottom(y)* is contained in *top(x)*.
2. Is *under* the thing x when *top(y)* is contained in *bottom(x)*.
3. *Touches x on the left* when *right-side(y)* is contained in *left-side(x)*
4. *Touches x on the right* when (*left-side(y)* is contained in *right-side(x)*).

This modus of orientation can be merged with mereological shape theory: one can say that a thing x *constitutes a dent on top/under/ on the left/on the right of the thing* y when, respectively,

1. $dent_{top}(x, y) \Leftrightarrow \exists z.TP(x, z) \wedge top(x) \subseteq top(z) \wedge comp(y, x, z)$.
2. $dent_{bottom}(x, y) \Leftrightarrow \exists z.TP(x, z) \wedge bottom(x) \subseteq bottom(z) \wedge comp(y, x, z)$.
3. $dent_{left}(x, y) \Leftrightarrow \exists z.TP(x, z) \wedge left - side(x) \subseteq left - side(z) \wedge comp(y, x, z)$.
4. $dent_{right}(x, y) \Leftrightarrow \exists z.TP(x, z) \wedge right - side(x) \subseteq right - side(z) \wedge comp(y, x, z)$.

These notions in turn allow for more precise definitions of fitting and filling; we restrict ourselves to filling as fitting is processed along same lines: we say that *a thing x fills a thing y on top/bottom/on the left-side/on the right-side*,

$$fills_{\alpha}(x, y) \Leftrightarrow \exists z.dent_{\alpha}(z, y) \wedge z = x \cdot y$$

where α is, respectively, *top, bottom, left, right*.

This bears on the notion of a join which can be made more precise: we say that *a thing x* (α, β)*–joins things y and z*

$$joins_{\alpha,\beta}(x, y, z) \Leftrightarrow \exists w.w = x + y + z \wedge fills_{\alpha}(x, y) \wedge fills_{\beta}(x, z)$$

where α, β=*top, bottom, left, right*.

A very extensive discussion of those aspects is given in Casati and Varzi (1999).

10.6.1.1 Qualitative Spatial Reasoning

With this analysis we enter the realm of Qualitative Spatial Reasoning. Qualitative spatial reasoning abstracts from qualitative details, cf., Cohn (1996); it is related to design, cf., Booch (1994) and planning, cf., Glasgow (1995).

Spatial reasoning employing mereology is a basis for analysis of semantics of orientational lexemes and semantics of motion, cf., Asher et al. (1995). It is basis for representation, and mapping of environments in behavioral robotics, cf., Kuipers (1994) and Arkin (1998). It is especially important for Geographic Information Systems (Frank and Campari 1993; Frank and Kuhn 1995; Hirtle and Frank 1997; Egenhofer and Golledge 1997).

Any formal approach to Spatial Reasoning requires Ontology, cf., Guarino (1994), Smith (1989), and Casati et al. (1998). In reasoning with spatial things, of primary importance is to develop an ontology of spatial things, taking into account complexity of these things.

10.6.1.2 A Case of Spatial Analysis of Limiting Things

We give two examples of spatial reasoning based on merology. In the first, we attempt, cf. Polkowski and Semeniuk–Polkowska (2010) at giving descriptions of various notions of boundary, or limiting, things like *separator, border, fence, hedge, confine*, involving in this discussion various models of mereology.

We introduce the notion of a *separator* $Sepr(x,z,y)$ for a triple x,z,y such that $\pi(x,y)$, $\pi(z,y)$, $extr(x,z)$ as,

$$Sepr(x,z,y) = Cls\{v : \pi(v,y), extr(v,x), extr(v,z)\} \tag{10.73}$$

Then,

1. $extr(Sepr(x,z,y),x)$.
2. $extr(Sepr(x,z,y),z)$.
3. $\pi(Sepr(x,z,y),y)$.
4. $x = Sepr(Sepr(x,z),z,y)$.
5. $z = Sepr(Sepr(x,z,y),x)$.

The notion of a separator comes close to the notion of a *border*: assume that Warszawa, the river Vistula, the left–bank part of Warszawa and its right–bank part are things in a Geographic Information System. Then the river Vistula is the separator between left– and right–bank parts of Warszawa, and it can justly be called the *border* of either, contrary to the topological boundaries of those parts which are left and right banks of the river.

The notion of connection, in particular the predicate of *external connectedness* EC allows for more detailed spatial analysis; in our example of Warszawa and the river Vistula, where things are Vistula and left and right banks of Warszawa, we have $EC(Vistula, left - bankside)$, $EC(Vistula, right - bankside)$. The connection relation in this case is defined in the ROM model. External connection leads to division of ingredients of any thing into two categories: tangential and non–tangential. In order to make this distinction, one introduces the complement of an entity, $-x$ as the class of all entities external to x. The tangential ingredient of x is z such that it is externally connected to an ingredient of $-x$. In the example of Warszawa and Vistula, the connection boundary of either bank of Warszawa is this bank itself; it is different from the idea of geographic boundary and any reasonable idea of a boundary. In order to rectify the idea, we can introduce a richer universe of parts, e.g., by declaring a part any region contained either in a bank of Warszawa or in the river, but not intersecting any two of these entities. Then the idea of Tarski (1929) may be applied of defining *ideal things* (*'points'*) as limits of ultrafilters of regions. Limits of ultrafilters of regions being parts of a bank of Warszawa constitute the geographic boundary of this bank.

For the notion of *confine* or *extent* we can apply rough mereogeometry and the notion of betweenness. Assume for simplicity, that entities are rectangles with sides parallel to coordinate axes. Given rectangles R_1, R_2, as proved above, the extent of R_1, R_2 is the smallest rectangle spanned by R_1, R_2.

10.6.1.3 A Digression on Time in Mereology

To analyze notions of a *fence* and a *hedge*, we resort to the property of *passability*: by a fence we understand a structure of iron wire made to be impassable, e.g. to small animals whereas a hedge is a structure usually of plants which we regard as passable. To express this difference, we introduce a new aspect of mereology, viz., timed mereology due to Tarski (1937) and Woodger (1937, 1939).

The time component is introduced into the framework of mereology with a set of notions and postulates (axioms) concerning aspects of time like *momentariness, coincidence in time, time slices*. Things are considered as spatial only and their relevance to time is expressed as momentary or as spatial and extended in time and then the predicate of part is understood as a global descriptor covering spatio–temporal extent of things whereas the temporal extension is described by the predicate *Temp, T* with the intended meaning that $T(u, v)$ means that the thing u *precedes* in time the thing v (in terminology of Leśniewski, Tarski and Woodger: *u wholly precedes v*) meaning that, e.g., when u and v have some temporal extent, then u ends before or at the precise moment when v begins.

The property (predicate) *Mom* meaning *momentary being* is introduced to denote things having only spatial aspect. This predicate is introduced by means of the following postulate,

(MOM) $Mom(x) \Leftrightarrow T(x, x)$

Thus, x begins and ends at the same time, so its time aspect is like a spike in time; it renders the phrase 'to exist in a moment of time'.

The predicate T is required to satisfy postulates

1. TM1 $T(x, y) \wedge T(y, z) \Rightarrow T(x, z)$.
2. TM2 $Mom(x) \wedge Mom(y) \Rightarrow T(x, y) \vee T(y, x)$.
3. TM3 $T(x, y) \Leftrightarrow \forall u, v.ingr(u, x) \wedge ingr(v, y) \Rightarrow T(u, v)$.

Postulate TM1 states that T is transitive, TM2 does state that of two momentary things, one precedes the other and TM3 relates T to the class operator, i.e., x precedes y if and only if each ingredient of x precedes each ingredient of y. Postulate TM3 provides a link between the part based mereology and the timed mereology, bonding spatial and temporal properties of things.

The notion of a *coincidence in time*, *CT* in symbols, is

$$CT(x,y) \Leftrightarrow T(x,y) \wedge T(y,x) \tag{10.74}$$

and it implies in turn a notion of a *time–slice*, $Slice(x,y)$, as

$$Slc(x,y) \Leftrightarrow Mom(x) \wedge ingr(x,y) \wedge \forall z.[ingr(z,y) \wedge C(z,x) \Rightarrow ingr(z,x)] \tag{10.75}$$

and thus a time–slice of an thing y is an ingredient of y which is spatially so arranged that any ingredient of y coinciding with it in time is also its ingredient.Time slices are unique up to coincidence in time: *if x, y are time–slices of z, then x, y coincide in time if and only if $x = y$.*

We use these notions in order to make a distinction between passable and non–passable boundaries, i.e., between hedges and fences. We say that a time–slice x of an entity y is a *time–front boundary* of y if and only if for each entity z it follows from $ingr(z,y)$ and $T(z,x)$ that $ingr(z,x)$; similarly, a time–slice w of y is a *time–rear boundary* of y if and only if for each entity z it follows from $ingr(z,y)$ and $T(w,z)$ that $ingr(z,w)$. The Boolean sum $x + w$ of x and w, is the *time–boundary* of y.

The front time boundary x of y is *passable* (is a *front time–hedge of* y) if and only if there is an entity z such that $T(z,x)$ and not $ingr(z,x)$; otherwise x is the *front time–fence of* y. Analogous definitions concern rear time–hedges and rear time–fences. Smith and Varzi (1997), make a distinction between *fiat boundaries* and *bona–fide boundaries*, the former defined as material boundaries of real entities whereas the latter understood as mental boundaries; time boundaries may serve as an example of the latter.

10.6.1.4 RCC: Region Connection Calculus. ROM Revisited

As an important example of mereological spatial reasoning we introduce here the RCC Calculus (Region Connection Calculus), cf. Randell et al. (1992), Cohn et al. (1993, 1996), Cohn (1996), Cohn and Gotts (1996) and Cohn and Varzi (1998). It is a calculus on closed regular sets (regions) in a regular topological space, i.e, in the frame of ROM. RCC admits Clarke's connection postulates CN1–CN3 and follows same lines in defining basic predicates. To preserve the flavor of this theory we give these predicates in the RCC notation

Table 10.1 Transition table for RCC8 calculus

–	DC	EC	PO	TPP	NTPP	TPPi	NTPPi
DC	–	DR,PO,PP	DR,PO,PP	DR,PO,PP	DC	DC	
EC	DR,PO,PPi	DR,PO,TPP,TPi	DR,PO,P	EC,PO,PP	PO,PP	DR	DC
PO	DR,PO,PPi	DR,PO,PPi	–	PO,PP	PO,P	DR,PO,PPi	DR,PO,PPi
TPP	DC	DR	DR,PO,PP	PP	NTPP	DR,PO,PP	–
NTPP	DC	DC	DR,O,PP	NTPP	NTPP	DR,PO,PP	–
TPPi	DR,PO,PPi	EC,PO,PPi	PO,PPi	PO,TPP,TPi	PO,PP	PPi	NTPPi
NTPPi	DR,PO,PPi	PO,PPi	PO,PPi	PO, PPi	0	NTPPi	NTPPi

1. *DISCONNECTED FROM*$(x)(y)$ $DC(x,y) \Leftrightarrow \neg C(x,y)$.
2. *IMPROPER PART OF*$(x)(y)$: $P(x,y) \Leftrightarrow \forall z.[C(z,y) \rightarrow C(z,x)]$.
3. *PROPER PART OF*$(x)(y)$: $PP(x,y) \Leftrightarrow P(x,y) \wedge \neg P(y,x)$.
4. *EQUAL*$(x)(y)$: $EQ(x,y) \Leftrightarrow P(x,y) \wedge P(y,x)$.
5. *OVERLAP*$(x)(y)$: $Ov(x,y) \Leftrightarrow \exists.z.P(x,z) \wedge P(y,z)$.
6. *DISCRETE FROM*$(x)(y)$: $DR(x,y) \Leftrightarrow \neg Ov(x,y)$.
7. *PARTIAL OVERLAP*$(x)(y)$: $POv(x,y) \Leftrightarrow Ov(x,y) \wedge \neg P(x,y) \wedge \neg P(y,x)$.
8. *EXTERNAL CONNECTED*$(x)(y)$: $EC(x,y) \Leftrightarrow C(x,y) \wedge \neg Ov(x,y)$.
9. *TANGENTIAL PART OF*$(x)(y)$: $TPP(x,y) \Leftrightarrow PP(x,y) \wedge \exists z.EC(x,z) \wedge EC(y,z)$.
10. *NON – TANGENTIAL PART OF*$(x)(y)$: $NTPP(x,y) \Leftrightarrow PP(x,y) \wedge \neg TPP(x,y)$.

To each non–symmetric predicate X RCC adds the inverse Xi (e.g., to $TPP(x,y)$ it adds $TPPi(y,x)$). The eight predicates: *DC, EC, PO, EQ, TPP, NTPP, TPPi, NTPPi* show the *JEPD property* (Jointly Exclusive and Pairwise Disjoint) and they form the fragment of RCC called RCC8.

Due to topological assumptions, RCC has some stronger properties than Clarke's calculus of C, where connection is simply the set intersection. Witness, the two properties, see.

1. If $\forall z.Ov(x,z) \leftrightarrow Ov(y,z)$, then $x = y$ (extensionality of overlapping). (If $x \neq y$, then, e.g., there is $z \in x - y$ and regularity of the space yields us an open neighborhood V of z such that $ClV \cap y = \emptyset$ and $Ov(V,x)$ negating the premise).
2. If $PP(x,y)$, then $\exists z.P(x,z) \wedge DR(y,z)$.
3. $\forall x.EC(x,-x)$.

RCC8 allows for additional predicates characterizing shape, connectivity, see Gotts et al. (1996) and regions with vague boundaries ("the egg–yolk" approach), see Gotts and Cohn (1995).

RCC8 is presented in the form of the *transition table*: a table in which for entries $R_1(x,y)$ and $R_2(y,z)$ a result $R_3(x,z)$ is given, see Egenhofer (1991). The transition table for RCC8 is shown in Table 10.1.

10.7 Mereology in Intelligent Planning and Navigation: The Case of Behavioral Robotics

We have stressed that by its nature, rough mereology does address concepts, relations among which are expressed by partial containment rendered as the predicate of a part to a degree. Behavioral robotics falls into this province, as usually robots as well as obstacles and other environmental things are modeled as figures or solids. We show applications of mereology to planning and navigation of autonomous mobile robots and their formations. First, we introduce the subject of planning in robotics.

10.7.1 Planning with Emphasis on Behavioral Robotics

Planning is concerned with setting a trajectory for a robot endowed with some sensing devices which allow it to perceive the environment in order to reach by the robot a goal in the environment at the same time bypassing obstacles.

Planning methods, cf., e.g., Choset et al. (2005), vary depending on the robot abilities, features of the environment and chosen methodology. Among them are simple geometric methods designed for a robot endowed with sensors detecting obstacles, e.g., touch sensors or range sensors and able to detect distance between any pair of points. These methods are called 'contour following', as for such a robot, the idea can be implemented of moving to goal in a straight line segment and in case of meeting with an obstacle to bypass it by circumnavigating its boundary until the straight line to goal is encountered anew. Typically, the robot performs a heuristic search of A^* type, see, e.g., Russell and Norvig (2009) or Choset et al. (2005) with the heuristic function $h(x) = \rho(x, O) + \rho(O, goal)$ where x is the current position of the robot, and the point O is selected as an end–point of the continuity interval of ρ – the distance function, whose values are bound by a constant R.When the distance measured by range sensors exceeds R the value of ρ is set to infinity. The graph of ρ against the position x exhibits then discontinuities and continuity intervals clearly outline boundaries of obstacles, hence, the idea of selecting O as a boundary continuity point. Minimization of h leads to optimization of the chosen safe trajectory.

A method of *potential field*, see Khatib (1986) consists in constructing a potential field composed of attractive potentials for goals and repulsive potentials for obstacles.

An example may be taken as the quadratic potential function

$$U_{attractive}(x) = \frac{1}{2} \cdot ||x - x_{goal}||^2 \tag{10.76}$$

which induces the gradient

$$\nabla U_{attractive}(x) = x - x_{goal} \tag{10.77}$$

which assures that the force (the gradient) exerted on the robot is greater when the robot is far from the goal and diminishes to zero as the robot is approaching the goal.

A repulsive potential should have opposite properties: it should exert a force tending to ∞ with the distance to the obstacle reaching 0. Denoting the distance from a point x to the closest obstacle with $s(x)$, the repulsive potential can be defined as in

$$U_{repulsive}(x) = \frac{1}{2} \cdot [\frac{1}{s(x)}] \tag{10.78}$$

with the gradient

$$\nabla U_{repulsive}(x) = -\frac{1}{s(x)^2} \cdot \nabla s(x) \tag{10.79}$$

The global potential function U is the sum of the attractive and repulsive parts:

$$U(x) = U_{attractive}(x) + U_{repulsive}(x)$$

Given U, the robot performs a well–known *gradient descent* : it does follow the direction of the gradient in small steps : the $(i+1)$–th position is given from the i-th position and the gradient therein as

$$x_{i+1} = x_i + \xi_i \cdot \nabla U(x_i) \tag{10.80}$$

In Polkowski and Ośmiałowski (2008, 2010) and Ośmiałowski (2009a) a mereological potential field planning method was proposed.

10.7.2 Mereological Planning via Potential Fields

Classical methodology of potential fields works with integrable force field given by formulas of Coulomb or Newton which prescribe force at a given point as inversely proportional to the squared distance from the target; in consequence, the potential is inversely proportional to the distance from the target. The basic property of the potential is that its density (=force) increases in the direction toward the target. We observe this property in our construction.

We apply the geometric rough inclusion

$$\mu^G(x, y, r) \Leftrightarrow \frac{||x \cap y||}{||x||} \tag{10.81}$$

where $||x||$ is the area of the region x. In our construction of the potential field, region will be squares: robots are represented by squares circumscribed on them (simulations were performed with disk–shaped Roomba robots, the intellectual property of iRobot. Inc.).

Geometry induced by means of a rough inclusion can be used to define a generalized potential field: the force field in this construction can be interpreted as the density of squares that fill the workspace and the potential is the integral of the density. We present now the details of this construction. We construct the potential field by a discrete construction. The idea is to fill the free workspace of a robot with squares of fixed size in such a way that the density of the square field (measured, e.g., as the number of squares intersecting the disc of a given radius r centered at the target) increases toward the target.

To ensure this property, we fix a real number – the `field growth step` in the interval $(0, \text{square edge length})$; in our exemplary case the parameter `field growth step` is set to 0.01.

The collection of squares grows recursively with the distance from the target by adding to a given square in the $(k + 1)$-th step all squares obtained from it by translating it by $k \times$ `field growth step` (with respect to Euclidean distance) in basic eight directions: N, S, W, E, NE, NW, SE, SW (in the implementation of this idea, the *floodfill algorithm* with a queue has been used, see Ośmiałowski (2009a,b)). Once the square field is constructed, the path for a robot from a given starting point toward the target is searched for.

The idea of this search consists in finding a sequence of *way–points* which delineate the path to the target. Way–points are found recursively as centroids of unions of squares mereologically closest to the square of the recently found way–point. We recall that the mereological distance between squares x, y is defined by means of

$$k(x, y) = min\{\, max\, r : \mu(x, y, r), max\, s : \mu(y, x, s)\} \tag{10.82}$$

We also remind that the mereological distance $k(x, y)$ takes on the value 1 when $x = y$ and the minimal value of 0 means that $x \cap y \subseteq Bd(x) \cap Bd(y)$. In order do define a "potential" of the rough mereological field, let us consider how many

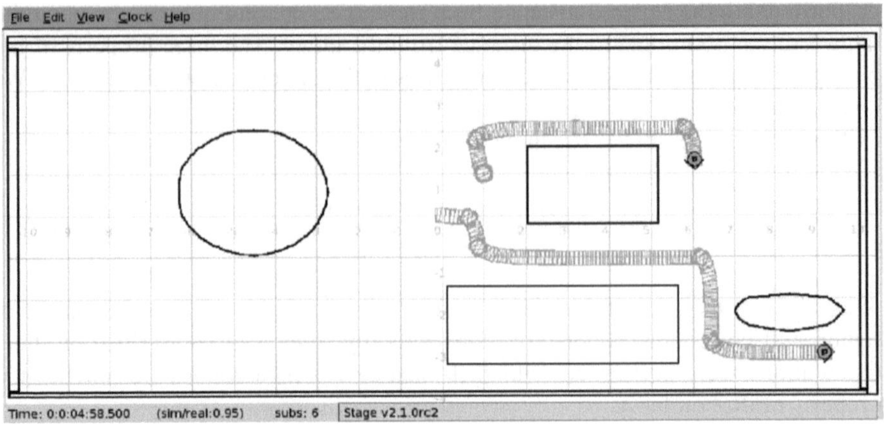

Fig. 10.1 Planned paths of Roomba robots to their targets

generations of squares will be centered within the distance r from the target. Clearly, we have

$$d + 2d + \ldots + kd \leq r \tag{10.83}$$

where d is the field growth step, k is the number of generations. Hence,

$$k^2 d \leq \frac{k(k+1)}{2} d \leq r \tag{10.84}$$

and thus

$$k \leq (\frac{r}{d})^{\frac{1}{2}} \tag{10.85}$$

The potential $V(r)$ can be taken as $\sim r^{\frac{1}{2}}$. The force field $F(r)$ is the negative gradient of $V(r)$,

$$F(r) = -\frac{d}{dr} V(r) \sim -\frac{1}{r^{\frac{1}{2}}} \tag{10.86}$$

Hence, the force decreases with the distance r from the target slower than traditional Coulomb force. It has advantages of slowing the robot down when it is closing on the target. Parameters of this procedure are: the `field growth step` set to 0.01, and the size of squares which in our case is 1.5 times the diameter of the Roomba robot.

A robot should follow the path proposed by planner shown in Fig. 10.1.

10.7.3 Planning for Teams of Robots

Problems of planning paths for teams of robots present an intellectual challenge due to aspects of cooperation, communication, task–sharing and division, and planning non–collision paths for robots. These problems require studies of cognitive theories, biology, ethology, organization and management. They can also lead to new solutions to problems of artificial intelligence. Passing from a single robot to teams of robots can be motivated also by pragmatic reasons, cf., Cao et al. (1997), as tasks for robots can be too complex for a single robot, or many robots can do the task easier at a lesser cost, or many robots can perform the task more reliably.

Practical studies along these lines were concerned with moving large things of irregular shapes by groups of robots, see Kube and Zhang (1996), search and rescue, see Jennings et al. (2001), formations of planetary outposts of mobile robots, see Huntsberger et al. (2007), multi–target inspection Parker (1997). Simulations of systems a few robots were studied, e.g., in CEBOT, see Fukuda and Nakagawa (1987), ACTRESS, see Asama et al. (1989), GOFER, see Caloud et al. (1990), cf., the ALLIANCE architecture in Parker (1998).

Many authors attacked these problems by extending methods elaborated for a single robot; helpful in those attempts were studies of behavior of migrating birds flying in 'boids', cf., Reynolds (1987) which brought forth elementary behaviors like collision–avoidance, velocity adjustment, leader–following, flock–centering, transferred into robot milieu, e.g., in Matarić (1993, 1994, 1997), Fredslund and Matarić (2002), Agah (1996), and Agah and Bekey (1997), which provided elementary robot behaviors like wandering, homing, following, avoidance, aggregation, dispersion.

In Balch and Arkin (1998) basic principles of *behavioral approach*, were formulated: it is vital to keep all robots within a certain distance from one another (e.g., to ensure mutual visibility), to move away when the distance becomes too close (to avoid congestion, collision, or resource conflict), to adapt own movement to movement of neighbors (e.g., by adjusting velocity of motion), to orient oneself on a leader, or a specific location, e.g., the gravity center of the group. They proposed that robots in a team obey rigid geometric constraints by means of references to the center of the group or to the assigned leader, or to the assigned neighbor.

10.7.4 Mereological Approach to Robot Formations

We recall that on the basis of the rough inclusion μ, and mereological distance κ geometric predicates of *nearness* and *betweenness*, are redefined in mereological terms.

Given two robots a, b as discs of same radii, and their safety regions as circumscribed regularly positioned rectangles A, B, we search for a proper choice of a region X containing A, and B with the property that a robot C contained in X can be said to be between A and B.

For two (possibly but not necessarily) disjoint rectangles A, B, we define the *extent*, $ext(A, B)$ of A and B as the smallest rectangle containing the union $A \cup B$. We know that in this setting, given two disjoint rectangles C, D, the only thing between C and D in the sense of the predicate T_B is the *extent* $ext(C, D)$ of C, D,, i.e., the minimal rectangle containing the union $C \cup D$.

For details of the exposition which we give now, please consult Ośmiałowski (2011) and Ośmiałowski and Polkowski (2009).

For robots a, b, c, we say that a robot b is *between robots a and c*, in symbols

$$\text{(between } b\ a\ c) \tag{10.87}$$

in case the rectangle $ext(b)$ is contained in the extent of rectangles $ext(a)$, $ext(c)$, i.e.,

$$\mu_0(ext(b), ext(ext(a), ext(c)), 1) \tag{10.88}$$

This can be generalized to the notion of *partial betweenness* which models in a more realistic manner spatial relations among a, b, c; we say in this case that robot b is *between robots a and c to a degree of at least r*, in symbols,

$$\text{(between–degr } r\ b\ a\ c\ ) \tag{10.89}$$

if and only if

$$\mu_0(ext(b), ext[ext(a), ext(c)], r), \tag{10.90}$$

i.e.,

$$\frac{||ext(b) \cap ext(ext(a), ext(c))||}{||ext(b)||} \geq r$$

For a team of robots, $T(r_1, r_2, \ldots, r_n) = \{r_1, r_2, \ldots, r_n\}$, an *ideal formation IF* on $T(r_1, r_2, \ldots, r_n)$ is a betweenness relation (between...) on the set $T(r_1, r_2, \ldots, r_n)$.

In implementations, ideal formations are represented as lists of expressions of the form

$$\text{(between } r_0\ r_1\ r_2) \tag{10.91}$$

indicating that the thing r_0 is between r_1, r_2, for all such triples, along with a list of expressions of the form

$$\text{(not–between } r_0\ r_1\ r_2) \tag{10.92}$$

indicating triples which are not in the given betweenness relation.

To account for dynamic nature of the real world, in which due to sensory perception inadequacies, dynamic nature of the environment etc., we allow for some deviations from ideal formations by allowing that the robot which is between two neighbors can be between them to a degree in the sense of (10.89). This leads to the notion of a real formation.

For a team of robots, $T(r_1, r_2, \ldots, r_n) = \{r_1, r_2, \ldots, r_n\}$, a *real formation* RF on $T(r_1, r_2, \ldots, r_n)$ is a betweenness to degree relation (between–deg ....) on the set $T(r_1, r_2, \ldots, r_n)$ of robots.

In practice, real formations will be given as a list of expressions of the form,

$$\text{(between–deg } \delta\ r_0\ r_1\ r_2), \tag{10.93}$$

indicating that the thing r_0 is to degree of δ in the extent of r_1, r_2, for all triples in the relation (between–deg ....), along with a list of expressions of the form,

$$\text{(not–between } r_0\ r_1\ r_2), \tag{10.94}$$

indicating triples which are not in the given betweenness relation.

Description of formations, as proposed above, can be a list of relation instances of large cardinality, cf., examples below. The problem can be posed of finding a minimal set of instances sufficient for describing a given formation, i.e., implying the full list of instances of the relation (between...). This problem turns out to be NP–hard, see Ośmiałowski and Polkowski (2009).

To describe formations we propose a language derived from LISP–like s–expressions: a formation is a list in LISP meaning with some restrictions that formulates our language. We will call elements of the list *things*. Typically, LISP lists are hierarchical structures that can be traversed using recursive algorithms. We restrict that top–level list (a root of whole structure) contains only two elements where the first element is always a formation identifier (a name). For instance

Example 1. (formation1 (some_predicate *param*1 ... *paramN*))

For each thing on a list (and for a formation as a whole) an extent can be derived and in facts, in most cases only extents of those things are considered. We have defined two possible types of things

1. *Identifier: robot or formation name (where formation name can only occur at top–level list as the first element)*;

2. *Predicate: a list in LISP meaning where first element is the name of given predicate and other elements are parameters; number and types of parameters depend on given predicate.*

Minimal formation should contain at least one robot. For example

Example 2. (formation2 roomba0)

To help understand how predicates are evaluated, we need to explain how extents are used for computing relations between things. Suppose we have three robots (*roomba*0, *roomba*1, *roomba*2) with *roomba*0 between *roomba*1 and *roomba*2 (so the *between* predicate is fulfilled). We can draw an extent of this situation as the smallest rectangle containing the union *roomba*1 ∪ *roomba*2 oriented as a regular rectangle, i.e., with edges parallel to coordinate axes. This extent can be embedded into bigger structure: it can be treated as an thing that can be given as a parameter to predicate of higher level in the list hierarchy. For example:

Example 3. (formation3 (between (between roomba0 roomba1 roomba2) roomba3 roomba4))

We can easily find more than one situation of robots that fulfill this example description. That is one of the features of our approach: one s–expression can describe many situations. This however makes very hard to find minimal s–expression that would describe already given arrangement of robots formation (as stated earlier in this chapter, the problem is NP–hard).

Typical formation description may look like below, see Ośmiałowski (2011)

Example 4.
```
(cross
    (set
        (max–dist 0.25 roomba0 (between roomba0 roomba1 roomba2))
        (max–dist 0.25 roomba0 (between roomba0 roomba3 roomba4))
        (not–between roomba1 roomba3 roomba4)
        (not–between roomba2 roomba3 roomba4)
        (not–between roomba3 roomba1 roomba2)
        (not–between roomba4 roomba1 roomba2)
    )
)
```

This is a description of a formation of five Roomba robots arranged in a cross shape. The *max–dist* relation is used to bound formation in space by keeping all robots close one to another.

We show a screen–shot of robots in the initial formation of cross–shape in a crowded environment, see Figs. 10.2 and 10.3. These behaviors witness the flexibility of our definition of a robot formation: first, robots can change formation, next, as the definition of a formation is relational, without metric constraints on robots,the formation can manage an obstacle without losing the prescribed formation (though, this feature is not illustrated in figures in this chapter).

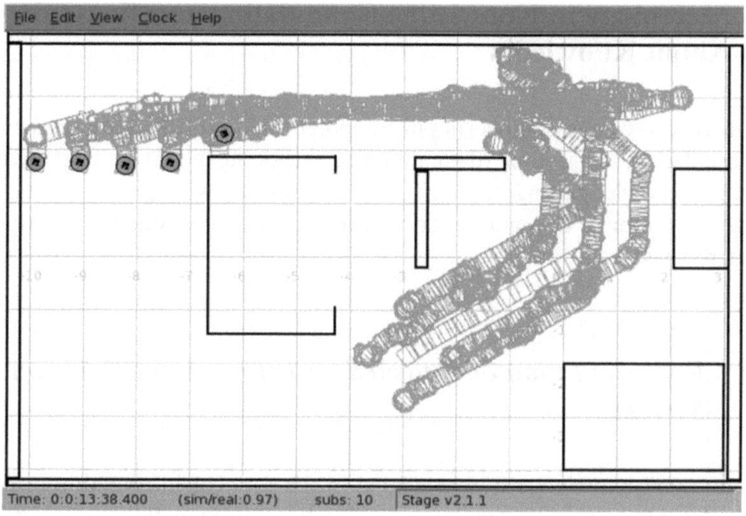

Fig. 10.2 Trails of robots moving in the line formation through the passage (From Polkowski (2011))

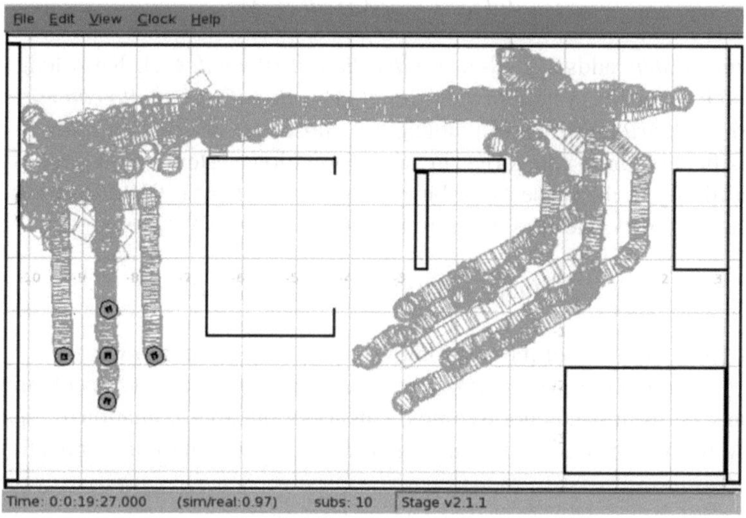

Fig. 10.3 Trails of robots in the restored cross formation in the free workspace after passing through the passage (From Polkowski (2011))

10.8 Mereology in Knowledge Granulation and Reasoning About Knowledge

The topic of knowledge engineering in computer science does encompass problems of representation, extraction (data mining), reasoning about and application of knowledge (knowledge engineering). We represent knowledge as annotated data expressed in symbolic or numeric, or hybrid form which encode information about a considered case.

10.8.1 Representation of Knowledge: Information/Decision Systems

We assume that knowledge is represented in the form of information or decision systems. An *information system*, cf., e.g., Pawlak (1991) or Polkowski (2002) is a pair (U, A) where U is a finite set of things and A is a finite set of *attributes*; each attribute a is a mapping $a : U \to V$ from the set U into a set V of *attribute values*. For each thing $u \in U$, the *information vector of u* is the set

$$Inf_A(u) = \{a(u) : a \in A\}.$$

A *decision system* adds to the set A a *decision attribute* $d \notin A$. Knowledge can be extracted from either system in the form of (1) a *classification into categories* or (2) a *decision algorithm* which is a judiciously chosen set of *decision rules*.

Classification into categories in an information system (U, A) relies on the indiscernibility in the sense of Leibniz (1969) and Forrest (2010): for a set $B \subseteq A$ of attributes, one defines the *B–indiscernibility relation* IND_B as

$$IND_B(u, v) \Leftrightarrow a(u) = a(v) \textit{ for each } a \in B \tag{10.95}$$

Classes $\{[u]_B : u \in U\}$ of the relation IND_B form $B-$ *categories.*

In a decision system (U, A, d), perceived as a window on a (possibly unknown) function $f_{A,d}$ from A-categories onto d-categories, an approximation to $f_{A,d}$ can be searched for in a form of a set of *decision rules* of the form of an implication

$$\bigwedge_{a \in B} (a, w_a) \Rightarrow (d, w_d) \tag{10.96}$$

where the *descriptor* (a, w_a) is a logical formula interpreted as $[(a, w_a)] = \{u \in U : a(u) = w_a\}$, extended recursively as $[\alpha \wedge \beta] = [\alpha] \cap [\beta]$, $[\alpha \vee \beta] = [\alpha] \cup [\beta]$, $[\neg\alpha] = U \setminus [\alpha]$. The implication in the formula (10.96) is satisfied to a degree r with respect to a set rough inclusion μ^S in case $\mu^S([\bigwedge_{a \in B}(a, w_a)], [(d, w_d)], r)$. In case $r = 1$ the rule is *true*.

10.8.2 Decision Rules

A *decision algorithm, classifier* is a judiciously chosen set of decision rules, approximating possibly most closely the real decision function $f_{A,d}$. This comes down to a search in the space of possible descriptors in order to find their successful combinations. In order to judge the quality, or, degree of approximation, decision rules are learned on a part of the decision system, the *training set* and then the decision algorithm is *tested* on the remaining part of the decision system, called the *test set*. Degree of approximation is measured by some coefficients of varied character. Simple measures of statistical character are found from the *contingency table*, see Arkin and Colton (1970). This table is built for each decision rule r and a decision value v, by counting the number n_t of training things, the number n_r of things satisfying the premise of the rule r (caught by the rule), $n_r(v)$ is the number of things counted in n_r and with the decision v, and $n_r(\neg v)$ is the number of things counted in n_r but with decision value distinct from v. To these factors, we add n_v, the number of training things with decision v and $n_{\neg v}$, the number of remaining things, i.e, $n_{\neg v} = n_t - n_v$.

For these values, *accuracy of the rule r relative to v* is the quotient

$$acc(r, v) = \frac{n_r(v)}{n_r} \tag{10.97}$$

and *coverage of the rule r relative to v* is

$$cov(r, v) = \frac{n_r(v)}{n_v} \tag{10.98}$$

These values are useful as indicators of a *rule strength* which is taken into account when classification of a test thing is under way: to assign the value of decision, a rule pointing to a decision with a maximal value of accuracy, or coverage, or combination of both can be taken; methods for combining accuracy and coverage into a single criterion are discussed, e.g., in Michalski (1990). Accuracy and coverage can, however, be defined in other ways; for a decision algorithm D, trained on a training set Tr, and a test set Tst, the *accuracy* of D is measured by its efficiency on the test set and it is defined as the quotient

$$accuracy(D) = \frac{n_{corr}}{n_{caught}} \tag{10.99}$$

where n_{corr} is the number of test things correctly classified by D and n_{caught} is the number of test things classified.

Similarly, *coverage* of D is defined as

$$coverage(D) = \frac{n_{caught}}{n_{test}} \tag{10.100}$$

where n_{test} is the number of test things. Thus, the product $accuracy(D) \cdot coverage(D)$ gives the measure of the fraction of test things correctly classified by D.

We have already mentioned that accuracy and coverage are often advised to be combined in order to better express the trade–off between the two: one may have a high accuracy on a relatively small set of caught things, or a lesser accuracy on a larger set of caught by the classifier things. Michalski (1990) proposes a combination rule of the form

$$MI = \frac{1}{2} \cdot A + \frac{1}{4} \cdot A^2 + \frac{1}{2} \cdot C - \frac{1}{4} \cdot A \cdot C \tag{10.101}$$

where A stands for accuracy and C for coverage. With the symbol MI, we denote the *Michalski index* as defined in (10.101). Other rule quality measures can be found, e.g., in Bruning and Kintz (1997), Bazan (1998), and Grzymala–Busse and Hu (2000).

Whereas indiscernibility classes are computationally feasible, cf., Skowron and Rauszer (1992), decision rules in optimal form are not, cf., op. cit. Methods for generation of rules with minimal set of descriptors, optimal rules, true rules, minimal sets of rules, strong (association) rules, etc., can be found in Pawlak and Skowron (1993), Skowron (1993), Skowron and Rauszer (1992), Grzymala–Busse (1992), and Agrawal et al. (1993).

10.8.3 Mereology as Similarity: Granulation of Knowledge

The creator of Fuzzy Set Theory Lotfi A. Zadeh (1979) proposed to compute with *granules of knowledge*. It was posed by L. A. Zadeh that the process of extraction of knowledge can be factored through the stage of *granulation* in which things are aggregated into *granules of knowledge* understood as collections or classes of things similar with respect to a chosen measure of similarity. Resulting *granular computing*, i.e., processing granules of knowledge promises lesser complexity as well as noise filtering.

In case discussed here, as similarity measure we choose a rough inclusion; it provides a similarity to a degree relation which is reflexive but not always symmetric, e.g., a set or geometric rough inclusion is not whereas (ari) or (airi) type rough inclusion is symmetric therefore inducing a hierarchy of *tolerance to a degree relations*; for a theory of tolerance relations, see, e.g., Shreider (1960).

The idea of mereological granulation of knowledge, see Polkowski (2004a, 2005a), cf., surveys Polkowski (2008, 2009a), presented here finds an effective application in problems of synthesis of classifiers from data tables. This application consists in granulation of data at preprocessing stage in the process of synthesis: after granulation, a new data set is constructed, called a *granular reflection*, to which various strategies for rule synthesis can be applied. This application can be regarded as a process of *filtration* of data, aimed at reducing noise immanent to data. Application of rough inclusions leads to a formal theory of granules of various *radii* allowing for various choices of coarseness degree in data.

For a given rough inclusion μ, the *granule* $g_\mu(u, r)$ *of the radius* r *about the center* u is defined as the class of property $\Phi^{\mu}_{u,r}$

$$g_\mu(u, r) = Cls\Phi^{\mu}_{u,r} \tag{10.102}$$

where

$$\Phi^{\mu}_{u,r}(v) \Leftrightarrow \mu(v, u, r) \tag{10.103}$$

Properties of granules depend, obviously, on the type of rough inclusion used in their definitions. In case of a symmetric and transitive rough inclusion μ, for each pair u, v of things, and $r \in [0, 1]$,

$$ingr(v, g_\mu(u, r)) \Leftrightarrow \mu(v, u, r)$$

holds which follows directly from the inference rule M3.

In effect, the granule $g_\mu(u, r)$ can be represented as the set $\{v : \mu(v, u, r)\}$. To justify this claim, assume that $ingr(v, g_\mu(u, r))$ holds. Thus, there exists z such that $Ov(z, v)$ and $\mu(z, u, r)$. There is x with $ingr(x, v)$, $ingr(x, z)$, hence, by transitivity of μ, also $\mu(x, u, r)$ holds. By symmetry of μ, $ingr(v, x)$, hence, $\mu(v, x, r)$ holds.

A more complicated case of other types of rough inclusions is discussed in Polkowski (2011).

Our idea of augmenting existing strategies for rule induction consists in using granules of knowledge. The principal assumption we can make is that the nature acts in a continuous way: if things are similar with respect to judiciously and correctly chosen attributes, then decisions on them should also be similar. A granule collecting similar things should then expose the most typical decision value for things in it while suppressing outlying values of decision, reducing noise in data, hence, leading to a better classifier.

In Polkowski and Artiemjew (2007) and in Artiemjew (2007) the theoretical analysis was confirmed as to its application merits. We proceed with a summary of methods and results of these verification.

10.8.4 The Idea of Granular Mereological Classifiers

We assume that we are given a decision system (U, A, d) from which a classifier is to be constructed; on the universe U, a rough inclusion μ is given, and a radius $r \in [0, 1]$ is chosen, see Polkowski (2004a, 2005a). We can find granules $g_\mu(u, r)$ for all $u \in U$, and make them into the set $G(\mu, r)$.

Table 10.2 Best results for Australian credit by some rough set based algorithms

Source	Method	Accuracy	Coverage	MI
Bazan (1998)	$SNAPM(0.9)$	$error = 0.130$	–	–
Nguyen SH (2000)	*simple.templates*	0.929	0.623	0.847
Nguyen SH (2000)	*general.templates*	0.886	0.905	0.891
Nguyen SH (2000)	*tolerance.gen.templ.*	0.875	1.0	0.891
Wróblewski (2004)	*adaptive.classifier*	0.863	–	–

From this set, a covering $Cov(\mu, r)$ of the universe U can be selected by means of a chosen strategy $\mathcal{G}$, i.e.,

$$Cov(\mu, r) = \mathcal{G}(G(\mu, r)) \tag{10.104}$$

We intend that $Cov(\mu, r)$ becomes a new universe of the decision system whose name will be the *granular reflection* of the original decision system. It remains to define new attributes for this decision system.

Each granule g in $Cov(\mu, r)$ is a collection of things; attributes in the set $A \cup \{d\}$ can be factored through the granule g by means of a chosen strategy $\mathcal{S}$, i.e., for each attribute $q \in A \cup \{d\}$, the new factored attribute $\overline{q}$ is defined by means of the formula

$$\overline{q}(g) = \mathcal{S}(\{a(v) : ingr(v, g_{\mu}(u, r))\}) \tag{10.105}$$

In effect, a new decision system $(Cov(\mu, r), \{\overline{a} : a \in A\}, \overline{d})$ is defined. The thing v with

$$Inf(v) = \{(\overline{a} = \overline{a}(g)) : a \in A\} \tag{10.106}$$

is called the *granular reflection of g*.

Granular reflections of granules need not be things found in data set; yet, the results show that they mediate very well between the training and test sets. In order to demonstrate the merits of this approach, we consider a standard data set *the Australian Credit Data Set* from Repository at UC Irvine (2012) and we collect the best results for this data set by various rough set based methods in Table 10.2. For a comparison we include in Table 10.3 results obtained by some other methods, as given in Statlog. In Table 10.4, we give a comparison of performance of rough set classifiers: exhaustive, covering and LEM (Grzymala–Busse 1992) implemented in RSES (2012) public domain system. We begin in the next section with granular classifiers in which granules are induced from the training set.

Table 10.3 A comparison of errors in classification by rough set and other paradigms

Paradigm	System/method	Austr. credit
Stat.Methods	*Logdisc*	0.141
Stat.Methods	*SMART*	0.158
Neural Nets	*Backpropagation2*	0.154
Neural Networks	*RBF*	0.145
Decision Trees	*CART*	0.145
Decision Trees	*C4.5*	0.155
Decision Trees	*ITrule*	0.137
Decision Rules	*CN2*	0.204

Table 10.4 Train and test (trn = 345 things, tst = 345 things) ; Australian credit; comparison of RSES implemented algorithms exhaustive, covering and LEM

Algorytm	Accuracy	Coverage	Rule number	MI
$covering(p = 0.1)$	0.670	0.783	589	0.707
$covering(p = 0.5)$	0.670	0.783	589	0.707
$covering(p = 1.0)$	0.670	0.783	589	0.707
$LEM2(p = 0.1)$	0.810	0.061	6	0.587
$LEM2(p = 0.5)$	0.906	0.368	39	0.759
$LEM2(p = 1.0)$	0.869	0.643	126	0.804

10.8.5 Classification by Granules of Training Things

We begin with a classifier in which granules computed by means of the rough inclusion μ_L form a granular reflection of the data set and then to this new data set the exhaustive classifier, see RSES (2012), is applied.

10.8.5.1 Procedure of the Test

1. *The data set (U, A, d) is input*;
2. *The training set is chosen at random. On the training set, decision rules are induced by means of exhaustive, covering and LEM algorithms implemented in the RSES system*;
3. *Classification is performed on the test set by means of classifiers of pt. 2*;
4. *For consecutive granulation radii r, granule sets $G(\mu, r)$ are found*;
5. *Coverings $Cov(\mu, r)$ are found by a random irreducible choice*;
6. *For granules in $Cov(\mu, r)$, for each r, we determine the granular reflection by factoring attributes on granules by means of majority voting with random resolution of ties*;
7. *For found granular reflections, classifiers are induced by means of algorithms in pt. 2*;
8. *Classifiers found in pt. 7, are applied to the test set*;
9. *Quality measures: accuracy and coverage for classifiers are applied in order to compare results obtained, respectively, in pts. 3 and 8.*

Table 10.5 Train–and–test; Australian Credit; Granulation for radii r; RSES exhaustive classifier; r = granule radius, tst = test set size, trn = train set size, rulex = rule number, aex = accuracy, cex = coverage

r	tst	trn	rulex	aex	cex	MI
Nil	345	345	5,597	0.872	0.994	0.907
0.0	345	1	0	0.0	0.0	0.0
0.0714286	345	1	0	0.0	0.0	0.0
0.142857	345	2	0	0.0	0.0	0.0
0.214286	345	3	7	0.641	1.0	0.762
0.285714	345	4	10	0.812	1.0	0.867
0.357143	345	8	23	0.786	1.0	0.849
0.428571	345	20	96	0.791	1.0	0.850
0.5	345	51	293	0.838	1.0	0.915
0.571429	345	105	933	0.855	1.0	0.896
0.642857	345	205	3,157	0.867	1.0	0.904
0.714286	345	309	5,271	0.875	1.0	0.891
0.785714	345	340	5,563	0.870	1.0	0.890
0.857143	345	340	5,574	0.864	1.0	0.902
0.928571	345	342	5,595	0.867	1.0	0.904

Table 10.6 Train–and–test; Australian credit;(layered–granulation)

r	acc	cov
0.500000	0.436	1.000
0.571429	0.783	1.000
0.642857	0.894	1.000
0.714286	0.957	1.000

In Table 10.5, the results are collected of results obtained after the procedure described above is applied. We can compare results expressed in terms of the Michalski index *MI* as a measure of the trade–off between accuracy and coverage; for template based methods, the best *MI* is 0.891, for covering or LEM algorithms the best value of *MI* is 0.804, for exhaustive classifier (r = nil) *MI* is equal to 0.907 and for granular reflections, the best *MI* value is 0.915 with few other values exceeding 0.900.

What seems worthy of a moment's reflection is the number of rules in the classifier. Whereas for the exhaustive classifier (r = nil) in non–granular case, the number of rules is equal to 5,597, in granular case the number of rules can be surprisingly small with a good *MI* value, e.g., at $r = 0.5$, the number of rules is 293, i.e., 5 % of the exhaustive classifier size, with the best *MI* at all of 0.915. This compression of classifier seems to be the most impressive feature of granular classifiers.

It is an obvious idea that this procedure can be repeated until a stable system is obtained to which further granulation causes no change; it is the procedure of *layered granulation*, see Artiemjew (2007). Table 10.6 shows some best results of this procedure for selected granulation radii. As coverage in all reported cases is equal to 1.0, the Michalski index MI is equal to accuracy. This initial, simple granulation, suggests further ramifications. For instance, one can consider, for a chosen value of $\varepsilon \in [0, 1]$, granules of the form

Table 10.7 ε_{opt} = optimal value of ε, acc = accuracy, cov = coverage. Best $r_{catch} = 0.1428$, $\varepsilon_{opt} = 0.35$: accuracy = 0.8681, coverage = 1.0

r_catch	optimal eps	acc	cov
Nil	*Nil*	0.845	1.0
0	0	0.555073	1.0
0.071428	0	0.83913	1.0
0.142857	0.35	0.868116	1.0
0.214286	0.5	0.863768	1.0
0.285714	0.52	0.831884	1.0
0.357143	0.93	0.801449	1.0
0.428571	1.0	0.514493	1.0
0.500000	1.0	0.465217	1.0
0.571429	1.0	0.115942	1.0

$$g_\mu(u, r, \varepsilon) = \{v \in U : \forall a \in A.|a(u) - a(v)| \leq \varepsilon\} \tag{10.107}$$

and repeat with these granules the procedure of creating a granular reflection and building from it a classifier. Another yet variation consists in mimicking the performance of the Łukasiewicz based rough inclusion and introducing a counterpart of the granulation radius in the form of the *catch radius*, r_{catch}. The granule is then dependent on two parameters: ε and r_{catch}, and its form is

$$g_\mu(u, \varepsilon, r_{catch}) = \{v \in U : \frac{|\{a \in A : |a(u) - a(v)| \leq \varepsilon}{|A|} \geq r_{catch}\} \tag{10.108}$$

Results of classification by granular classifier induced from the granular reflection obtained by means of granules (10.108) are shown in Table 10.7.

10.8.6 A Treatment of Missing Values

A particular but important problem in data analysis is the treatment of missing values. In many data, some values of some attributes are not recorded due to many factors, like omissions, inability to take them, loss due to some events etc.

Analysis of systems with missing values requires a decision on how to treat missing values; Grzymala–Busse and Hu (2000) analyze nine such methods, among them, (1) *most common attribute value*, (2) *concept restricted most common attribute value*, (3) *assigning all possible values to the missing location*, (4) *treating the unknown value as a new valid value*, etc. Their results indicate that methods (3), (4) perform very well and in a sense stand out among all nine methods.

We adopt and consider two methods, i.e., (3), (4) from the above mentioned. As usual, the question on how to use granular structures in analysis of incomplete systems, should be answered first.

The idea is to embed the missing value into a granule: by averaging the attribute value over the granule in the way already explained, it is hoped the average value would fit in a satisfactory way into the position of the missing value.

We will use the symbol $*$, commonly used for denoting the missing value; we will use two methods (3), (4) for treating $*$, i.e, either $*$ is a *don't care* symbol meaning that any value of the respective attribute can be substituted for $*$, hence, $* = v$ for each value v of the attribute, or $*$ is a new value on its own, i.e., if $* = v$ then v can only be $*$.

Our procedure for treating missing values is based on the granular structure $(G(\mu, r), \mathcal{G}, \mathcal{S}, \{a* : a \in A\})$; the strategy $\mathcal{S}$ is the majority voting, i.e., for each attribute a, the value $a^*(g)$ is the most frequent of values in $\{a(u) : u \in g\}$. The strategy $\mathcal{G}$ consists in random selection of granules for a covering.

For a thing u with the value of $*$ at an attribute a,, and a granule $g = g(v, r) \in G(\mu, r)$, the question whether u is included in g is resolved according to the adopted strategy of treating $*$: in case $* =$ *don't care*, the value of $*$ is regarded as identical with any value of a hence $|IND(u, v)|$ is automatically increased by 1, which increases the granule; in case $* = *$, the granule size is decreased. Assuming that $*$ is sparse in data, majority voting on g would produce values of a^* distinct from $*$ in most cases; nevertheless the value of $*$ may appear in new things g^*, and then in the process of classification, such value is repaired by means of the granule closest to g^* with respect to the rough inclusion μ_L, in accordance with the chosen method for treating $*$.

In plain words, things with missing values are in a sense absorbed by close to them granules and missing values are replaced with most frequent values in things collected in the granule; in this way the method (*3*) or (*4*) in Grzymala–Busse and Hu (2000) is combined with the idea of a frequent value, in a novel way.

We have thus four possible strategies:

1. *Strategy A: in building granules* $* =$ *don't care, in repairing values of* $*$, $* =$ *don't care*;
2. *Strategy B: in building granules* $* =$ *don't care, in repairing values of* $*$, $* = *$;
3. *Strategy C: in building granules* $* = *$, *in repairing values of* $*$, $* =$ *don't care*;
4. *Strategy D: in building granules* $* = *$, *in repairing values of* $*$, $* = *$.

We show how effective are these strategies, see Polkowski and Artiemjew (2007) by perturbing the data set *Pima Indians Diabetes*, from UC Irvine Repository (2012). First, in Table 10.8 we show results of granular classifier on the non–perturbed (i.e., without missing values) Pima Indians Diabetes data set. We now perturb this data set by randomly replacing 10 % of attribute values in the data set with missing $*$ values. Results of granular treatment in case of Strategies A, B, C, D in terms of accuracy are reported in Table 10.9. As algorithm for rule induction, the exhaustive algorithm of the RSES system has been selected. 10-fold cross validation (CV–10) has been applied.

Strategy A reaches the accuracy value for data with missing values within 94 % of the value of accuracy without missing values (0.9407–1.0) at the radius of 0.875. With Strategy B, accuracy is within 94 % from the radius of 0.875 on. Strategy C is

Table 10.8 10-fold CV; Pima; exhaustive algorithm, r = radius, macc = mean accuracy, mcov = mean coverage

r	macc	mcov
0.0	0.0	0.0
0.125	0.0	0.0
0.250	0.6835	0.9956
0.375	0.7953	0.9997
0.500	0.9265	1.0
0.625	0.9940	1.0
0.750	1.0	1.0
0.875	1.0	1.0

Table 10.9 Accuracies of strategies A, B, C, D. 10-fold CV; Pima Indians; exhaustive algorithm; r = radius, maccA = mean accuracy of A, maccB = mean accuracy of B, maccC = mean accuracy of C, maccD = mean accuracy of D

r	maccA	maccB	maccC	maccD
0.250	0.0	0.0	0.0	0.645
0.375	0.0	0.0	0.0	0.7779
0.500	0.0	0.0	0.0	0.9215
0.625	0.5211	0.5831	0.5211	0.9444
0.750	0.7705	0.7769	0.7705	0.9994
0.875	0.9407	0.9407	0.9407	0.9987

much better: accuracy with missing values reaches 99 % of accuracy in no missing values case from the radius of 0.625 on. Strategy D gives results slightly better than C with the same radii.

We conclude that the essential for results of classification is the strategy of treating the missing value of $*$ as $* = *$ in both strategies C and D; the repairing strategy has almost no effect: C and D differ very slightly with respect to this strategy.

10.8.7 Granular Rough Mereological Classifiers Using Residuals

Rough inclusions used in Sects. 10.8.4–10.8.6 in order to build classifiers do take, to a certain degree, into account the distribution of values of attributes among things, by means of the parameters ε and the catch radius r_{catch}. The idea that metrics used in classifier construction should depend locally on the training set is, e.g., present in classifiers based on the idea of nearest neighbor, see, e.g., a survey in Polkowski (2009b). In order to construct a measure of similarity based on distribution of attribute values among things, we resort to residual implications–induced rough inclusions. This rough inclusion can be transferred to the universe U of a decision system; to this end, first, for given things u, v, and $\varepsilon \in [0, 1]$, factors

$$dis_{\varepsilon}(u, v) = \frac{|\{a \in A : |a(u) - a(v)| \geq \varepsilon\}|}{|A|} \tag{10.109}$$

and

$$ind_{\varepsilon}(u, v) = \frac{|\{a \in A : |a(u) - a(v)| < \varepsilon\}|}{|A|} \tag{10.110}$$

are introduced. The weak variant of rough inclusion $\mu_{\rightarrow_T}$ is defined, see Polkowski (2007), as

$$\mu^*_T(u, v, r) \text{ if and only if } dis_{\varepsilon}(u, v) \rightarrow_T \; ind_{\varepsilon}(u, v) \geq r \tag{10.111}$$

These similarity measures will be applied in building granules and then in data classification.

Tests are done with the Australian credit data set; the results are validated by means of the 5–fold cross validation (CV–5). For each of t–norms: M, P, L, three cases of granulation are considered

1. *Granules of training things* (GT);
2. *Granules of rules induced from the training set* (GRT);
3. *Granules of granular things induced from the training set* (GGT).

In this approach, training things are made into granules for a given ε. Things in each granule g about a test thing u, vote for decision value at u as follows: for each decision class c, the value

$$p(c) = \frac{\sum_{\text{training thing v in g falling in c}} w(u, v)}{\text{size of c in training set}} \tag{10.112}$$

is computed where the weight $w(u, v)$ is computed for a given t–norm T as

$$w(u, v) = dis_{\varepsilon}(u, v) \rightarrow_T \; ind_{\varepsilon}(u, v) \tag{10.113}$$

The class c* assigned to u is the one with the largest value of p.

Weighted voting of rules in a given granule g for decision at test thing u goes according to the formula $d(u) = argmax p(c)$, where

$$p(c) = \frac{\sum_{\text{rule in g pointing to c}} w(u, r) \cdot support(r)}{\text{size of c in training set}} \tag{10.114}$$

where weight $w(u, r)$ is computed as

$$dis_{\varepsilon}(u, r) \rightarrow_T \; ind_{\varepsilon}(u, r) \tag{10.115}$$

The optimal (best) results in terms of accuracy of classification are collected in Table 10.10.

Table 10.10 5-fold CV; Australian; residual metrics. met = method of granulation, T = t–norm, ε_{opt} = optimal ε, macc = mean accuracy, mcov = mean coverage

met	T	ε_{opt}	macc	mcov
GT	*M*	0.04	0.848	1.0
GT	*P*	0.06	0.848	1.0
GT	*L*	0.05	0.846	1.0
GRT	*M*	0.02	0.861	1.0
GRT	*P*	0.01	0.851	1.0
GGT	*M*	0.05	0.855	1.0
GRT	*P*	0.01	0.852	1.0

These results show that rough mereological granulation provides better or at least on par with best results by other methods accuracy of classification at the radically smaller classifier size measured as the number of decision rules in it.

10.9 Mereology in Artificial Intelligence

Though the topics relegated to this section may be as well assigned to Knowledge Engineering, yet we relate them to Artificial Intelligence as tools which may be helpful in reasoning about complex systems and hard decision problems.

10.9.1 Cognitive Reasoning

We focus here on cognitive methods known also as network related methods. Of those, *neural networks* are well–known as a tool useful in pattern recognition, classification and machine learning. Based on the structure of the physiological neuron, discovered by Ramón y Cajal (1889), artificial neuron was defined in McCulloch and Pitts (1943) as the structure composed of a finite set of inputs labeled $x_1, x_2, \ldots, x_n$, a body with a *threshold* Θ and the *output*, y; according to the physiological archetype, this neuron computes by the rule

$$y = 1 \Leftrightarrow \sum_i x_i \geq \Theta \text{ else } y = 0 \tag{10.116}$$

Later developments include a *perceptron* defined in Rosenblatt (1958). A simplified perceptron adds to McCulloch–Pitts neuron *weights* on inputs, and an additional input with constant value of 1 and a weight b, called *bias*. Thus, the computation rule has the form

$$y = 1 \Leftrightarrow \sum_i w_i \cdot x_i + b \geq \Theta \text{ else } y = 0 \tag{10.117}$$

which allows for a greater flexibility. Either type of neuron is able to classify binary concepts by means of the *separating hyperplane*, H, which separates the space E^n of possible input vectors into two semi–planes, and it is defined, e.g., in case of perceptrons, as

$$\sum_i w_i \cdot x_i + b = \Theta \tag{10.118}$$

The idea of networks of neurons was advocated by Alan Turing (1948) who proposed a learning scheme for networks of neurons connected through *modifiers*, and it was revived in Grossberg (1973) with networks of perceptrons. Such networks of connected perceptrons produce the intersection of respective semi–planes which cuts the space of input vectors into convex closed regions in ideal case assigning input vectors representing distinct categories of things to distinct regions. Finally, the back–propagation learning, see Werbos (1994), added in place of Θ's differentiable sigmoid *transfer functions*.

Our model of perceptron, see Polkowski (2005b), differs from the standard model, as its neurons are perceived as intelligent agents working with knowledge represented in information systems. An essential feature of network perceptrons from the point of view of learning is differentiability of transfer functions; hence, we introduced a special type of rough inclusions, called *gaussian* in Polkowski (2005b) because of their form, by letting

$$\mu_G(x, y, r) \text{ iff } e^{-|\sum_{a \in DIS(x,y)} w_a|^2} \geq r \tag{10.119}$$

where $w_a \in (0, +\infty)$ is a weight associated with the attribute a for each attribute $a \in A$; cf. (10.23) for *DIS*. Computation by this perceptron is directed by the gradient of the function

$$f(x, y) = e^{-|\sum_{a \in DIS(x,y)} w_a|^2} \tag{10.120}$$

whose w_a component is

$$\frac{\partial f}{\partial w_a} = f \cdot (-2 \cdot \sum w_a) \tag{10.121}$$

It follows from the last equation that gradient search would go in direction of minimizing the value of $\sum_a w_a$.

We denote the perceptron by the agent symbol *ag*; it is endowed with an information system $I_{ag} = (U_{ag}, A_{ag})$. The input to *ag* is in the form of a thing x.

The rough mereological perceptron is endowed with a set of *target concepts* $t = g_{\mu_G}(T \in U_{ag}/IND(A_{ag}), r_t)$. The result of computation with a dedicated target t for a training thing x is a granule $g = g_{\mu_G}(x, r(res))$ such that $ingr(g, t)$.

During computation, weights are incremented by the *learning rule*

$$w_a \leftarrow w_a + \Delta \cdot \frac{\partial E}{\partial w_a} \tag{10.122}$$

where Δ is the *learning rate*.

At a stage *current* of computing, where $\gamma = |r_{current} - r|$, for a natural number k, the value of $\Delta_{current}$ which should be taken at the step *current* in order to achieve the target in at most k steps should be taken as, see Polkowski (2005b)

$$\Delta_{current} \simeq \frac{\gamma}{2 \cdot k \cdot f^2 \cdot (\sum_a w_a)^2} \tag{10.123}$$

10.9.2 MAS Reasoning: Many–Agent Systems

Reasoning in artificial intelligence is often concerned with 'complex cases' like, e.g., robotic soccer, in which performing successfully tasks requires participation of a number of 'agents' bound to cooperate, and in which a task is performed with a number of steps, see, e.g., Stone (2000); other areas where such approach seems necessary concern assembling and design, see Amarel (1991), fusion of knowledge, e.g., in robotics, fusion of information from sensors, see, e.g., Canny (1988), Choset et al. (2005), or, Stone (2000), as well as in machine learning and fusion of classifiers, see, e.g., Dietterich (2000).

Rough mereological approach to these problems was initiated with Polkowski and Skowron (1998, 1999a,b, 2001); here, we propose a logically oriented formulation.

Rough inclusions and granular intensional logics based on them can be applied in describing workings of a collection of intelligent agents which are called here *granular agents.*

A granular agent a will be represented as a tuple

$$(U_a, \mu_a, L_a, prop_a, synt_a, aggr_a)$$

where

1. *U_a is a collection of objects available to the agent a.*
2. *μ_a is a rough inclusion on objects in U_a.*
3. *L_a is a set of unary predicates in first–order open calculus, interpretable in U_a.*
4. *$prop_a$ is the propagation function that describes how uncertainty expressed by rough inclusions at agents connected to a propagates to a.*
5. *$synt_a$ is the logic propagation functor which expresses how formulas of logics at agents connected to the agent a are made into a formula at a.*
6. *$aggr_a$ is the synthesis function which describes how objects at agents connected to a are made into an object at a.*

We assume for simplicity that agents are arranged into a rooted tree; for each agent a distinct from any leaf agent, we denote by B_a the children of a, understood as agents connected to a and directly sending to a objects, logical formulas describing them, and uncertainty coefficients like values of rough inclusions.

For $b \in B_a$, the symbol x_b will denote an object in U_b; similarly, ϕ_b will denote a formula of L_b, and μ_b will be a rough inclusion at b with values r_b. The same convention will be obeyed by objects at a.

Postulates governing the working of the scheme are

MA1 *If* $ingr_b(x'_b, x_b)$ *for each* $b \in B_a$, *then* $ingr_a(aggr(\{x'_b\}), aggr(\{x_b\}))$.
This postulate does assure that ingredient relations are in agreement with aggregate operator of forming complex objects: ingredients of composed objects form an ingredient of a complex object. We can say that $aggr \circ ingr = ingr \circ aggr$, i.e, the resulting diagram commutes.

MA2 *If* $x_b \models \phi_b$, *then* $aggr(\{x_b\}) \models synt(\{\phi_b\})$.
This postulate is about agreement between aggregation of objects and their logical descriptions: descriptions of composed objects merge into a description of the resulting complex object.

MA3 *If* $\mu_b(x_b, y_b, r_b)$ *for* $b \in B_a$, *then* $\mu_a(aggr(\{x_b\}), aggr(\{y_b\}), prop\{r_b\})$.
This postulate introduces the propagation function, which does express how uncertainty at connected agents is propagated to the agent a. One may observe the uniformity of *prop*, which in the setting of MA3 depends only on values of r_b's; this is undoubtedly a simplifying assumption, but we want to avoid unnecessary and obscuring the general view complications, which of course can be multiplied at will.

MA4 *For* $b \in B_a$, $ingr_b(x_b, g_{mu_r}(u_b, r_b))$ *implies*

$$ingr_a(aggr(\{x_b\}), g_{\mu_a}(aggr(\{u_b\}), prop(\{r_b\}))).$$

Admitting MA4, we may also postulate that in case agents have at their disposal variants of rough mereological granular logics, intensions propagate according to the *prop* functor

MA5 *If* $I^{\vee}_{\nu_b}(g_b)(\phi_b) \geq r_b$ *for each* $b \in B_a$, *then*

$$I^{\vee}_{\nu_a}(aggr(\{g_b\}))(synt(\{\phi_b\})) \geq prop(\{r_b\}).$$

A simple exemplary case of knowledge fusion was examined in Polkowski (2008). We consider an agent $a \in Ag$ with two incoming connections from agents b, c, i.e., $B_a = \{b, c\}$. Each agent is applying the rough inclusion $\mu = \mu^I_L$, see (10.26), to an information system (U_a, A_a), (U_b, A_b), (U_c, A_c). Each agent is also applying the rough inclusion on sets of the form (10.27) in evaluations related to extensions of formulae intensions.

We consider a simple fusion scheme in which information systems at b, c are combined thing–wise to make the information system at a; thus, $aggr_a(x, y) = (x, y)$. Such case may happen, e.g., when an object is described with help of a camera image by some features and at the same time it is perceived and recognized with range sensors like infrared or laser sensors and some localization means like GPS.

Then: *uncertainty propagation and granule propagation are described by the Łukasiewicz t–norm L and extensions of logical intensions propagate according to the product t–norm P*.

10.9.3 Granular Logics: Reasoning in Information Systems

The idea of a granular rough mereological logic, see Polkowski (2004b) and Polkowski and Semeniuk–Polkowska (2005), consists in measuring the meaning of a unary predicate in the model which is a universe of an information system against a granule defined by means of a rough inclusion. The result can be regarded as the degree of truth (the logical value) of the predicate with respect to the given granule. The obtained logics are intensional as they can be regarded as mappings from the set of granules (possible worlds) to the set of logical values in the interval [0, 1], the value at a given granule regarded as the extension at that granule of the intension. A discussion of intensional logics can be found, e.g., in Gallin (1975), Van Benthem (1988), Hughes and Creswell (1996) and Fitting (2004).

For an information/decision system (U, A, d), with a rough inclusion ν, e.g., (10.27), on subsets of U and for a collection of unary predicates Pr, interpreted in the universe U (meaning that for each predicate $\phi \in Pr$ the meaning $[[\phi]]$ is a subset of U), we define the intensional logic GRM_{ν} by assigning to each predicate ϕ in Pr its intension $I_{\nu}(\phi)$ defined by its extension $I_{\nu}^{\vee}(g)$ at each particular granule g, as

$$I_{\nu}^{\vee}(g)(\phi) \geq r \;\Leftrightarrow\; \nu(g, [[\phi]], r) \tag{10.124}$$

With respect to the rough inclusion (10.27) the formula (10.124) becomes

$$I_{\nu_L}^{\vee}(g)(\phi) \geq r \;\Leftrightarrow\; \frac{|g \cap [[\phi]]|}{|g|} \geq r \tag{10.125}$$

A formula ϕ interpreted in the universe U of a system (U, A, d) is *true at a granule* g with respect to a rough inclusion ν if and only if $I_{\nu}^{\vee}(g)(\phi) = 1$ and ϕ is *true* if and only if it is true at each granule g. A rough inclusion ν is *regular* when $\nu(X, Y, 1)$ holds if and only if $X \subseteq Y$. Hence, for a regular ν, a formula ϕ is true if and only if for $g \subseteq [[\phi]]$ for each granule g.

A particularly important case of a formula is that of decision rules; clearly, for a decision rule $r : p \Rightarrow q$ in the descriptor logic, the rule r is true at a granule g with respect to a regular rough inclusion ν if and only if $g \cap [[p]] \subseteq [[q]]$.

Analysis of decision rules in the system (U, A, d) can be given in a more general setting of *dependencies*. For two sets $C, D \subseteq A \cup \{d\}$ of attributes, one says that

D depends functionally on C when $IND(C) \subseteq IND(D)$, symbolically denoted $C \mapsto D$. Functional dependence can be represented locally by means of functional dependency rules of the form

$$\phi_C(\{v_a : a \in C\}) \Rightarrow \phi_D(\{w_a : a \in D\}) \tag{10.126}$$

where $\phi_C(\{v_a : a \in C\})$ is the formula $\bigwedge_{a \in C}(a = v_a)$, and $[[\phi_C]] \subseteq [[\phi_D]]$.

We introduce a regular rough inclusion on sets ν_3 defined as

$$\nu_3(X, Y, 1) \Leftrightarrow X \subseteq Y \text{ else } \nu_3(X, Y, \frac{1}{2}) \Leftrightarrow X \cap Y \neq \emptyset \text{ else } \nu_3(X, Y, 0) \tag{10.127}$$

Then one proves that $\alpha : \phi_C \Rightarrow \phi_D$ *is a functional dependency rule if and only if* α *is true in the logic induced by* ν_3. A specialization of this statement holds for decision rules. Further applications to modalities in decision systems and the Perception Calculus in the sense of Zadeh (2004) can be found in Polkowski (2011).

Acknowledgements The author would like to extend thanks to Dr Piotr Artiemjew and Dr Paul Ośmiałowski for submitting experimental material in research on, respectively, granular classifiers and behavioral robotics.

References

Agah, A. (1996). Robot teams, human workgroups and animal sociobiology: A review of research on natural and artificial multi–agent autonomous systems. *Advanced Robotics*, *10*, 523–545.

Agah, A., & Bekey, G. A. (1997). Tropism based cognition: A novel software architecture for agents in colonies. *Journal of Experimental and Theoretical Artificial Intelligence*, *9*(2–3), 393–404.

Agrawal, R., Imieliński, T., & Swami, A. (1993). Mining association rules between sets of items in large databases. In *Proceedings ACM SIGMOD conference*, Washington, DC (pp. 207–216).

Amarel, S. (1991). Panel on AI and design. In *Proceedings of 12th international conference on AI*, Sydney (pp. 563–565).

Aristotle. (1989). *Prior analytics*. Indianapolis: Hackett.

Aristotle. (2004). *Metaphysics* (W. Ross, Trans.). Book Delta, 1203b. Classical Library. Available: http://www.classicallibrary.org/aristotle/metaphysics/

Arkin, R. C. (1998). *Behavior-based robotics*. Cambridge: MIT.

Arkin, H., & Colton, R. R. (1970). *Statistical methods*. New York: Barnes and Noble.

Artiemjew, P. (2007). Classifiers from granulated data sets: Concept dependent and layered granulation. In *Proceedings RSKD'07. Workshop at ECML/PKDD'07*, Warsaw (pp. 1–9). Warsaw: Warsaw University Press.

Asama, H., Matsumoto, A., & Ishida, Y. (1989). Design of an autonomous and distributed robot system: ACTRESS. In *Proceedings IEEE/RSJ IROS*, Tsukuba (pp. 283–290).

Asher, N., Aurnague, M., Bras, M., Sablayrolles, P., & Vieu, L. (1995) *De l'espace-temps dans l'analyse du discours* (Rapport interne IRIT/95-08-R). Institut de Recherche en Informatique, Univ. Paul Sabatier, Toulouse.

Asher, N., & Vieu, L. (1995). Toward a geometry of common sense: A semantics and a complete axiomatization of mereotopology. In *Proceedings IJCAI'95*, Montreal. San Mateo: Morgan Kaufmann.

Aurnague, M., & Vieu, L. (1995). A theory of space-time for natural language semantics. In K. Korta & J. M. Larrazábal (Eds.), *Semantics and pragmatics of natural language: Logical and computational aspects* (ILCLI series, pp. 69–126). San Sebastian: I. University of Pais Vasco.

Aurnague, M., Vieu, L., & Borillo, A. (1997). Representation formelle des concepts spatiaux dans la langue. In M. Denis (Ed.), *Langage et Cognition Spatiale* (pp. 69–102). Paris: Masson.

Baker, L. R. (2004). The ontology of artifacts. *Philosophical Explorations, 7*(2), 99–111.

Balcar, B., & Štěpánek, P. (1986). *Teorie Množin*. Praha: Academia.

Balch, T., & Arkin, R. C. (1998). Behavior–based formation control for multiagent robot teams. *IEEE Transactions on Robotics and Automation, 14*(12), 926–931.

Bazan, J. G. (1998). A comparison of dynamic and non–dynamic rough set methods for extracting laws from decision tables. In L. Polkowski & A. Skowron (Eds.), *Rough sets in knowledge discovery 1* (pp. 321–365). Heidelberg: Physica Verlag.

Björner, D., & Eir, A. (2010). Compositionality: Ontology and mereology of domains. In D. Dams, U. Hannemann, & M. Steffen (Eds.), *Concurrency, compositionality, and correctness* (Lecture notes in computer science, Vol. 5930, pp. 22–60). Berlin: Springer.

Bocheński, I. M. (1954). *Die Zeitgenössichen Denkmethoden*. Bern: A. Francke A. G.

Booch, G. (1994). *Object–oriented analysis and design with applications*. Menlo Park: Addison–Wesley.

Boothroyd, G. (2005). *Assembly automation and product design* (2nd ed.). Boca Raton: Taylor and Francis.

Boothroyd, G., Dewhurst, P., & Knight, W. (2002). *Product design for manufacture and assembly* (2nd ed.). New York: Marcel Dekker.

Borgo, S., & Vieu, L. (2009). Artefacts in formal ontology. In A. Meijers (Ed.), *Handbook of philosophy of technology and engineering sciences* (pp. 273–308). Amsterdam/London/Boston: Elsevier.

Bruning, J. L., & Kintz, B. L. (1997). *Computational handbook of statistics* (4th ed.). Columbus: Allyn and Bacon.

Caloud, P., Choi, W., Latombe, J.-C., Le Pape, C., & Yin, M. (1990). Indoor automation with many mobile robots. In *Proceedings IEEE/RSJ IROS*, Tsuchiura (pp. 67–72).

Canny, J. F. (1988). *The complexity of robot motion planning*. Cambridge: MIT.

Casati, R., Smith, B., & Varzi, A. C. (1998). Ontological tools for geographic representation. In N. Guarino (Ed.), *Formal ontology in information systems*. Amsterdam: IOS.

Casati, R., & Varzi, A. C. (1999). Parts and places: The structures of spatial representations. Cambridge: MIT.

Choset, H., Lynch, K. M., Hutchinson, S., Kantor, G., Burgard, W., Kavraki, L. E., & Thrun, S. (2005). *Principles of robot motion: Theory, algorithms, and implementations*. Cambridge: MIT.

Clarke, B. L. (1981). A calculus of individuals based on connection. *Notre Dame Journal of Formal Logic, 22*(2), 204–218.

Clay, R. (1974). Relation of Leśniewski's mereology to Boolean algebra. *The Journal of Symbolic Logic, 39*, 638–648.

Cohn, A. G. (1996). Calculi for qualitative spatial reasoning. In J. Calmet, J. A. Campbell, & J. Pfalzgraf (Eds.), *Artificial intelligence and symbolic mathematical computation* (Lecture notes in computer science, Vol. 1138, pp. 124–143) Berlin: Springer.

Cohn, A. G., Gooday, J. M., Bennett, B., & Gotts, N. M. (1996). A logical approach to representing and reasoning about space. In J. Calmet, J. A. Campbell, & J. Pfalzgraf (Eds.), *Artificial intelligence and symbolic mathematical computation* (Lecture notes in computer science, Vol. 1138, pp. 124–143). Berlin: Springer.

Cohn, A. G., & Gotts, N. M. (1996). Representing spatial vagueness: A mereological approach. In *Principles of knowledge representation and reasoning. Proceedings of the 5th international conference KR'96*, Boston (pp. 230–241). San Francisco: Morgan Kaufmann.

Cohn, A. G., Randell, D., Cui, Z., & Bennett B. (1993). Qualitative spatial reasoning and representation. In N. Carrete & M. Singh (Eds.), *Qualitative reasoning and decision technologies* (pp. 513–522). CIMNE: Barcelona.

Cohn, A. G., & Varzi, A. C. (1998). Connections relations in mereotopology. In H. Prade (Ed.), *Proceedings ECAI'98: 13th European conference on artificial intelligence*, Brighton (pp. 150–154). Chichester: Wiley.

de Laguna, Th. (1922). Point, line and surface as sets of solids. *The Journal of Philosophy*, *19*, 449–461.

Dietterich, T. G. (2000). Ensemble methods in machine learning. In: *1st international workshop on multiple classifier systems* (Lecture notes in computer science, Vol. 1857, pp. 1–15). Berlin: Springer.

Egenhofer, M. J. (1991). Reasoning about binary topological relations. In O. Gunther, H. & Schek (Eds.), *Advances in spatial databases: SSD'91 proceedings*, Zurich (pp. 143–160). Berlin.

Egenhofer, M. J., & Golledge, R. G. (Eds.). (1997). *Spatial and temporal reasoning in geographic information systems*. Oxford: Oxford University Press.

Fitting, M. C. (2004). First–order intensional logic. *Annals of Pure and Applied Logic*, *127*, 171–193.

Forrest, P. (2010) The identity of indiscernibles. In E. N. Zalta (Ed.), *The Stanford encyclopedia of philosophy* (Winter 2012 ed.). http://plato.stanford.edu/archives/win2012/entries/identity-indiscernible/.

Frank, A. U., & Campari, I. (Eds.). *Spatial information theory: A theoretical basis for GIS* (Lecture notes in computer science, Vol. 716). Berlin: Springer.

Frank, A. U., & Kuhn, W. (Eds.). *Spatial information theory: A theoretical basis for GIS* (Lecture notes in computer science, Vol. 988). Berlin: Springer.

Fredslund, J., & Matarić, M. (2002). A general algorithm for robot formation using local sensing and minimal communication. *IEEE Transactions on Robotics and Automation*, *18*(5), 837–846.

Frege, G. (1903). *Grundgesetze der Arithmetic. II.* Jena: Verlag Hermann Pohle.

Fukuda, T., & Nakagawa, S. (1987). A dynamically reconfigurable robotic system (concept of a system and optimal configurations). In *International conference on industrial electronics, control, and instrumentation (IECON'87)*, Cambridge (pp. 588–595).

Gallin, D. (1975). *Intensional and higher–order modal logic*. Amsterdam: North Holland.

Glasgow, J. (1995). A formalism for model–based spatial planning. In: A. U. Frank & W. Kuhn (Eds.), *Spatial information theory – A theoretical basis for GIS* (Lecture notes in computer science, Vol. 988, pp. 501–518). Berlin: Springer.

Gödel, K. (1951). *The consistency of the axiom of choice and of the generalized continuum hypothesis with the axioms of set theory* (Annals of mathematics studies). Princeton: Princeton University Press.

Gotts, N. M., & Cohn, A. G. (1995). A mereological approach to representing spatial vagueness. In *Working papers. The ninth international workshop on qualitative reasoning, QR'95*, Amsterdam.

Gotts, N. M., Gooday, J. M., & Cohn, A. G. (1996). A connection based approach to commonsense topological description and reasoning. *The Monist*, *79*(1), 51–75.

Grossberg, S. (1973). Contour enhancement, short–term memory, and constancies in reverberating neural networks. *Studies in Applied Mathematics*, *52*, 213–257.

Grzymala–Busse, J. W. (1992). LERS – A system for learning from examples based on rough sets. In R. Słowiński (Ed.), *Intelligent decision support: Handbook of advances and applications of the rough sets theory* (pp. 3–18). Dordrecht: Kluwer.

Grzymala–Busse, J. W., & Hu, M. (2000). A comparison of several approaches to missing attribute values in data mining. In *Lecture notes in artificial intelligence (LNAI)* (Vol. 2005, pp. 378–385). Berlin: Springer.

Guarino, N. (1994). The ontological level. In R. Casati, B. Smith, G. White (Eds.), *Philosophy and the cognitive sciences*. Vienna: Hoelder-Pichler-Tempsky.

Hájek, P. (1998). *Metamathematics of fuzzy logic*. Dordrecht: Kluwer.

Hilpinen, R. (1995). Belief systems as artifacts. *The Monist*, *78*, 136–155.

Hirtle, S. C., & Frank, A. U. (Eds.). Spatial information theory: A theoretical basis for GIS (Lecture notes in computer science, Vol. 1329). Berlin: Springer.

Nguyen, S. H. (2000). Regularity analysis and its applications in data mining. In: L. Polkowski, S. Tsumoto, & T. Y. Lin (Eds.), *Rough set methods and applications: New developments in knowledge discovery in information systems* (pp. 289–378). Heidelberg: Physica Verlag.

Hughes, G. E., & Creswell, M. J. (1996). A new introduction to modal logic. London: Routledge.

Huntsberger, T., Stroupe, A., Aghazarian, H., Garrett, M., Younse, P., & Powell, M. (2007). TRESSA: Teamed robots for exploration and science of steep areas. *Journal of Field Robotics, 24*(11/12), 1015–1031.

Iwanuś, B. (1973). On Leśniewski's elementary ontology. *Studia Logica, XXXI*, 73–119.

Jennings, J. S., Whelan, G., & Evans, W. F. (2001). Cooperative search and rescue with a team of mobile robots. In *SPIE* (Vol. 4364). Orlando: Aerosense.

Khatib, O. (1986). Real–time obstacle avoidance for manipulators and mobile robots. *International Journal of Robotics Research, 5*, 90–98.

Kim, K.-Y., Yang, H., & Kim D.-W. (2008). Mereotopological assembly joint information representation for collaborativbe product design. *Robotics and Computer-Integrated Manufacturing, 24*(6), 744–754.

Kube, C. R., & Zhang, H. (1996). The use of perceptual cues in multi–robot box–pushing. In *Proceedings IEEE international conference on robotics and automation (ICRA'96)*, Minneapolis (pp. 2085–2090).

Kuipers, B. J. (1994). *Qualitative reasoning: Modeling and simulation with incomplete knowledge*. Cambridge: MIT.

Leibniz, G. W. (1969). Discourse on metaphysics. In L. Loemker, G. W. Leibniz (Eds.), *Philosophical papers and letters* (2nd ed.). Dordrecht: D. Reidel.

Lejewski, C. Z. (1958). On Leśniewski's ontology. *Ratio, I*(2), 150–176.

Leonard, H. S., & Goodman, N. (1940). The calculus of individuals and its uses. *Journal of Symbolic Logic, 5*, 45–55.

Leśniewski, S. (1916). Podstawy Ogólnej Teoryi Mnogości, I (Foundations of General Set Theory, I, in Polish). Prace Polskiego Koła Naukowego w Moskwie, Sekcya Matematyczno–przyrodnicza, No. 2, Moscow.

Leśniewski, S. (1927). O podstawach matematyki (On foundations of mathematics, in Polish). *Przegląd Filozoficzny, XXX*, 164–206; *Przegląd Filozoficzny, XXXI*, 261–291 (1928); *Przegląd Filozoficzny, XXXII*, 60–101 (1929); *Przegląd Filozoficzny, XXXIII*, 77–105 (1930); *Przegląd Filozoficzny, XXXIV*, 142–170 (1931).

Leśniewski, S. (1930). Über die Grundlagen der Ontologie. C.R. Soc. Sci. Lettr. Varsovie Cl. III, 23 Anneé, 111–132.

Leśniewski, S. (1982). On the foundations of mathematics. *Topoi, 2*, 7–52.

Ling, C.-H. (1965). Representation of associative functions. *Publicationes Mathematicae Debrecen, 12*, 189–212.

Łukasiewicz, J. (1939). On Aristotle's syllogistic. *CR Acad. Polon. Lettr. Cracoviae, 44*, 220–226.

Łukasiewicz, J. (1957). *Aristotle's syllogistic from the standpoint of modern formal logic* (2nd ed.). Oxford: Oxford University Press.

Łukasiewicz, J. (1970). Die Logischen Grundlagen der Warcheinlichtkeitsrechnung. Cracow (1913); cf. [engl. trans]: L. Borkowski (Ed.), *Jan Łukasiewicz: Selected works* (pp. 16–63). Amsterdam/Warsaw: North Holland/Polish Scientific Publishers.

Margolis, E., & Laurence, S. (Eds.). (2007). *Creations of the mind: Theories of artifacts and their representation*. Oxford/New York: Oxford University Press.

Matarić, M. (1993). Kin recognition, similarity and group behavior. In: *Proceedings of 15th annual cognitive society conference*, Boulder (pp. 705–710). Hillsdale: Lawrence Erlbaum Associates.

Matarić, M. (1994). Interaction and intelligent behavior. PhD dissertation. MIT EECS Department.

Matarić, M. (1997). Behavior–based control: Examples from navigation, learning, and group behavior. *Journal of Experimental and Theoretical Artificial Intelligence, 9*(2,3), 323–336.

McCulloch, W., & Pitts, W. (1943). A logical calculus of the ideas immanent in nervous activity. *Bulletin of Mathematical Biophysics, 7*, 115–133.

Michalski, R. (1990). Pattern recognition as rule–guided inductive inference. *IEEE Transactions on Pattern Analysis and Machine Intelligence PAMI, 2*(4), 349–361.

Miéville, D. (1984). *Un Developpement des Systemes Logiques de Stanislas Leśniewski. Prototétique-Ontologie-Méréologie*. Bern/Frankfurt: Peter Lang.
Ośmiałowski, P. (2009a). On path planning for mobile robots: Introducing the mereological potential field method in the framework of mereological spatial reasoning. *Journal of Automation, Mobile Robotics and Intelligent Systems (JAMRIS)*, *3*(2), 24–33.
Ośmiałowski, P. (2009b). A case of planning and programming of a concurrent behavior: Planning and navigating with formations of robots. In *Proceedings of CSP 2009. Concurrency, specification, programming*, Kraków. Warsaw University Press.
Ośmiałowski, P. (2011). *Planning and navigation for mobile autonomous robots*. Warszawa: PJIIT Publishers.
Ośmiałowski, P., & Polkowski, L. (2009). *Spatial reasoning based on rough mereology: Path planning problem for autonomous mobile robots. Transactions on rough sets XII* (Lecture notes in computer science, Vol. 6190, pp. 143–169). Berlin: Springer.
Parker, L. E. (1997). Cooperative motion control for multi–target observations. In *IEEE/JRS international conference on intelligent robots and systems (JRS IROS'97)*, Grenoble (pp. 1591–1598).
Parker, L. E. (1998). ALLIANCE: An architecture for fault tolerant multirobot cooperation. *IEEE Transactions on Robotics and Automation*, *14*(2), 220–240.
Pawlak, Z. (1991). *Rough sets: Theoretical aspects of reasoning about data*. Dordrecht: Kluwer.
Pawlak, Z., Skowron, A. (1993). A rough set approach for decision rules generation. In *Proceedings of IJCAI'93 workshop W12*, Chambéry. The Management of Uncertainty in AI, France; and ICS Research Report 23/93. Warsaw University of Technology, Institute of Computer Science.
Plutarch. (1936). The E at Delphi. In *Moralia* (Vol. 5). Cambridge: Harvard University Press.
Polkowski, L. (2002). *Rough sets: Mathematical foundations*. Heidelberg: Physica Verlag/ Springer.
Polkowski, L. (2003). A rough set paradigm for unifying rough set theory and fuzzy set theory. *Fundamenta Informaticae*, *54*, 67–88; and: *Proceedings RSFDGrC03*, Chongqing (Lecture notes in artificial intelligence, Vol. 2639, pp. 70–78). Berlin: Springer.
Polkowski, L. (2004a). Toward rough set foundations. Mereological approach. In *Proceedings RSCTC04*, Uppsala (Lecture notes in artificial intelligence, Vol. 3066, pp. 8–25). Berlin: Springer.
Polkowski, L. (2004b). A note on 3–valued rough logic accepting decision rules. *Fundamenta Informaticae*, *61*, 37–45.
Polkowski, L. (2005a). Formal granular calculi based on rough inclusions. In *Proceedings of IEEE 2005 conference on granular computing GrC05*, Beijing (pp. 57–62). IEEE.
Polkowski, L. (2005b). Rough–fuzzy–neurocomputing based on rough mereological calculus of granules. *International Journal of Hybrid Intelligent Systems*, *2*, 91–108.
Polkowski, L. (2007). Granulation of knowledge in decision systems: The approach based on rough inclusions. The method and its applications. In *Proceedings RSEISP 07*, Warsaw, June 2007 (Lecture notes in artificial intelligence, Vol. 4585, pp. 271–279). Berlin: Springer.
Polkowski, L. (2008). A unified approach to granulation of knowledge and granular computing based on rough mereology: A survey. In W. Pedrycz, A. Skowron, & V. Kreinovich (Eds.), *Handbook of granular computing* (pp. 375–400). Chichester: Wiley.
Polkowski, L. (2009a). Granulation of knowledge: Similarity based approach in information and decision systems. In R. A. Meyers (Ed.), *Springer encyclopedia of complexity and system sciences*. Berlin: Springer. Article 00 788.
Polkowski, L. (2009b). Data-mining and knowledge discovery: Case based reasoning, nearest neighbor and rough sets. In R. A. Meyers (Ed.), *Encyclopedia of complexity and system sciences*. Berlin: Springer. Article 00 391.
Polkowski, L. (2011). *Approximate reasoning by parts. An introduction to rough mereology*. Berlin: Springer.
Polkowski, L., & Artiemjew, P. (2007). On granular rough computing with missing values. In *Lecture notes in artificial intelligence (LNAI)* (Vol. 4585, pp. 271–279). Berlin: Springer.

Polkowski, L., & Ośmiałowski, P. (2008). A framework for multi–agent mobile robotics: Spatial reasoning based on rough mereology in player/stage system. In *Lecture notes in artificial intelligence (LNAI)* (Vol. 5306, pp. 142–149). Berlin: Springer.

Polkowski, L., & Ośmiałowski, P. (2010). Navigation for mobile autonomous robots and their formations: An application of spatial reasoning induced from rough mereological geometry. In A. Barrera (Ed.), *Mobile robots navigation* (pp. 329–354). Zagreb: InTech.

Polkowski, L., & Semeniuk–Polkowska, M. (2005). On rough set logics based on similarity relations. *Fundamenta Informaticae*, *64*, 379–390.

Polkowski, L., & Semeniuk–Polkowska, M. (2010). Boundaries, borders, fences, hedges. In *Proceedings CS&P 2010 (concurrency, specification and programming)*, Humboldt University, Berlin.

Polkowski, L., & Skowron, A. (1994). Rough mereology. In *Proceedings of ISMIS'94* (Lecture notes in artificial intelligence, Vol. 869, pp. 85–94). Berlin: Springer.

Polkowski, L., & Skowron, A. (1997). Rough mereology: A new paradigm for approximate reasoning. *International Journal of Approximate Reasoning*, *15*(4), 333–365.

Polkowski, L., & Skowron, A. (1998). Rough mereological foundations for design, analysis, synthesis and control in distributed systems. *An International Journal of Information Sciences*, *104*(1–2), 129–156.

Polkowski, L., & Skowron, A. (1999a). Grammar systems for distributed synthesis of approximate solutions extracted from experience. In Gh. Paun & A. Salomaa (Eds.), *Grammatical models of multi-agent systems* (pp. 316–333). Amsterdam: Gordon and Breach.

Polkowski, L., & Skowron, A. (1999b). Towards adaptive calculus of granules. In L. A. Zadeh & J. Kacprzyk (Eds.), *Computing with words in information/intelligent systems 1. Foundations*. Heidelberg: Physica Verlag.

Polkowski, L., & Skowron, A. (2001). Rough mereological calculi of granules: A rough set approach to computation. *An International Journal of Computational Intelligence*, *17*(3), 472–492.

Ramón y Cajal, S. (1889). Sur la morphologie et les connexions des elements de la retine des oiseaux. *Anatomisches Anzeiger*, *4*, 111–121.

Randell, D., Cui, Z., & Cohn, A. G. (1992). A spatial logic based on regions and connection. In *Proceedings of the 3rd international conference on principles of knowledge representation and reasoning KR'92*, Cambridge (pp. 165–176). San Mateo: Morgan Kaufmann.

Reynolds, C. (1987). Flocks, herds and schools. A distributed behavioral model. *Computer Graphics*, *21*(4), 25–34.

Rosenblatt, F. (1958). The perceptron: A probabilistic model for information storage and organization in the brain. *Psychological Review*, *65*(6), 386–408.

RSES: Rough Set Exploration System. (2012). A system for data analysis; available at http://logic.mimuw.edu.pl/~rses/; last entered 25. 03. 2014.

Russell, S. J., & Norvig, P. (2009). *Artificial intelligence: Modern approach* (3rd ed.). Upper Saddle River: Prentice Hall.

Salustri, F. A. (2002). Mereotopology for product modelling. A new framework for product modelling based on logic. *Journal of Design Research*, *2*, 22.

Seibt, J. (2009). Forms of emergent interaction in general process theory. *Synthese*, *1666*, 479–512.

SEP (Stanford Encyclopedia of Philosophy). (2012). Artifact; available http://plato.stanford.edu/entries/artifact; last entered 25. 03. 2014.

Shreider, Yu. (1960). *Equality, resemblance, order*. Moscow: Mir Publishers.

Simons, P. (2003). *Parts. A sudy in ontology*. Oxford: Clarendon.

Skowron, A. (1993). Boolean reasoning for decision rules generation. In *Proceedings of ISMIS'93*, Trondheim (Lecture notes in artificial intelligence, Vol. 689, pp. 295–305). Berlin: Springer.

Skowron, A., Rauszer C. (1992). The discernibility matrices and functions in decision systems. In R. Słowiński (Ed.) *Intelligent decision support. Handbook of applications and advances of the Rough sets theory* (pp. 311–362). Dordrecht: Kluwer.

Słupecki, J. (1949–1950). On Aristotle's syllogistic. *Studia philosophica (Poznań)*, *4*, 275–300.
Słupecki, J. (1955). S. Leśniewski's calculus of names. *Studia Logica*, *III*, 7–72.
Smith, B. (1989). Logic and formal ontology. In J. N. Mohanty, & W. McKenna, (Eds.), *Husserl's phenomenology: A textbook* (pp. 29–67). Lanham: University Press of America.
Smith, B. (1996). Mereotopology: A theory of parts and boundaries. *Data and Knowledge Engineering, 20*, 287–303.
Smith, B., & Varzi, A. C. (1997). Fiat and bona fide boundaries. *Philosophy and Phenomenological Research*, *60*(2), 401–420.
Sobociński, B. (1949). L'analyse de l'antinomie Russellienne par Leśniewski. *Methodos*, *I*, 94–107, 220–228, 308–316; *Methodos*, *II*, 237–257 (1950).
Sobociński, B. (1954–1955). Studies in Leśniewski's mereology. Yearbook for 1954–55 of the polish society of art and sciences abroad, Vol. V (1955), London (pp. 34–43).
Srzednicki, J., Surma, S. J., Barnett, D., & Rickey, V. F. (Eds.). (1992). *Collected works of Stanisław Leśniewski*. Dordrecht: Kluwer.
Stone, P. (2000). *Layered learning in multiagent systems: A winning approach to robotic soccer*. Cambridge: MIT.
Tarski, A. (1929). Les fondements de la géométrie des corps. *Supplement to Annales de la Société Polonaise de Mathématique*, *7*, 29–33.
Tarski, A. (1935). Zur Grundlegung der Booleschen Algebra. I. *Fundamenta Mathematicae*, *24*, 177–198.
Tarski, A. (1937). Appendix E. In J. H. Woodger (Ed.), *The axiomatic method in biology* (p. 160). Cambridge: Cambridge University Press.
Tarski, A. (1959). What is elementary geometry? In L. Henkin, P. Suppes, & A. Tarski (Eds.), *The axiomatic method with special reference to geometry and physics* (pp. 16–29). Amsterdam: North-Holland.
Turing, A. M. (1948). *Intelligent machinery. A report*. National Physical Laboratory, Mathematical Division.
UC Irvine Repository. (2012). Available at http://archive.uci.edu/ml/datasets; last entered 25. 03. 2014.
Uny, C. Y., Fukunaga, A. S., & Kahng, A. B. (1997). Cooperative mobile robotics: Antecedents and directions. *Autonomous Robots*, *4*, 7–27.
Van Benthem, J. (1983). *The logic of time*. Dordrecht: Reidel.
Van Benthem, J. (1988). *A manual of intensional logic*. Stanford: CSLI Stanford University.
Varzi, A. C. (2008). The extensionality of parthood and composition. *The Philosophical Quarterly*, *58*, 108–133.
Werbos, P. J. (1994). The Roots of backpropagation: From ordered derivatives to neural networks and political forecasting. New York: Wiley.
Whitehead, A. N. (1916). La théorie relationniste de l'espace. *Revue de Métaphysique et de Morale*, *23*, 423–454.
Whitehead, A. N. (1919). *An enquiry concerning the principles of human knowledge*. Cambridge: Cambridge University Press.
Whitehead, A. N. (1920). *The concept of nature*. Cambridge: Cambridge University Press.
Whitehead, A. N. (1929). *Process and reality: An essay in cosmology*. New York: Macmillan.
Woodger, J. H. (1937). *The axiomatic method in biology*. Cambridge: Cambridge University Press.
Woodger, J. H. (1939). The technique of theory construction. In R. Carnap et al. (Eds.), *International encyclopedia of unified science, II* (Vol. 5, pp. 1–81). Chicago: Chicago University Press.
Wróblewski, J. (2004). Adaptive aspects of combining approximation spaces. In: S. K. Pal, L. Polkowski, & A. Skowron (Eds.), *Rough neural computing: Techniques for computing with words* (pp. 139–156). Berlin: Springer
Zadeh, L. A. (1965). Fuzzy sets. *Information and Control*, *8*, 338–353.

Zadeh, L. A. (1979). Fuzzy sets and information granularity. In M. Gupta, R. Ragade, & R. R. Yager (Eds.), *Advances in fuzzy set theory and applications*, (pp. 3–18). Amsterdam: North–Holland.
Zadeh, L. A. (2004). Toward a unified theory of uncertainty. In *Proceedings of IPMU 2004*, Perugia (Vol. 1, pp. 3–4).
Zermelo, E. (1908). Untersuchungen über die Grundlgen der Mengenlehre I. *Mathematische Annalen*, *65*, 261–281.

Chapter 11
Discrete Mereotopology

Antony Galton

11.1 From Mereology to Mereotopology

Mereology, as the theory of parts and wholes, leads to a set of five jointly exhaustive and pairwise disjoint (JEPD) relations that may hold between any pair of entities X and Y that come under its purview, namely

X is a proper part of Y	$\mathsf{PP}(x, y)$
X coincides with Y	$\mathsf{EQ}(x, y)$
X partially overlaps Y	$\mathsf{PO}(x, y)$
X contains Y as a proper part	$\mathsf{PPI}(x, y)$
X is disjoint from Y	$\mathsf{DR}(x, y)$

For the logical development, we first stipulate that the primitive relation of parthood (P) is reflexive and transitive[1]:

$$\mathsf{P}(x, x)\ , \tag{A1}$$

$$\mathsf{P}(x, y) \wedge \mathsf{P}(y, z) \rightarrow \mathsf{P}(x, z)\ . \tag{A2}$$

[1] We adopt the usual convention in presenting first-order theories that free variables in formulae presented as axioms or theorems are understood to be universally quantified, so that, e.g., $\mathsf{P}(x, y) \wedge \mathsf{P}(y, z) \rightarrow \mathsf{P}(x, z)$ is to be read as if it were written $\forall x \forall y \forall z(\mathsf{P}(x, y) \wedge \mathsf{P}(y, z) \rightarrow \mathsf{P}(x, z))$.

A. Galton (✉)
University of Exeter, Exeter, UK
e-mail: apgalton@ex.ac.uk

C. Calosi and P. Graziani (eds.), *Mereology and the Sciences*, Synthese Library 371, DOI 10.1007/978-3-319-05356-1_11,

We then define overlap as possession of a common part:

$$\mathsf{O}(x, y) =_{\text{def}} \exists z(\mathsf{P}(z, x) \wedge \mathsf{P}(z, y)) \tag{D1}$$

and go on to define the five relations listed above as follows:

$$\mathsf{PP}(x, y) =_{\text{def}} \mathsf{P}(x, y) \wedge \neg\mathsf{P}(y, x) \ , \tag{D2}$$

$$\mathsf{EQ}(x, y) =_{\text{def}} \mathsf{P}(x, y) \wedge \mathsf{P}(y, x) \ , \tag{D3}$$

$$\mathsf{PO}(x, y) =_{\text{def}} \mathsf{O}(x, y) \wedge \neg\mathsf{P}(x, y) \wedge \neg\mathsf{P}(y, x) \ , \tag{D4}$$

$$\mathsf{PPI}(x, y) =_{\text{def}} \mathsf{PP}(y, x) \ , \tag{D5}$$

$$\mathsf{DR}(x, y) =_{\text{def}} \neg\mathsf{O}(x, y) \ . \tag{D6}$$

This system of relations is known in the Qualitative Spatial Reasoning (QSR) community as RCC5, the five-element Region Connection Calculus.[2]

If the terms of the formal language are interpreted as referring to *spatial* entities, which we here call *regions*, it is generally felt that mereology alone does not provide sufficient expressive power to be useful for QSR. In addition to parthood and the relations derived from it, we need also to be able to distinguish between, on the one hand, *internal* and *peripheral* parts, and on the other, between *contact* and *separation*. To express these, a primitive relation C (for contact, or *connection*) is introduced, and stipulated to be reflexive and symmetric:

$$\mathsf{C}(x, x) \ , \tag{A3}$$

$$\mathsf{C}(x, y) \rightarrow \mathsf{C}(y, x) \ . \tag{A4}$$

The minimal relationship between P and C is that anything connected to an entity is automatically connected to anything that entity is part of:

$$\mathsf{P}(x, y) \rightarrow \forall z(\mathsf{C}(z, x) \rightarrow \mathsf{C}(z, y)) \ . \tag{A5}$$

Using P and C as primitives, we can now define a number of important additional relations as follows:

- X is *disconnected* from Y:

$$\mathsf{DC}(x, y) =_{\text{def}} \neg\mathsf{C}(x, y) \ . \tag{D7}$$

---

[2]The Region Connection Calculus was introduced, though not under that name, in Randell et al. (1992). The version there presented is RCC8, described below; explicit recognition of RCC5 in QSR came later. Strictly speaking, this set of mereological relations should only be called a connection calculus if they are defined in terms of connection rather than, as here, in terms of parthood.

- X is *externally connected* to Y:

$$\mathsf{EC}(x,y) =_{\text{def}} \mathsf{C}(x,y) \wedge \neg\mathsf{O}(x,y)\ . \tag{D8}$$

- X is a *tangential part* of Y:

$$\mathsf{TP}(x,y) =_{\text{def}} \mathsf{P}(x,y) \wedge \exists z(\mathsf{C}(x,z) \wedge \neg\mathsf{O}(z,y))\ . \tag{D9}$$

  (i.e., X is a part of Y that is connected to something disjoint from Y).
- X is a *non-tangential part* of Y:

$$\mathsf{NTP}(x,y) =_{\text{def}} \mathsf{P}(x,y) \wedge \forall z(\mathsf{C}(x,z) \rightarrow \mathsf{O}(z,y))\ . \tag{D10}$$

  (i.e., X is a part of Y that is only connected to things that overlap Y).

Note that any part of a region must be either a tangential part or a non-tangential part of it, but not both. In particular, a region is a non-tangential part of itself if and only if it is not connected to any region disjoint from it and is therefore a union of one or more connected components of the whole space.

The system of eight JEPD relations known as RCC8 comprises DC, EC, PO, EQ, TPP (defined as the conjunction of PP and TP), NTPP (the conjunction of PP and NTP), and the inverses of TPP and NTPP.

The logical language here is denoted $\mathscr{L}_{\mathsf{P},\mathsf{C}}$, and comprises all first-order formulae in which the non-logical language is restricted to the two binary predicates P and C—all formulae containing the other RCC8 relations being reducible to formulae containing just P and C, via the definitions given above. Systems of this kind, which combine the mereological notion of parthood with the topological notion of connection, are called *mereotopologies*.

Mereotopologies are normally interpreted as referring to regions which can be indefinitely subdivided. This is expressed by positing the formula

$$\exists y \mathsf{PP}(y,x) \tag{N1}$$

as an axiom. The domain of such an interpretation is usually taken to be some collection of non-empty subsets of $\mathbb{R}^n$ for some positive integer n (typically either 2 or 3). In order to ensure infinite subdivisibility, only infinite subsets should be considered as possible domain elements, but this still leaves open many different possible such collections, for example

- All infinite subsets of $\mathbb{R}^n$
- All non-empty open subsets of $\mathbb{R}^n$
- All non-empty regular open[3] subsets of $\mathbb{R}^n$

[3] A regular open set is a set that is equal to the interior of its closure; a regular closed set is equal to the closure of its interior.

- All non-empty regular closed subsets of $\mathbb{R}^n$ (Gotts 1996)
- All open polygonal (polyhedral, etc.) subsets of $\mathbb{R}^n$ (Pratt and Lemon 1997; Pratt and Schoop 1997; Pratt-Hartmann and Schoop 2002)

In these interpretations, it is usual to interpret the predicates P and C as standing for the following relations:

- X is part of Y if and only if $X \subseteq Y$.
- X is connected to Y if and only if $\overline{X} \cap \overline{Y} \neq \emptyset$, i.e., the topological closures of X and Y have at least one point in common.

The interpretation of parthood in terms of the subset relation explains why the domain has to be restricted to non-empty sets: if the empty set were allowed, then any pair of regions would overlap, since they would have the empty set as a common part.

It should be noted that under any of these interpretations, parthood is necessarily *antisymmetric*, satisfying the formula

$$\mathsf{P}(\mathsf{x},\mathsf{y}) \wedge \mathsf{P}(\mathsf{y},\mathsf{x}) \rightarrow \mathsf{x} = \mathsf{y}\ , \tag{A6}$$

and thus *extensional*, meaning that two distinct entities cannot have exactly the same parts:

$$\forall \mathsf{z}(\mathsf{P}(\mathsf{z},\mathsf{x}) \leftrightarrow \mathsf{P}(\mathsf{z},\mathsf{y})) \rightarrow \mathsf{x} = \mathsf{y}\ . \tag{T1}$$

From now on we shall assume that P denotes an antisymmetric relation; a consequence of this is that $\mathsf{EQ}(\mathsf{x},\mathsf{y})$ becomes equivalent to $\mathsf{x} = \mathsf{y}$, meaning that the symbol EQ can be dropped.

Connection, on the other hand, need not be extensional: that is, it does not necessarily follow that two entities are identical if they are connected to exactly the same things. In the first two models above, connection is not extensional; for example, if the domain of discourse consists of all infinite subsets of $\mathbb{R}^n$, an open set and its closure are connected to exactly the same sets, yet they are not identical. In the last three models listed above, connection *is* extensional, that is, they satisfy

$$\forall \mathsf{z}(\mathsf{C}(\mathsf{x},\mathsf{z}) \leftrightarrow \mathsf{C}(\mathsf{y},\mathsf{z})) \rightarrow \mathsf{x} = \mathsf{y}\ . \tag{N2}$$

In such models, if X is connected to everything Y is connected to, then X is part of Y, which means that the converse of (A5) holds. In this case, parthood can be characterised exactly in terms of connection, as follows:

$$\mathsf{P}(\mathsf{x},\mathsf{y}) \leftrightarrow \forall \mathsf{z}(\mathsf{C}(\mathsf{x},\mathsf{z}) \rightarrow \mathsf{C}(\mathsf{y},\mathsf{z}))\ . \tag{N3}$$

It is easy to see that (N2) and (N3) together imply (A6) and hence (T1). If we have (N3), we can use it to *define* P, leaving just the one primitive predicate C. In this case the logical language can be reduced to $\mathscr{L}_C$.

It should be noted that although individual terms in $\mathscr{L}_{\mathsf{P,C}}$ refer to regions, by interpreting them as denoting subsets of $\mathbb{R}^n$ we are implicitly postulating a universe of points, even though these cannot be referred to in the language. To avoid this unsatisfactory situation, Stell (2000a) showed that the terms of RCC could be interpreted as referring to elements of structures called Boolean Connection Algebras (BCAs), which do not presuppose that regions are collections of points. This notion was generalised by Li and Ying (2004) to Generalized Boolean Connection Algebras, which as well as subsuming Stell's BCAs can provide models for the discrete versions of RCC which we turn to next.

11.2 Discrete Mereotopology and Adjacency Spaces

If we wish to interpret the mereotopological predicates over *discrete* domains, in which entities are *not* indefinitely subdivisible, it is no longer possible to define parthood in terms of connection. In a discrete domain, every entity decomposes into *atoms*, which have no proper parts. We define

$$\mathsf{Atom}(\mathsf{x}) =_{\text{def}} \neg\exists \mathsf{y}\mathsf{PP}(\mathsf{y},\mathsf{x}) \tag{D11}$$

Then it can be seen that (N1) is equivalent to $\neg\exists\mathsf{x}\mathsf{Atom}(\mathsf{x})$.

If A is an atom which is connected to its complement A^c, then anything connected to A must either be A itself or overlap A^c, and hence in either case must be connected to A^c; but on the other hand A is obviously *not* part of its complement,[4] so we have

$$\forall z(\mathsf{C}(\mathsf{a},\mathsf{z}) \rightarrow \mathsf{C}(\mathsf{b},\mathsf{z})) \wedge \neg\mathsf{P}(\mathsf{a},\mathsf{b})$$

(where a and b denote A and A^c respectively), contradicting (N3).

For *discrete* mereotopology, then, both P and C are needed as primitive predicates in the logical language. How should they be interpreted? The subsets of $\mathbb{R}^n$ listed above are no longer appropriate, and an obvious substitute here would be to use subsets of $\mathbb{Z}^n$, i.e., sets of points with integer coordinates. How should C be interpreted in this case? Since overlap is a form of connection,[5] we need only concern ourselves with the interpretation of non-overlapping connection, i.e., EC.

An example is illustrated in Fig. 11.1, where the atomic regions are shown as unit squares, which can be mapped in the obvious way to elements of $\mathbb{Z}^2$. The external

[4] At least, this is obvious so long as "part" and "complement" are understood in the usual sense; however, Roy and Stell (2002) showed that by replacing the ordinary set-theoretical complement operation by a weaker operation, the dual pseudo-complement, defined over a class of structures called dual p-algebras, one obtains a model of discrete space in which (N3) holds.

[5] It follows from (A5) that $\forall\mathsf{x}\forall\mathsf{y}(\mathsf{O}(\mathsf{x},\mathsf{y}) \rightarrow \mathsf{C}(\mathsf{x},\mathsf{y}))$.

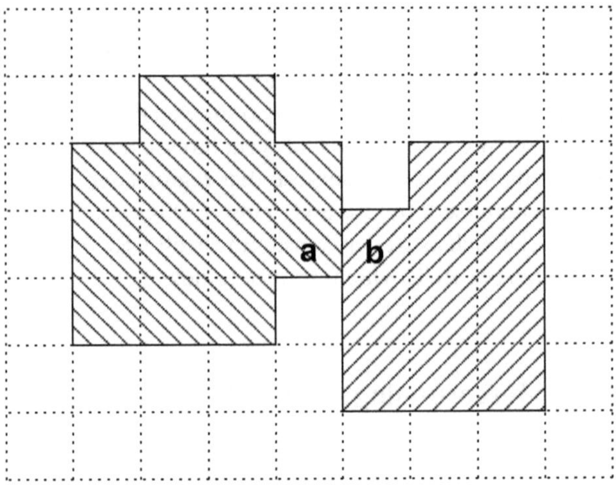

Fig. 11.1 External connection in discrete space

connection between the two differently shaded regions depends on the fact that the squares labelled 'a' and 'b', one from each region, are *adjacent* to each other, and it is this notion of adjacency which forms the basis for a general way of interpreting the connection predicate in discrete mereotopologies. We do not confine ourselves to subsets of $\mathbb{Z}^n$ but rather to regions defined over a more general class which we define as follows:

Definition. An *adjacency space* is a non-empty set U of entities called *cells* together with a reflexive, symmetric relation $\sim \subseteq U \times U$, called *adjacency*.

An adjacency space can be regarded as a graph; but this does not mean that the theory of adjacency spaces is identical to graph theory. An important difference emerges when we consider substructures. A graph is specified by a set V of vertices and a set E of edges, where each edge joins an element of V to an element of V. A subgraph is specified by a subset $V' \subseteq V$ of the vertices and a subset $E' \subseteq E$ of the edges, with the proviso that each edge in E' joins an element of V' to an element of V'. There is no requirement that an edge in E which happens to join an element of V' to an element of V' must be in E'. Thus there can be many different subgraphs of (V, E) all of which have the vertex-set V'. In adjacency spaces, the adjacency or otherwise of two cells is fixed by the *space* and is automatically inherited by the substructures. Thus a substructure of an adjacency space can be specified by giving its cells alone, without reference to adjacency.

These substructures, which may be thought of as aggregates of cells, are called *regions*, and it is these entities that discrete mereotopology is primarily concerned with, not the cells themselves. A cell might, indeed, be considered to be the aggregate which is composed of precisely that cell and nothing else; but for theoretical purposes it is convenient to specify a region in terms of the (non-empty) *set* of cells which make it up, and in that case a one-cell region is conceptually distinct from its only cell, the former being, in fact, the singleton set of the latter. Therefore an interpretation I of the logical language $\mathscr{L}_{\mathsf{P,C}}$ over an adjacency space $(U, \sim)$ is specified as follows:

- Each individual term t of the logical language denotes a non-empty subset $t^I \subseteq U$.
- A formula $\mathsf{P}(t_1, t_2)$ is interpreted to mean that $t_1^I \subseteq t_2^I$.
- A formula $\mathsf{C}(t_1, t_2)$ is interpreted to mean that there are cells $x \in t_1^I$ and $y \in t_2^I$ such that $x \sim y$.

Thus two regions are regarded as connected so long as some cell in one is adjacent to some cell in the other.[6]

The theory of Discrete Mereotopology (DM), as we shall understand it in this paper, comprises all and only those formulae of the language $\mathscr{L}_{\mathsf{P,C}}$ which are satisfied by every adjacency space under the scheme of interpretation just specified. It is easy to see that DM includes all the formulae thus far introduced with labels beginning with either A or T: the formulae (A1), (A2), (A6), and (T1) characterising parthood, (A3) and (A4) characterising connection, and (A5) relating parthood and connection.[7,8]

DM does *not* include the converse of (A5), meaning that the definitional reduction of P to C given by (N3) is not available here. The other important non-theorem is (N1), which expresses the infinite subdivisibility of regions that is characteristic of non-discrete (dense or continuous) models; instead, DM includes the formula

$$\forall x \exists y (\mathsf{P}(y, x) \wedge \mathsf{Atom}(y)) , \qquad \text{(A7)}$$

which says that every region has an atomic region as part. The predicate Atom is clearly satisfied by just the singleton subsets of the universe U; and *every* subset of

[6] These ideas were presented, without explicit use of the term "adjacency space", in Galton (1999). The term "adjacency space" was used in Galton (2000).

[7] Of course, as written, not all of these are $\mathscr{L}_{\mathsf{P,C}}$ formulae; they become so when the predicates other than P and C are expanded in accordance with their definitions, given by the formulae whose labels begin with D.

[8] Formulae whose labels beginning with T logically follow from those with labels beginning with A; thus the latter can be regarded as *axioms* and the former as *theorems*. However, the distinction is somewhat arbitrary (since there are in principle many different ways of assigning "A" and "T" labels) and only comes into its own when we wish to consider to what extent reasoning about adjacency spaces can be accomplished purely by means of symbolic manipulation of $\mathscr{L}_{\mathsf{P,C}}$ formulae, without reference to any interpretation. In that case it becomes of interest whether or not there is a finitely-specifiable set of $\mathscr{L}_{\mathsf{P,C}}$ formulae whose logical consequences comprise all and only the true formulae of discrete mereotopology—in short, whether this theory can be completely axiomatised.

U has at least one singleton subset; under the interpretation, these two sets count as a region and an atomic part of that region.

It will be convenient to introduce a predicate AP to say that one region is an atomic part of another; this is straightforwardly defined as follows:

$$\mathsf{AP(x,y)} =_{\text{def}} \mathsf{Atom(x)} \wedge \mathsf{P(x,y)}\ , \tag{D12}$$

enabling us to rewrite (A7) as $\forall x \exists y \mathsf{AP}(y, x)$.

The mereotopological relations of atoms are much simpler than those of general regions. In particular, if A overlaps B, where A is an atom, then the common part of A and B cannot be a proper part of A and must therefore be A itself. We thus have

$$\mathsf{Atom(x)} \rightarrow (\mathsf{O(x,y)} \rightarrow \mathsf{P(x,y)}). \tag{T2}$$

An important mereological principle, which forms part of General Extensional Mereology, is the *Strong Supplementation Principle* (Simons 1987, p. 29), which states that any region that is *not* part of a given region must have a part that does not overlap that region:

$$\neg \mathsf{P(y,x)} \rightarrow \exists z(\mathsf{P(z,y)} \wedge \neg \mathsf{O(z,x)}) \tag{A8}$$

In combination with (A7), this leads to a powerful extensionality principle for DM, namely

$$\forall z(\mathsf{AP(z,x)} \leftrightarrow \mathsf{AP(z,y)})) \rightarrow x = y. \tag{T3}$$

Whereas, in order to show that two regions are the same, (T1) requires us check that they agree in *all* their parts, with (T3) it suffices to check that they agree in just their atomic parts.

To see how this follows from (A8), suppose we have

$$\forall z(\mathsf{AP(z,x)} \leftrightarrow \mathsf{AP(z,y)}). \tag{*}$$

We must show that $x = y$. Suppose not; then by (A6), either $\neg \mathsf{P}(x, y)$ or $\neg \mathsf{P}(y, x)$. Without loss of generality we may assume the former. By (A8), this means there is a region u such that

$$\mathsf{P(u,x)} \wedge \neg \mathsf{O(u,y)}\ . \tag{**}$$

By (A7), there is a region v such that $\mathsf{AP}(v, u)$, and by transitivity therefore $\mathsf{AP}(v, x)$. By $(*)$ this means that $\mathsf{AP}(v, y)$. We now have $\mathsf{P}(v, u) \wedge \mathsf{P}(v, y)$, so $\mathsf{O}(u, y)$, which contradicts $(**)$.

It is clear that the Strong Supplementation Principle is satisfied when regions are interpreted as subsets of an adjacency space, so both (A8) and (T3) belong to

DM. The upshot of this is that we can define a region uniquely by characterising its atomic parts. We will use this in the following fashion: if a predicate ϕ is defined by a rule of the form

$$\phi(\mathsf{x}) =_{\text{def}} \forall \mathsf{z}(\mathsf{Atom}(\mathsf{z}) \rightarrow (\mathsf{P}(\mathsf{z},\mathsf{x}) \leftrightarrow \psi(\mathsf{z},\mathsf{x})))$$

then it follows that any two regions with the property ϕ are identical.

The formulae (A1), (A2), (A6), and (A8) together constitute the axiomatic basis for the system designated **EM** (for Extensional Mereology) in Varzi (1996). The addition of (A7) yields Atomic Extensional Mereology **AEM**.

The study of discrete mereotopology can be pursued on two levels, which we may loosely characterise as "set-theoretical" and "logical". At the set-theoretical level, the class of adjacency spaces are treated as mathematical objects in their own right, independently of any particular logical language chosen for describing them. At the logical level, on the other hand, one focusses on the particular first-order language $\mathscr{L}_{\mathsf{P,C}}$, which is the common language in which to express mereotopological theses, regardless of whether discrete or continuous spaces are intended. The set-theoretical level provides a metalanguage within which one can specify interpretations of the logical level. While much of what is said at one level can be transposed easily to the other, it is important to maintain a clear conceptual distinction between them. This is supported here by a typographical distinction: formulae at the logical level are always printed in a sanserif font.

Moving back and forth between the levels we can investigate what formulae of $\mathscr{L}_{\mathsf{P,C}}$ are satisfied in adjacency spaces (these formulae constituting the theory of DM), and conversely which properties of adjacency spaces can be expressed in the language. We can then ask whether the theory of adjacency spaces, insofar as it can be expressed in $\mathscr{L}_{\mathsf{P,C}}$, is axiomatisable, i.e., whether there is a finitely specifiable set of $\mathscr{L}_{\mathsf{P,C}}$ formulae from which all and only the true theorems of DM follow as logical consequences.

In the remainder of this paper we present a few of the most important features of DM, briefly discuss its relation to some other approaches, and describe an area in which it is being applied.

11.3 Examples of Adjacency Spaces

The adjacency space in Fig. 11.1 is underdetermined, in that we did not specify how the relation $\sim$ was to be defined. It was assumed that the reader would naturally understand that the cells labelled 'a' and 'b' were to count as adjacent. In fact there are (at least) two different, and equally natural, ways of understanding adjacency in $\mathbb{Z}^2$. Under *orthogonal adjacency*, only cells which share an edge count as adjacent; thus each cell is adjacent to four cells other than itself. We denote this relation $\sim_4$. *Orthodiagonal adjacency* is where cells count as adjacent so long as they share at

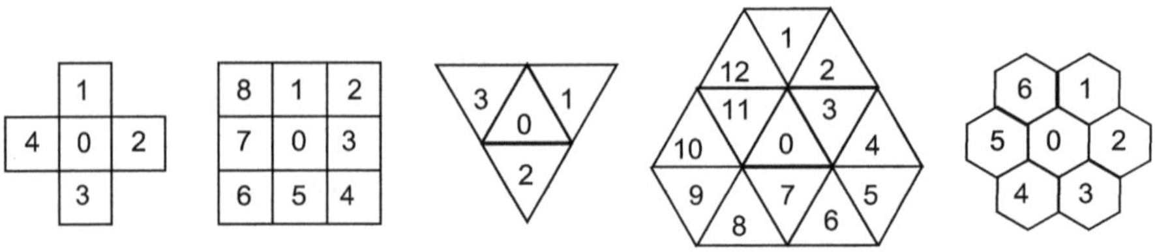

Fig. 11.2 Adjacency relations in regular planar tessellations

least one boundary point—either along an edge or at a corner; each cell is adjacent to eight others, and hence we denote this relation $\sim_8$. These two adjacency relations are defined as follows:

- $(x, y) \sim_4 (x', y')$ iff $|x - x'| + |y - y'| \leq 1$.
- $(x, y) \sim_8 (x', y')$ iff $|x - x'| \leq 1$ and $|y - y'| \leq 1$.

Both of these spaces are *homogeneous* in the sense that all cells "look the same"; more formally, for any $x, y \in U$ there is a bijective adjacency-preserving function from U to U which maps x onto y.

The adjacency spaces $(\mathbb{Z}^2, \sim_4)$ and $(\mathbb{Z}^2, \sim_8)$ have played a prominent part in work on discrete spaces, mainly because they are the most natural spaces within which to model *digital pictures*, as seen, for example, on a computer screen in which the display is produced by assigning colour values to each element in a rectangular array of pixels (see, e.g., Rosenfeld 1979; Kong and Rosenfeld 1989).

Other homogeneous adjacency spaces correspond to tessellations of triangles or hexagons. In the triangular case, there are two possible adjacency relations: $X \sim_3 Y$ if triangles X and Y share an edge, and $X \sim_{12} Y$ if they share at least one boundary point. With the hexagonal lattice there is only one natural adjacency relation, $\sim_6$, which holds between hexagons that share an edge. All these cases are illustrated in Fig. 11.2.

Homogeneous adjacency spaces do not need to be infinite. Familiar examples of finite spaces are provided by the five platonic solids. The faces of a dodecahedron, for example, can be thought of as a 12-element adjacency space, where adjacency is interpreted as edge-sharing between the pentagonal faces.

Beyond these examples, adjacency spaces do not have to be homogeneous. Non-homogeneous tessellations include the triangulated irregular networks (TIN) used in Geographical Information Science. An example is shown in Fig. 11.3a. As with the homogeneous tessellation of squares, two kinds of adjacency can be defined on a TIN: either adjacency along edges only, or adjacency at edges and vertices. The advantage of a hexagonal tessellation is that this ambiguity does not arise, and for this reason we will use such tessellations for our illustrative examples in what follows—though for practical convenience, the hexagonal tessellation will be represented in the form of an isomorphic "staggered squares" grid, as shown in Fig. 11.3b.

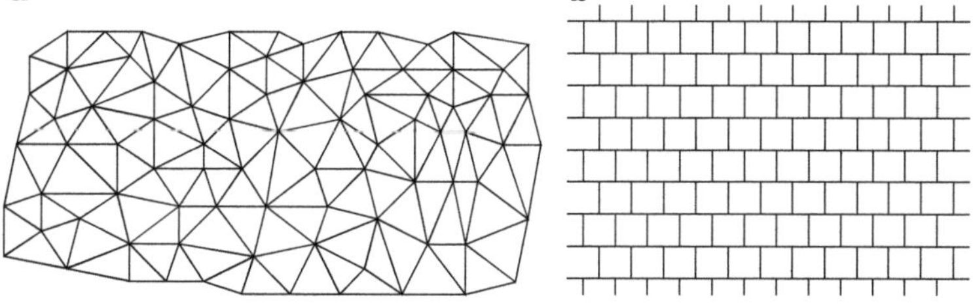

Fig. 11.3 (**a**) A triangular irregular network (TIN). (**b**) A grid of staggered squares, isomorphic to the regular hexagonal tessellation

11.4 Mereotopological Relations on Adjacency Spaces

Given the interpretation of the relations P and C over an adjacency space, the interpretations of all the relations defined in terms of P and C, such as the RCC8 relations, become fixed. Following the standard convention in model theory, we write $(U, \sim) \models R[X, Y]$ to mean that the relation denoted by a predicate R (defined in $\mathscr{L}_{\mathsf{P,C}}$) holds between the elements $X, Y \in U$. Thus for example we have, for regions X, Y:

- $(U, \sim) \models \mathsf{DC}[X, Y]$ iff there are no cells $x \in X$ and $y \in Y$ such that $x \sim y$.
- $(U, \sim) \models \mathsf{EC}[X, Y]$ iff $X \cap Y = \emptyset$ and there are cells $x \in X$ and $y \in Y$ such that $x \sim y$.
- $(U, \sim) \models \mathsf{TP}[X, Y]$ iff $X \subseteq Y$ and there are cells $x \in X$ and $y \notin Y$ such that $x \sim y$.
- $(U, \sim) \models \mathsf{NTP}[X, Y]$ iff for all cells $x \in X$, if $x \sim y$ then $y \in Y$.
- $(U, \sim) \models \mathsf{EQ}[X, Y]$ iff $X = Y$.

Examples of all the RCC8 relations are illustrated in Fig. 11.4, using regions defined on the "staggered squares" grid.

It should be noted that *every* region is a non-tangential part of U since the consequent of the defining condition, $y \in Y$, is always true when $Y = U$.

The set of subsets of U forms a Boolean algebra under the usual set-theoretic operations of union, intersection, and complement, with U itself acting as the top element (generally notated 1 or $\top$) and $\emptyset$ as the bottom element (notated 0 or $\bot$). Since $\emptyset$ is not a region, the regions just fall short of being a Boolean algebra: they form a *quasi-Boolean algebra*. In mereotopology, therefore, the Boolean operations are appropriately restricted so that neither their range nor their domain contains the empty set, as we show below.

The universe U can be characterised in $\mathscr{L}_{\mathsf{P,C}}$ as the region which every other region is part of. This can be expressed by the predicate U, defined by

$$\mathsf{U}(\mathsf{x}) =_{\text{def}} \forall \mathsf{y} \mathsf{P}(\mathsf{y}, \mathsf{x}) \tag{D13}$$

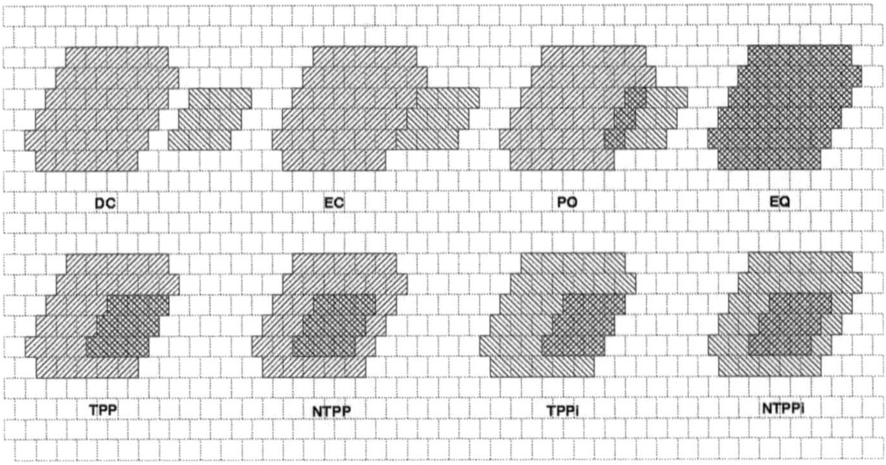

Fig. 11.4 RCC8 relations in an adjacency space

That such a region exists is stated by the formula

$$\exists x \mathsf{U}(x), \tag{A9}$$

which is generally accepted as an axiom in RCC, whether in a discrete or continuous setting. By antisymmetry of P, it straightforwardly follows from this that there can be at most one universal region; in terms of adjacency structures we have

$$(U, \sim) \models \mathsf{U}[X] \text{ if and only if } X = U.$$

We can therefore introduce a constant symbol U to denote the unique universal region, defined contextually as follows:

$$\phi(\mathsf{U}) =_{\text{def}} \forall x(\mathsf{U}(x) \to \phi(x)) , \tag{D14}$$

where ϕ stands for any open formula with one free variable.

Since not all pairs of regions have a Boolean product (intersection), we cannot represent it by a function symbol in $\mathscr{L}_{\mathsf{P,C}}$; instead we define a relational predicate Prod, the intended meaning of $\mathsf{Prod}(x, y, z)$ being that z is the intersection of x and y, defined as follows:

$$\mathsf{Prod}(x, y, z) =_{\text{def}} \forall v(\mathsf{P}(v, z) \leftrightarrow \mathsf{P}(v, x) \wedge \mathsf{P}(v, y)) . \tag{D15}$$

Similarly, the Boolean sum (union) is defined by

$$\mathsf{Sum}(x, y, z) =_{\text{def}} \forall v(\mathsf{O}(v, z) \leftrightarrow \mathsf{O}(v, x) \vee \mathsf{O}(v, y)) , \tag{D16}$$

and the Boolean difference by

$$\mathsf{Diff}(x, y, z) =_{\text{def}} \forall v(\mathsf{P}(v, z) \leftrightarrow \mathsf{P}(v, x) \wedge \neg\mathsf{O}(v, y)) \; . \tag{D17}$$

The existence of regions playing the role of z in these formulae is stated by the following formulae, which also specify the conditions on x and y for such a z to exist:

$$\mathsf{O}(x, y) \rightarrow \exists z \mathsf{Prod}(x, y, z) \; , \tag{A10}$$

$$\exists z \mathsf{Sum}(x, y, z) \; , \tag{A11}$$

$$\neg\mathsf{P}(x, y) \rightarrow \exists z \mathsf{Diff}(x, y, z) \; . \tag{A12}$$

As with (D13), it follows from (D15), (D16), and (D17) that products, sums, and differences, where they exist, are unique. Note that (A12), in conjunction with (A1), logically implies (A8), meaning that the latter could be relegated to the status of a theorem (with a 'T' label) rather than an axiom; however, we shall retain the designation (A8) to avoid confusion.

The complement of a non-universal region can be defined as its difference from U, i.e.,

$$\mathsf{Compl}(x, y) =_{\text{def}} \mathsf{Diff}(U, x, y) \; .$$

Since we always have $\mathsf{P}(v, U)$, this may be expanded as

$$\mathsf{Compl}(x, y) =_{\text{def}} \forall v(\mathsf{P}(v, y) \leftrightarrow \neg\mathsf{O}(v, x)) \; . \tag{D18}$$

It is easy to show that, for non-empty sets $X, Y \subseteq U$, $(U, \sim) \models \mathsf{Compl}[X, Y]$ if and only if $X = Y^c$, thus ensuring that the $\mathscr{L}_{\mathsf{P,C}}$-definable predicate Compl captures the set-theoretic relation of complementation insofar as it applies to regions in adjacency spaces. It does not *immediately* follow from this, of course, that Compl behaves like complementation in arbitrary models of DM, but that this is so is shown by the following theorem:

$$\mathsf{Compl}(x, y) \leftrightarrow \neg\mathsf{O}(x, y) \wedge \mathsf{Sum}(x, y, U) \; , \tag{T4}$$

which says that one region is the complement of another if and only if they are disjoint regions whose sum is the universe. A corollary of this is that complementarity is mutual:

$$\mathsf{Compl}(x, y) \rightarrow \mathsf{Compl}(y, x) \tag{T5}$$

The proofs of these theorems are given in Appendix 1.

With the addition of (A9), (A10), (A11), and (A12) to **EM** we obtain the system of Closed Extensional Mereology designated **CEM** by Varzi (1996). Li and Ying (2004) show that every model of **CEM** is isomorphic to a complete quasi-Boolean algebra, and therefore that **ACEM**, the atomic variant of **CEM** with the additional axiom (A7) is isomorphic to an atomic complete quasi-Boolean algebra,

In standard mereotopology, where there are no atomic regions, and parthood is defined in terms of connection, the definition of these Boolean operations in $\mathscr{L}_{\mathsf{P,C}}$ has proved somewhat problematic, it being difficult to demonstrate that the required interrelationships hold when connection is taken into account. In particular, (D18) does not suffice to captured the desired behaviour and needs to be supplemented by an additional axiom, represented here by the formula

$$\mathsf{Compl}(\mathsf{x},\mathsf{y}) \rightarrow \forall \mathsf{z}(\mathsf{C}(\mathsf{z},\mathsf{y}) \leftrightarrow \neg\mathsf{NTP}(\mathsf{z},\mathsf{x})) \qquad \text{(T6)}$$

In DM, however, it can be proved that (T6) follows from (D18) and the existing axioms, as demonstrated in Appendix 2.

11.5 Quasi-topological Operators

The relation NTP picks out those subregions of a given region which are disconnected from the complement of the region: the neighbours of each cell in the subregion are all in the region itself. The union of all the non-tangential parts of a region thus consists of *all* the cells in the region whose neighbours are also all in the region:

$$\bigcup\{X \mid (U,\sim) \models \mathsf{NTP}[X,R]\} = \{x \in U \mid \forall y(x \sim y \rightarrow y \in R)\}$$

This set is called the *(discrete) interior* of region R and is denoted $int_D(R)$. So long as it is non-empty, it is of course a region itself.

While int_D, considered as an operator on sets of cells, is a total function, when considered as an operator on regions it is only a partial function, since if $int_D(R)$ is empty, R does not have an interior region. In the language $\mathscr{L}_{\mathsf{P,C}}$, therefore, the notion of interior is expressed by means of a relational predicate $\mathsf{Int}(\mathsf{x},\mathsf{y})$, meaning that y is *an* interior of x, defined by:

$$\mathsf{Int}(\mathsf{x},\mathsf{y}) =_{\text{def}} \forall \mathsf{z}(\mathsf{P}(\mathsf{z},\mathsf{y}) \leftrightarrow \mathsf{NTP}(\mathsf{z},\mathsf{x})) \qquad \text{(D19)}$$

Thus a region is an interior of R if and only if its parts are all and only the non-tangential parts of R. As discussed earlier, it suffices, in fact, to consider just atomic parts, so we have

$$\mathsf{Int}(\mathsf{x},\mathsf{y}) \leftrightarrow \forall \mathsf{z}(\mathsf{Atom}(\mathsf{z}) \rightarrow (\mathsf{P}(\mathsf{z},\mathsf{y}) \leftrightarrow \mathsf{NTP}(\mathsf{z},\mathsf{x}))). \qquad \text{(D19}'\text{)}$$

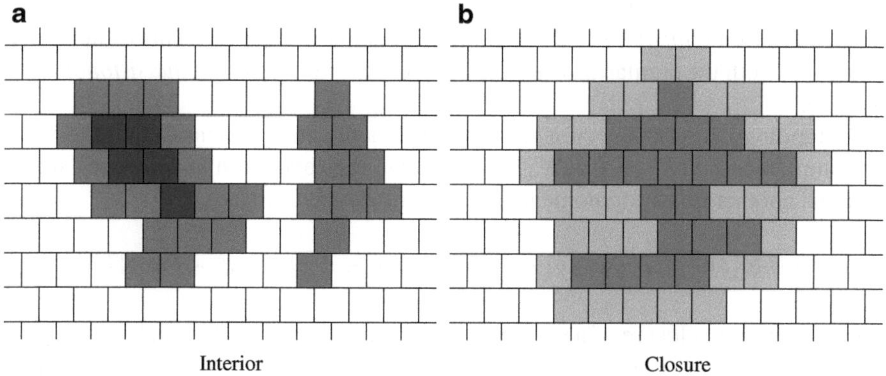

Fig. 11.5 Discrete interiors and closures. In (**a**), a two-part region (*all shading*) and its discrete interior (*dark grey*); in (**b**), a region (*mid-grey*) and its discrete closure (*all shading*)

It now follows that a region cannot have two distinct interiors, since if $\mathsf{Int}(\mathsf{x}, \mathsf{y}_1) \wedge \mathsf{Int}(\mathsf{x}, \mathsf{y}_2)$ then from (D19) we have $\forall \mathsf{z}(\mathsf{P}(\mathsf{z}, \mathsf{y}_1) \leftrightarrow \mathsf{P}(\mathsf{z}, \mathsf{y}_2))$ so in particular $\mathsf{P}(\mathsf{y}_2, \mathsf{y}_1)$ and $\mathsf{P}(\mathsf{y}_1, \mathsf{y}_2)$, whence $\mathsf{y}_1 = \mathsf{y}_2$ by antisymmetry of P (A6). The appropriateness of (D19) follows from the easily demonstrated fact that, if $Y \neq \emptyset$, $(U, \sim) \models \mathsf{Int}[X, Y]$ if and only if $Y = int_D(X)$. Thus Int does capture in $\mathscr{L}_{\mathsf{P,C}}$ the relationship between a region and its discrete interior, so long as the latter is non-empty.

Any connected component of U, and any union of such connected components (including U itself), is its own interior, since it is a non-tangential part of itself. If a region has no non-tangential parts, and hence no interior, we describe it as "thin". In Fig. 11.5a, the left-hand region has its interior shaded a darker grey; the right-hand region is thin, since all of its cells have at least one neighbour outside the region. Note that U. and any of its connected components, cannot be thin since it is its own interior; in particular, therefore, a single cell is a thin region so long as it is connected to at least one other cell, but if it is a connected component of U (and thus an isolated cell, disconnected from the rest of the space), it is not thin.

We refer to the *discrete* interior in order to distinguish int_D from the topological interior operator int, which does not apply to adjacency spaces since they are not defined as topologies. The two operators share a number of common features, notably:

- $int_{(D)}(U) = U$,
- $\forall X(int_{(D)}(X) \subseteq X)$,
- $\forall X \forall Y(X \subseteq Y \rightarrow int_{(D)}(X) \subseteq int_{(D)}(Y))$.

The most important *difference* between the discrete and topological interior operators is that whereas the latter is idempotent, i.e.,

- $\forall X(int(int(X)) = int(X))$,

this is not, in general true for the former, since, in an adjacency space, the only regions for which $int_D(X) = X$ are unions of connected components of the space. In view of both the similarities and the differences, we call int_D a *quasi-topological* operator.

In topology, the *closure* of a set is defined as the complement of the interior of its complement: $cl(X) = (int(X^c))^c$. The analogous operation in adjacency spaces gives us another quasi-topological operator, the *discrete closure*,

$$cl_D(R) = (int_D(R^c))^c = \{x \mid \exists y(x \sim y \wedge y \in R)\} = \bigcap\{X \mid (U, \sim) \models \mathsf{NTP}[X, R]\}.$$

Thus the discrete closure of a region R consists of all those cells which are adjacent to an element of R; it is the intersection of all regions which R is a non-tangential part of.

As with the interiors, the discrete closure shares some properties with the topological closure, namely

- $cl_{(D)}(\emptyset) = \emptyset$,
- $\forall X(X \subseteq cl_{(D)}(X))$,
- $\forall X \forall Y(X \subseteq Y \rightarrow cl_{(D)}(X) \subseteq cl_{(D)}(Y))$,

but not:

- $\forall X(cl(cl(X)) = cl(X))$.

As with interiors, the only regions in adjacency space for which $cl_D(X) = X$ are the unions of connected components of the space.

Analogously to discrete interior, we can define the discrete closure relation in $\mathscr{L}_{\mathsf{P,C}}$ by

$$\mathsf{Cl(x, y)} =_{\text{def}} \forall \mathsf{z(P(y, z)} \leftrightarrow \mathsf{NTP(x, z))} , \tag{D20}$$

which says that y is the closure of x if and only if x is a non-tangential part of all and only those regions y is part of.

Our decision to develop mereology with a universal element but no null element leads to a certain asymmetry. In set-theoretical interpretations the asymmetry shows up as the fact that the universal set is recognised as determining a region but the empty set is not. A consequence of this is that, unlike the discrete interior, discrete closure is a total function on regions: not every region has a discrete interior, but every region has a discrete closure. This means that we can also define the closure *function* cl by

$$\phi(\mathsf{cl(x)}) =_{\text{def}} \forall \mathsf{y(Cl(x, y)} \rightarrow \phi(\mathsf{y})), \tag{D21}$$

where ϕ stands for any open formula with one free variable. Figure 11.5b shows a region (dark grey cells) and its discrete closure (the region plus the lighter grey cells).

Within $\mathscr{L}_{\mathsf{P,C}}$, in order to characterise the relationship between discrete closure and interior, we need to use the predicate Compl already defined in (D18). The relationship between discrete interior and discrete closure is then expressed in $\mathscr{L}_{\mathsf{P,C}}$ by the formula

$$\mathsf{Compl(x, y)} \wedge \mathsf{Int(y, z)} \wedge \mathsf{Compl(z, w)} \rightarrow \mathsf{Cl(x, w)}, \tag{T7}$$

a proof of which is given in Appendix 3.

From now on, we shall use the words 'closure' and 'interior' on their own to mean the *discrete* closure and interior; if we need to refer to the topological operators, we shall do so explicitly. A useful way of characterising the closure and interior in an adjacency space is to use the idea of the *neighbourhood* of a cell, defined as follows:

$$N(x) = \{y \in U \mid x \sim y\}$$

Thus neighbourhoods are in fact the closures of atoms, and the closure of any region $R \subseteq U$ is the union of the neighbourhoods of its constituent cells:

$$cl_D(R) = \bigcup\{N(x) \mid x \in R\}.$$

The interior comprises those cells whose neighbourhoods are parts of the region:

$$int_D(R) = \{x \in U \mid N(x) \subseteq R\} .$$

Useful operations result from combining closure and interior, in either order. If a set X includes some thin spikelike parts, these will disappear when the interior is taken, and will not be restored if closure is then applied. Thus $cl_D(int_D(X))$ is essentially like X but with any thin parts removed (Fig. 11.6, left). On the other hand, if the region has any thin holes or fissures, these will be filled in by the closure operation and not be opened out again when interior is applied. Thus $int_D(cl_D(X))$ is like X but with any thin holes or fissures filled in (Fig. 11.6, right). Regions which lack spikes or fissures, i.e., for which $X = cl_D(int_D(X)) = int_D(cl_D(X))$, may be called *regular*.

Referring back to the earlier discussion of the distinction between adjacency spaces and graphs, it should be noted that Stell (2000b) has reformulated these quasi-topological operators in terms of two kinds of complementation definable on graphs. Recall that a subgraph is specified by giving both its vertices and its edges. The *negation* $\neg G$ of a subgraph G consists of all the vertices of U that are not in G, and all the edges of U joining vertices in $\neg G$. The *supplement* $\sim G$ consists of all the edges of U that are not in G, and all the vertices of V that are incident with an edge in $\neg G$. Then the subgraphs $\neg\sim G$ and $\sim\neg G$ correspond to $int_D(G)$ and $cl_D(G)$ respectively. It is worth noting that Stell defines a *region* in a graph to be subgraph G such that $\neg\neg G = G$; thus a region, in this sense, includes all the edges

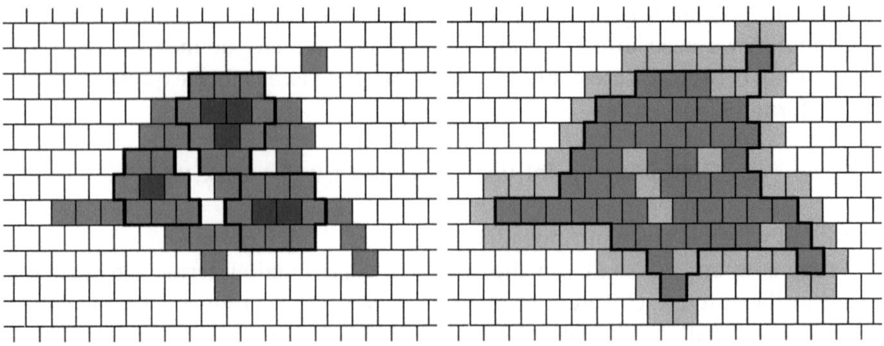

Fig. 11.6 *Left*: A region (*mid grey*) with its interior (*dark grey*) and the closure of the interior (*heavy outline*). *Right*: The same region, with its closure (*all shading*) and the interior of the closure (*heavy outline*)

of U that join vertices of G to a vertices of G; it can therefore can be specified just by giving its vertices, and thus corresponds to a region in adjacency space.

11.6 Measures of Size and Distance

From now on we assume that the universe is connected, meaning that every region (other than the universe itself) is connected to its complement:

$$\forall X \subset U((U, \sim) \models \mathsf{C}[X, X^c]) .$$

This means that the universe consists of a single connected component, itself, and is therefore the only region which is its own closure and interior (the empty set also has this property, but it is not a region).[9]

We have already characterised a *thin* region as one with empty interior. More generally we can define the *thickness* of a region as the number of successive interior operations required to reduce the region to nothing. For a region R of thickness n we have the sequence

$$R, int_D(R), int_D(int_D(R)), \ldots, int_D^n(R) = \emptyset .$$

Thus for $X \subseteq U$ we define

$$Thickness(X) =_{\text{def}} \begin{cases} 0 & (\text{if } X = \emptyset) \\ n+1 & (\text{if } X \neq \emptyset \text{ and } Thickness(int_D(X)) = n) \\ \infty & (\text{Otherwise}) \end{cases}$$

[9]In standard treatments of mereotopology this is posited as an axiom: $\mathsf{Compl(x, y)} \rightarrow \mathsf{C(x, y)}$.

The thin regions are then those with thickness 1. The universe, U, since it is its own interior, can never be reduced to nothing in this way, so its thickness is infinite—even if it is finite in the sense of containing only finitely many cells.[10] Regions other than the universe can only have infinite thickness if the universe is infinite. Thickness provides a measure of how 'substantial' a region is.

A region of thickness n is the union of a sequence of $n-1$ *shells* surrounding a central *core* of thickness 1. If $Thickness(R) = n$, then the shells of region R are

$$R \setminus int_D(R), int_D(R) \setminus int_D(int_D(R)), \ldots, int_D^{n-2}(R) \setminus int_D^{n-1}(R), int_D^{n-1}(R)\,,$$

where the final term in the sequence is the core, whose interior is empty.

We can perform an analogous construction, starting from a region R and repeatedly forming the closure, thus building up a sequence of regions:

$$R, cl_D(R), cl_D(cl_D(R)), cl_D(cl_D(cl_D(R))), \ldots\,.$$

If the universe is infinite, this process may go on for ever, depending on the starting point R, otherwise, given that the universe is connected, we will reach a point where $cl_D^n(R) = U$. Give the complementarity of closure and interior, this occurs when $Thickness(R^c) = n$. We can think of the sequence of closures being built up by the successive addition of outer shells,

$$cl_D(R) \setminus R, cl_D(cl_D(R)) \setminus cl_D(R), \ldots, cl_D^n(R) \setminus cl_D^{n-1}(R), \ldots\,.$$

The notion of *distance* is usually defined in terms of shortest paths; but as we shall see, in adjacency spaces it can also be defined in terms of closures.

A *path* of length n from cell x to cell y is a sequence $x_0, x_1, \ldots, x_n$ such that $x_0 = x$, $x_n = y$, and for $i = 1, \ldots, n$, $x_{i-1} \sim x_i$. We can prove that there is a path of length n from x to y if and only if $y \in cl_D^n(\{x\})$. We use induction on n:

Base case ($n = 0$). A path of length 0 from x to y consists of a single point x_0 which must be equal to both x and y. Clearly this exists if and only if $x = y$. Since $cl_D^0(\{x\}) = \{x\}$ this means that $y \in cl_D^0(\{x\})$ as required.

Induction step (from $n-1$ to n). Assume the result holds for $n-1$. If $x_0, x_1, \ldots, x_n$ is a path of length n from x to y, then $x_0, x_2, \ldots, x_{n-1}$ is a path of length $n-1$ from x to x_{n-1}. By hypothesis such a path exists if and only if $x_{n-1} \in cl_D^{n-1}(\{x\})$. Since $x_{n-1} \sim x_n = y$, this means that $y \in cl_D(cl_D^{n-1}(\{x\})) = cl_D^n(\{x\})$, as required.

[10] An alternative definition of thickness, which would allow the thickness of the universe to be finite, would be to restrict the definition given in the text to non-U regions, and define $Thickness(U) =_{\text{def}} \max_{X \subset U} Thickness(X) + 1$, it being understood that this expression evaluates to ∞ if there is no upper bound to the thickness of non-universal regions.

This gives us a natural measure of the distance between two cells:

$$d(x, y) = \min_{n \in \mathbb{N}}(y \in cl^n_D(\{x\})) \ .$$

From the above, this is the length of a shortest path from x to y, and it is easy to see that it is a true metric, i.e.,

- $d(x, y) = 0$ if and only if $x = y$,
- $d(x, y) = d(y, x)$
- $d(x, z) \leq d(x, y) + d(y, z)$ (triangle inequality).

Note that there will not, in general, be a unique shortest path between two cells. A familiar example is the adjacency space $(\mathbb{Z}^2, \sim_4)$. The points (m, n) and $(m+1, n+1)$ are linked by two minimal paths, of length 2, one going via $(m, n+1)$ and the other via $(m+1, n)$; in general, between the points (m, n) and $(m+h, n+k)$ there are ${}^{h+k}C_h$ minimal paths, of length $h+k$. This is the "Manhattan" or "city-block" distance.

An obvious generalisation of distance to regions gives us the *proximal distance* between two regions, defined as the smallest number of closure operations that can be applied to X to produce a region that overlaps Y; as before, this is equivalent to the more familiar definition as the shortest distance between a cell in one region and a cell in the other:

$$pd(X, Y) = \min_{n \in \mathbb{N}}(cl^n_D(X) \cap Y \neq \emptyset) = \min_{\substack{x \in X \\ y \in Y}} d(x, y).$$

Unfortunately proximal distance is not a true metric since (1) the proximal distance between distinct but overlapping regions is zero, and (2) the triangle inequality does not hold, i.e., we can have regions X, Y, Z such that $pd(X, Z) > pd(X, Y) + pd(Y, Z)$.

A more satisfactory measure of distance for regions is the *Hausdorff distance*, defined as the greatest distance between any point in one of the regions and the nearest point in the other:

$$hd(X, Y) = \max\left(\max_{x \in X}\min_{y \in Y} d(x, y), \max_{y \in Y}\min_{x \in X} d(x, y)\right).$$

If $\max_{x \in X}\min_{y \in Y} d(x, y) = n$, then for any $x \in X, y \in Y$ we have $y \in cl^n_D(\{x\}) \subseteq cl^n_D(X)$, so $Y \subseteq cl^n_D(X)$, and likewise with X and Y reversed. Thus an equivalent formulation in terms of closures is

$$hd(X, Y) = \min\{n \in \mathbb{N} \mid X \subseteq cl^n_D(Y) \wedge Y \subseteq cl^n_D(X)\}.$$

The Hausdorff distance between two regions in adjacency space is thus the smallest n such that each region is within the nth closure of the other. Unlike proximal distance, Hausdorff distance is a true metric.

It should be noticed that while the Hausdorff distance between a region and its closure is always 1, i.e., $hd(X, cl_D(X)) = 1$, this is not necessarily the case for a region and its interior. In fact

$$hd(X, int_D(X)) = 1 \text{ if and only if } cl_D(int_D(X)) = X.$$

Such an X is called a *regular closure set* in Smyth and Webster (2007).

11.7 Relation to Mathematical Morphology

Mathematical Morphology (MM) comprises a set of mathematical tools for manipulating images. Readers familiar with MM will recognise a clear similarity between our discrete interior and closure operations and the erosion and dilation operators of that theory. Here we make this relationship explicit. While the theory of MM may be developed both for continuous and discrete images, for our purposes we will consider only the discrete case: in this case we are working with $\mathbb{Z}^2$. An *image* is any subset of this set.

In Mathematical Morphology, there is no pre-defined adjacency relation. Instead, erosion and dilation may be performed with respect to an arbitrary *structuring element* which in effect determines which points are to count as adjacent. A structuring element is itself an image, typically small. Given an image X and a structuring element B, we define

- The *dilation* of X by B is the image

$$X \oplus B = \{x + b \mid x \in X, b \in B\}.$$

- The *erosion* of X by B is the image

$$X \ominus B = \{y \in \mathbb{Z}^2 \mid \forall b \in B(y + b \in X)\}.$$

Addition here is coordinate-wise, i.e., treating points as vectors. The dilation of a region by B expands B by replacing each of its points by a copy of B; the location of the copy depends on where B itself is with respect to the origin. It is usual to assume that the origin is one of the points of B, otherwise dilation will result in a displacement of the image as well as an expansion. If the structuring element is taken to be a 3×3 square centered on the origin, then the dilation of any image will be exactly its closure with respect to the adjacency relation $\sim_8$. If instead we take a cross-shaped structuring element consisting of the origin and the four points orthogonally adjacent to it (like the leftmost image in Fig. 11.2), dilation then gives closure with respect to the relation $\sim_4$.

Erosion removes the outer part of the image, retaining a point only if a copy of the structuring element anchored on that point would lie entirely within the image.

So long as the structuring element has central symmetry (i.e., $B = -B$), erosion and dilation are related exactly as interior and closure. More generally we have

$$X \oplus B = (X^c \ominus (-B))^c ,$$

where $-B = \{-x \mid x \in B\}$.

MM makes much use of operations called *opening* and *closing*, defined as follows:

Opening: $X \circ B = (X \ominus B) \oplus B$,
Closing: $X \bullet B = (X \oplus B) \ominus B$.

These correspond to the DM operations $cl_D(int_D(X))$ and $int_D(cl_D(X))$ illustrated in Fig. 11.6.

Mathematical Morphology may appear to be in some ways more general, and in other ways less so, than Discrete Mereotopology, but both these appearances are misleading.

The sense in which MM may appear to be more general than DM is that it allows arbitrary structuring elements; but this generality could be recovered in DM by defining a different adjacency relation on $\mathbb{Z}^2$ for each possible structuring element. For example, if the structuring element consists of just the points $\{(0, 0), (1, 0), (0, 1)\}$, then the corresponding adjacency relation would relate any point (x, y) to (x, y), $(x + 1, y)$, $(x, y + 1)$ and nothing else.

On the other hand, as usually presented in terms of structuring elements, MM would appear to presuppose spaces which are homogeneous in the sense that they allow a copy of the same structuring element to be located at each point in the space. As we have seen, however, DM is equally happy in non-homogeneous spaces where such arbitrary translation of structuring elements does not make sense; the discrete closure and interior operations of DM work just as well in this setting as with homogeneous spaces, but the most familiar forms of MM gain no purchase in this context.

This is not, however, the full story, since a number of researchers have investigated forms of MM which allow variable structuring elements—see for example Roerdink and Heijmans (1988) and Verly and Delanoy (1993). And it is certainly true more generally that the study of MM is a much more mature research area than that of DM, and as a result has been developed to a considerably greater degree of mathematical sophistication and generality—see for example Bloch et al. (2007).

11.8 Relation to Digital Topology

Adjacency spaces are examples of a general class of mathematical structures called *closure spaces* (Čech 1966). A closure space is a pair (U, cl), where U is any set, and cl is a function mapping each set $X \subseteq U$ to a set $cl(X) \subseteq U$ such that

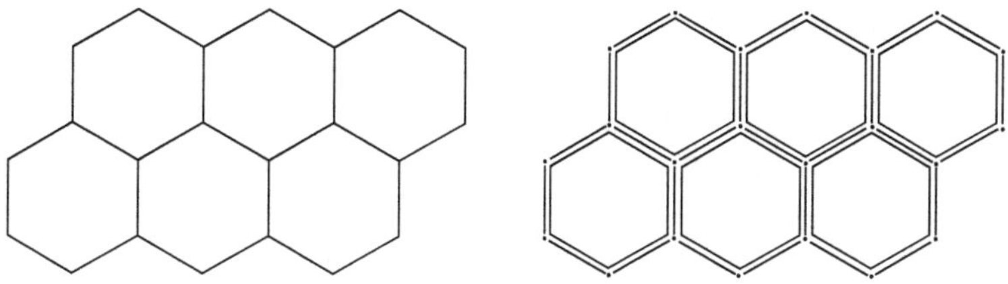

Fig. 11.7 An adjacency space in the form of a hexagonal grid (*left*), and the topological space obtained by the addition of bounding lines and points (*right*)

1. $cl(\emptyset) = \emptyset$,
2. $X \subseteq cl(X)$,
3. $cl(X \cup Y) = cl(X) \cup cl(Y)$.

This notion generalises topological closure, which in addition to satisfying conditions 1–3 also satisfies the idempotency rule

4. $cl(cl(X)) = cl(X)$,

which as we have noted is *not*, in general, satisfied by sets in an adjacency space.

Although the discrete closure operation in an adjacency space is not a topological closure (since it is not idempotent), one can define topological spaces associated with any adjacency space. A trivial way of doing this is to specify the *discrete topology* on U, that is, the topology under which $cl(X) = X$ for every $X \subseteq U$. Much more interesting is to extend U by including in the universe not just the original elements of U (conceived of as atomic regions) but also boundary elements where these atomic regions adjoin one another, and boundaries of those, etc, depending on the dimensionality one wishes to confer on the space. This is illustrated in Fig. 11.7, where, on the left is shown part of an adjacency space in the form of a regular hexagonal lattice, and on the right is shown a space which includes, in addition to the hexagons of the original lattice, a set of line segments representing the boundaries of the hexagons and a set of points representing the ends of the line segments. This space can be made into a topology by specifying that the closure of any set consists of the elements of that set together with all their bounding elements. Thus the closure of one hexagonal tile consists of the hexagon together with its six bounding lines and its six bounding points, and the closure of a line element is the line together with its two bounding points. It is easy to see that this closure operation satisfies the conditions (1)–(4) above, and therefore defines a topological space. Such topological spaces, if finite, are called *cellular complexes*, and these are investigated in the context of applications to image analysis in Kovalevsky (1989).

The topological spaces obtained in this way from rectangular grids of the form $\mathbb{Z}^n$ are called *Khalimsky spaces* (Khalimsky et al. 1990; Kong et al. 1991). See also Kong and Rosenfeld (1991) for a discussion of the relationship between these topological approaches and graph-based approaches such as DM.

11.9 An Application

Discrete Mereotopology has been applied to the analysis of histological images (Randell and Landini 2008; Randell et al. 2013). Real-world images invariably have imperfections which means that when standard segmentation algorithms are applied to them in order to extract information about the entities pictured, the resulting structures are not always in conformity with theoretical models—e.g., one might find cell nuclei overlapping the boundaries of their cytoplasm, or distinct tissue types wrongly labelled. DM can be used to identify maximally parsimonious ways of repairing such ill-formed images, using "conceptual neighbourhood diagrams" (Freksa 1992) to identify sequences of operations that can transform an existing inappropriate structure to one that is in conformity with expectation. This method works at the level of individual image pixels, and can therefore harness the power of mathematical morphology alongside DM; but by considering a coarser segmentation of the image one can exploit the ability of DM to allow reasoning about arbitrary adjacency spaces.

To illustrate, Fig. 11.8a shows a Haemotoxylin and Eosin stained section of an odontogenic keratocyst lining. Image-processing techniques are used to extract theoretical cell boundaries from this image, defining "virtual cells" or "v-cells" in the epithelial compartment separating the background free space at the top of the image from the connective tissue at the bottom. The segmentation into v-cells is shown in Fig. 11.8b. Individual v-cells, as well as the whole block of connective tissue and the background space, can be regarded as atomic regions of an adjacency space. The v-cells adjacent to the connective tissue form what is called the *basal layer*. If V is the region in the image consisting of the v-cells, and C is the region corresponding to the connective tissue, then the basal layer B can be identified as $V \cap cl_D(C)$. By taking successive closures of the basal layer we can segment the epithelium into layers as shown in Fig. 11.8c. This operation allows one to derive a more meaningful measure of tissue thickness, for example, than crude measures involving pixel counts or Euclidean distance. Such measures can provide important diagnostic criteria for histopathology. By taking a single target cell within the segmented image, one can similarly use the closure operation to generate nested rings of v-cells, as shown in Fig. 11.8d, which can again provide useful information, at the cellular level, on local tissue architecture.

Acknowledgements I would like to thank Professor Gabriel Landini and Dr David Randell of the School of Dentistry, University of Birmingham, for supplying the images used in Fig. 11.8 and for providing critical commentary on a draft of this chapter. Dr John Stell from the School of Computing, University of Leeds, also provided useful feedback on an earlier draft of this paper.

Appendix 1: Proof of (T4) and (T5)

The first theorem to be proved is

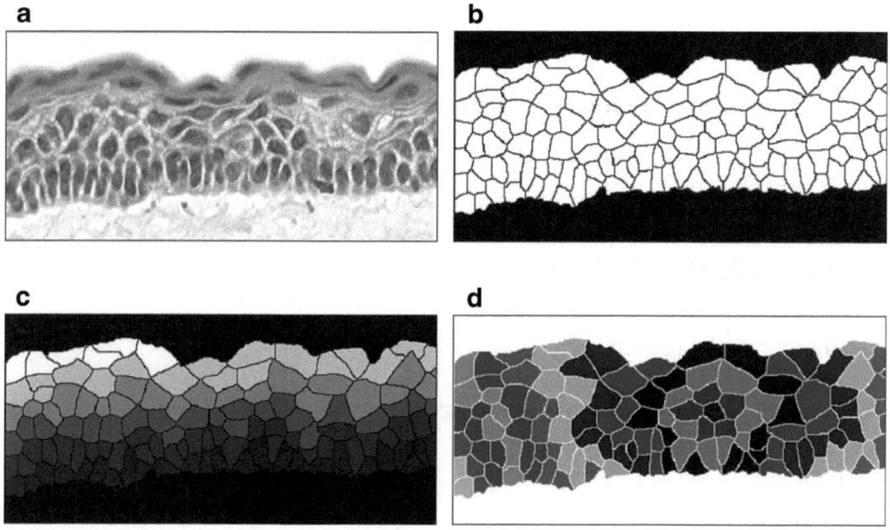

Fig. 11.8 Application of discrete closure operation to a histological image (Images courtesy of Prof. Gabriel Landini and Dr D. Randell). (**a**) Stained epithelium section. (**b**) Segmentation into "virtual cells" (v-cells). (**c**) Layering of epithelial v-cells. (**d**) Nested shells of v-cells around a target cell

$$\forall x \forall y(\mathsf{Compl}(x, y) \leftrightarrow \neg \mathsf{O}(x, y) \wedge \mathsf{Sum}(x, y, \mathsf{U})) \tag{T4}$$

First, suppose $\mathsf{Compl}(x, y)$, so since $\mathsf{P}(y, y)$ we have $\neg\mathsf{O}(y, x)$ by (D18) and therefore

$$\neg\mathsf{O}(x, y). \tag{1}$$

Let v be any region; if $\neg\mathsf{O}(v, x)$, then $\mathsf{P}(v, y)$ (since $\mathsf{Comp}(x, y)$), so $\mathsf{O}(v, y)$. Hence $\mathsf{O}(v, x) \vee \mathsf{O}(v, y)$. Since v is arbitrary, we always have $\mathsf{O}(v, \mathsf{U})$, and hence we have $\mathsf{O}(v, \mathsf{U}) \leftrightarrow \mathsf{O}(v, x) \vee \mathsf{O}(v, y)$, i.e.,

$$\mathsf{Sum}(x, y, \mathsf{U}) \tag{2}$$

Conversely, suppose we have $\neg\mathsf{O}(x, y) \wedge \mathsf{Sum}(x, y, \mathsf{U})$.

Let $\mathsf{P}(u, y)$, so that whenever $\mathsf{P}(w, u)$ we have also $\mathsf{P}(w, y)$. Therefore from $\neg\mathsf{O}(x, y)$, i.e. $\neg\exists w(\mathsf{P}(w, y) \wedge \mathsf{P}(w, x))$, we infer $\neg\exists w(\mathsf{P}(w, u) \wedge \mathsf{P}(w, x))$, i.e., $\neg\mathsf{O}(u, x)$. Hence we have shown $\forall u(\mathsf{P}(u, y) \rightarrow \neg\mathsf{O}(u, x))$.

Now suppose $\neg\mathsf{O}(u, x)$. We must show $\mathsf{P}(u, y)$. Suppose not; then from $\neg\mathsf{P}(u, y)$, by (A8), there is a region w such that $\mathsf{P}(w, u) \wedge \neg\mathsf{O}(w, y)$. From $\neg\mathsf{O}(w, y)$, since $\mathsf{Sum}(x, y, \mathsf{U})$ we have $\mathsf{O}(w, x)$ (since $\mathsf{O}(w, \mathsf{U})$ in any case). From $\mathsf{P}(w, u)$ and $\mathsf{O}(w, x)$ it is an easy deduction that $\mathsf{O}(u, x)$, contradicting our assumption. Hence we have shown $\forall u(\neg\mathsf{O}(u, x) \rightarrow \mathsf{P}(u, y))$.

We now have $\forall u(\mathsf{P}(u, y) \leftrightarrow \neg\mathsf{O}(u, x))$, i.e., $\mathsf{Compl}(x, y)$. □

The proof of (T5) is now straightforward:

$$\mathsf{Compl}(x,y) \Leftrightarrow \neg \mathsf{O}(x,y) \wedge \mathsf{Sum}(x,y,U) \Leftrightarrow \neg \mathsf{O}(y,x) \wedge \mathsf{Sum}(y,x,U) \Leftrightarrow \mathsf{Compl}(y,x).$$

Appendix 2: Proof of (T6)

The theorem to be proved is

$$\forall x \forall y (\mathsf{Compl}(x,y) \rightarrow \forall z (\mathsf{C}(z,y) \leftrightarrow \neg \mathsf{NTP}(z,x))) \quad \text{(T6)}$$

Assuming

$$\mathsf{Compl}(a,b), \quad (1)$$

suppose, first, that $\mathsf{C}(c,b)$. From (1), since $\mathsf{P}(b,b)$, we have $\neg\mathsf{O}(b,a)$, hence we have $\mathsf{C}(c,b) \wedge \neg\mathsf{O}(b,a)$ and therefore $\neg\mathsf{NTP}(c,a)$ by (D10). Hence we have $\forall z(\mathsf{C}(z,b) \rightarrow \neg\mathsf{NTP}(z,a))$.

Next, suppose we have $\neg\mathsf{NTP}(c,a)$. Then from (D10), either we have

$$\neg \mathsf{P}(c,a) \quad \text{(2a)}$$

or there is a region d such that

$$\mathsf{C}(c,d) \wedge \neg \mathsf{O}(d,a). \quad \text{(2b)}$$

In the former case, from (1) and (T5) we have $\mathsf{Compl}(b,a)$, so (2a) implies $\mathsf{O}(c,b)$, which implies $\mathsf{C}(c,b)$. In the latter case (2b), from (1) and $\neg\mathsf{O}(d,a)$ we have $\mathsf{P}(d,b)$. Then from $\mathsf{C}(c,d)$ and $\mathsf{P}(d,b)$ we have $\mathsf{C}(c,b)$ by (A5). Thus in either case we have $\mathsf{C}(c,b)$ and we have proved $\forall z(\neg\mathsf{NTP}(z,a) \rightarrow \mathsf{C}(z,b))$.

Combining the results and generalising give us (T6). □

Appendix 3: Proof of (T7): Relationship of Discrete Interior and Closure

The theorem to be proved is

$$\mathsf{Compl}(x,y) \wedge \mathsf{Int}(y,z) \wedge \mathsf{Compl}(z,w) \rightarrow \mathsf{Cl}(x,w) \quad \text{(T7)}$$

Using the definitions (D19, D20, D18), this means that from

$$\forall x(\mathsf{P}(x, b) \leftrightarrow \neg \mathsf{O}(x, a)) \tag{1}$$

$$\forall x(\mathsf{P}(x, c) \leftrightarrow \mathsf{NTP}(x, b)) \tag{2}$$

$$\forall x(\mathsf{P}(x, d) \leftrightarrow \neg \mathsf{O}(x, c))) \tag{3}$$

we must derive

$$\forall x(\mathsf{P}(d, x) \leftrightarrow \mathsf{NTP}(a, x)) \tag{4}$$

From (T5) we can rewrite (1) and (3) as

$$\forall x(\mathsf{P}(x, a) \leftrightarrow \neg \mathsf{O}(x, b)) \tag{5}$$

$$\forall x(\mathsf{P}(x, c) \leftrightarrow \neg \mathsf{O}(x, d)) \tag{6}$$

so from (2) and (6) we have

$$\forall x(\neg \mathsf{O}(x, d) \leftrightarrow \mathsf{NTP}(x, b)) \tag{7}$$

Then from (1) and (7) (since $\mathsf{NTP}(x, b)$ implies $\mathsf{P}(x, b)$) we have

$$\forall x(\neg \mathsf{O}(x, d) \rightarrow \neg \mathsf{O}(x, a)) \tag{8}$$

Suppose $\neg \mathsf{P}(a, d)$. Then by (A8) there must be a region e such that $\mathsf{P}(e, a) \wedge \neg \mathsf{O}(e, d)$. By (8) this would imply $\mathsf{P}(e, a) \wedge \neg \mathsf{O}(e, a)$, a contradiction. Therefore we have $\mathsf{P}(a, d)$, and therefore

$$\forall x(\mathsf{P}(d, x) \rightarrow \mathsf{P}(a, x)) \tag{9}$$

Suppose $\mathsf{P}(d, g)$. and let f be any region connected to a, i.e., $\mathsf{C}(a, f)$. Suppose $\neg \mathsf{O}(f, d)$. Then by (6) we have $\mathsf{P}(f, c)$ and so by (2), $\mathsf{NTP}(f, b)$. Therefore any region connected to f must overlap b. Since we have $\mathsf{C}(a, f)$ this means that $\mathsf{O}(a, b)$. But from (1) we know that $\neg \mathsf{O}(a, b)$ and we have a contradiction. Therefore $\mathsf{O}(f, d)$, and therefore, since $\mathsf{P}(d, g)$, we have $\mathsf{O}(f, g)$.

Thus we have

$$\forall x(\mathsf{P}(d, x) \rightarrow \forall y(\mathsf{C}(a, y) \rightarrow \mathsf{O}(y, x))) \tag{10}$$

Combining (9) and (10) (and using (D10)) we get

$$\forall x(\mathsf{P}(d, x) \rightarrow \mathsf{NTP}(a, x)) \tag{11}$$

which is one half of (4).

For the converse, let $\mathsf{NTP}(a, e)$; we must show that $\mathsf{P}(d, e)$. Assume on the contrary that

$$\neg\mathsf{P}(d, e) \tag{12}$$

By (A8), there is a region f such that $\mathsf{P}(f, d)$ and $\neg\mathsf{O}(f, e)$.

From $\mathsf{P}(f, d)$ we have, by (3), $\neg\mathsf{O}(f, c)$, and therefore

$$\neg\mathsf{P}(f, c) \tag{13}$$

From $\neg\mathsf{O}(f, e)$, we have, since $\mathsf{NTP}(a, e)$,

$$\neg\mathsf{C}(a, f) \tag{14}$$

Also from $\neg\mathsf{O}(f, e)$, given $\mathsf{NTP}(a, e)$, and therefore $\mathsf{P}(a, e)$, we have $\neg\mathsf{O}(f, a)$, and therefore, by (1), $\mathsf{P}(f, b)$.

We will show that in fact $\mathsf{NTP}(f, b)$.

To this end we must show that anything connected to f overlaps b. Suppose $\neg\mathsf{O}(z, b)$. By (5) this implies $\mathsf{P}(z, a)$, and therefore, from (14), (A4) and (A5) $\neg\mathsf{C}(z, f)$. Hence if $\mathsf{C}(z, f)$ it follows that $\mathsf{O}(z, b)$. Thus we have $\mathsf{NTP}(f, b)$.

By (2) this gives $\mathsf{P}(f, c)$, contradicting (13). Hence assumption (12) is false, and we conclude, as required, that $\mathsf{P}(d, e)$.

We have now shown that $\forall x(\mathsf{NTP}(a, x) \rightarrow \mathsf{P}(d, x))$, which, in combination with (11), gives us (4). □

References

Bloch, I., Heijmans, H., & Ronse, C. (2007). Mathematical morphology. In M. Aiello, I. Pratt-Hartmann, & J. van Benthem (Eds.), *Handbook of spatial logics* (pp. 857–944). Dordrecht: Springer.

Čech, E. (1966). *Topological spaces*. Chichester: Wiley. Revised edition by Zdeněk Frolík and Miroslav Katětov.

Freksa, C. (1992). Temporal reasoning based on semi-intervals. *Artificial Intelligence, 54*, 199–227.

Galton, A. P. (1999). The mereotopology of discrete space. In C. Freksa & D. M. Mark (Eds.), *Spatial information theory: Cognitive and computational foundations of geographic science* (pp. 251–66). New York: Springer.

Galton, A. P. (2000). Continuous motion in discrete space. In A. G. Cohn, F. Giunchiglia, & B. Selman (Eds.), *Proceedings of the seventh international conference on principles of knowledge representation and reasoning (KR2000)*, Breckenridge (pp. 26–37). San Francisco: Morgan Kaufmann.

Gotts, N. M. (1996). An axiomatic approach to topology for spatial information systems. Technical report 96.25, University of Leeds School of Computer Studies.

Khalimsky, E., Kopperman, R., & Meyer, P. R. (1990). Computer graphics and connected topologies on finite ordered sets. *Topology and Its Applications, 36*, 1–17.

Kong, T. Y., & Rosenfeld, A. (1989). Digital topology: Introduction and survey. *Computer Vision, Graphics and Image Processing, 48*, 357–93.

Kong, T. Y., & Rosenfeld, A. (1991). Digital topology: A comparison of the graph-based and topological approaches. In G. M. Reed, A. W. Roscoe, & R. F. Wachter (Eds.), *Topology and category theory in computer science* (pp. 273–89). Oxford: Clarendon.

Kong, T. Y., Kopperman, R., & Meyer, P. R. (1991). A topological approach to digital topology. *The American Mathematical Monthly, 98*(10), 901–17.

Kovalevsky, V. A. (1989). Finite topology as applied to image analysis. *Computer Vision, Graphics, and Image Processing, 46*, 141–61.

Li, S., & Ying, M. (2004). Generalized region connection calculus. *Artificial Intelligence, 160*, 1–34.

Pratt, I., & Lemon, O. (1997). Ontologies for plane, polygonal mereotopology. *Notre Dame Journal of Formal Logic, 38*(2), 225–45.

Pratt, I., & Schoop, D. (1997). A complete axiom system for polygonal mereotopology of the real plane. Technical report UMCS-97-2-2, University of Manchester.

Pratt-Hartmann, I., & Schoop, D. (2002). Elementary polyhedral mereotopology. *Journal of Philosophical Logic, 31*, 469–498.

Randell, D. A., & Landini, G. (2008). Discrete mereotopology in automated histological image analysis. In *Proceedings of the second ImageJ user and developer conference*, Luxembourg (pp. 151–156).

Randell, D. A., Cui, Z., & Cohn, A. G. (1992). A spatial logic based on regions and connection. In B. Nebel, C. Rich, & W. Swartout (Eds.), *Proceedings of the third international conference on principles of knowledge representation and reasoning (KR'92)*, Cambridge (pp. 165–76). San Mateo: Morgan Kaufmann.

Randell, D. A., Landini, G., & Galton, A. (2013). Discrete mereotopology for spatial reasoning in automated histological image analysis. *IEEE Transactions on Pattern Analysis and Machine Intelligence*, 35(3), 568–581.

Roerdink, J. B. T. M., & Heijmans, H. J. A. M. (1988). Mathematical morphology for structures without translation symmetry. *Signal Processing, 15*(3), 271–277.

Rosenfeld, A. (1979). Digital topology. *American Mathematical Monthly, 86*(8), 621–30.

Roy, A. J., & Stell, J. G. (2002). A qualitative account of discrete space. In M. J. Egenhofer & D. M. Mark (Eds.), *Proceedings of GIScience 2002*, Boulder (LNCS, Vol. 2478) (pp. 276–290). Springer.

Simons, P. (1987). *Parts: A study in ontology*. Oxford: Clarendon.

Smyth, M. B., & Webster, J. (2007). Discrete spatial models. In M. Aiello, I. Pratt-Hartmann, & J. van Benthem (Eds.), *Handbook of spatial logics* (pp. 713–798). Dordrecht: Springer.

Stell, J. G. (2000a). Boolean connection algebras: A new approach to the region-connection calculus. *Artificial Intelligence, 122*, 111–136.

Stell, J. G. (2000b). The representation of discrete multi-resolution spatial knowledge. In A. G. Cohn, F. Giunchiglia, & B. Selman (Eds.), *Proceedings of the seventh international conference on principles of knowledge representation and reasoning (KR2000)*, Breckenridge (pp. 38–47). San Francisco: Morgan Kaufmann.

Varzi, A. C. (1996). Parts, wholes, and part-whole relations: The prospects of mereotopology. *Data and Knowledge Engineering, 20*, 259–286.

Verly, J. G., & Delanoy, R. L. (1993). Adaptive mathematical morphology for range imagery. *IEEE Transactions on Image Processing, 2*(2), 272–275.

Chapter 12
A Rôle for Mereology in Domain Science and Engineering: To Every Mereology There Corresponds a λ–Expression

Dines Bjørner

In memory of Douglas T. Ross 1929–2007[1]

12.1 Introduction

The term 'mereology' is accredited to the Polish mathematician, philosopher and logician Stansław Leśniewski (1886–1939). In this contribution we shall be concerned with only certain aspects of mereology, namely those that appear most immediately relevant to domain science (a relatively new part of current computer science). Our knowledge of 'mereology' has been through studying, amongst others, Casati and Varzi (1999) and Lejewski (1983).

12.1.1 Computing Science Mereology

"Mereology (from the Greek $\mu\epsilon\rho o\varsigma$ 'part') is the theory of part-hood relations: of the relations of part to whole and the relations of part to part within a whole".[2] In this contribution we restrict 'parts' to be those that, firstly, are spatially distinguishable, then, secondly, while "being based" on such spatially distinguishable parts, are

[1] See Sect. 12.7.1.

[2] Achille Varzi: Mereology, http://plato.stanford.edu/entries/mereology/ 2009 and Casati and Varzi (1999)

D. Bjørner (✉)
Fredsvej 11, DK-2840 Holte, Denmark
e-mail: bjorner@gmail.com

C. Calosi and P. Graziani (eds.), *Mereology and the Sciences*, Synthese Library 371,
DOI 10.1007/978-3-319-05356-1_12,

Fig. 12.1 'Adjacent' and "embedded within" parts

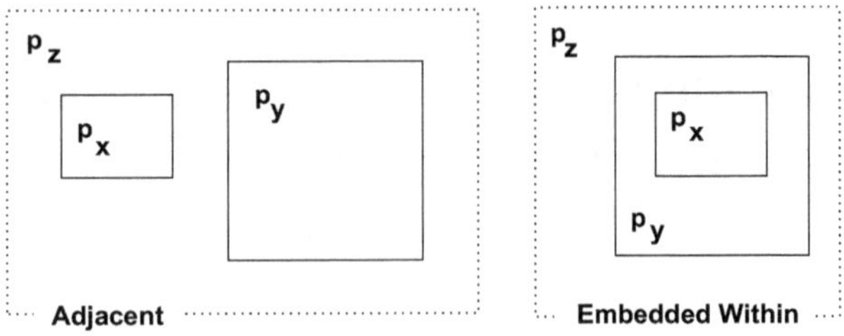

Fig. 12.2 'Adjacent' and "embedded within" parts

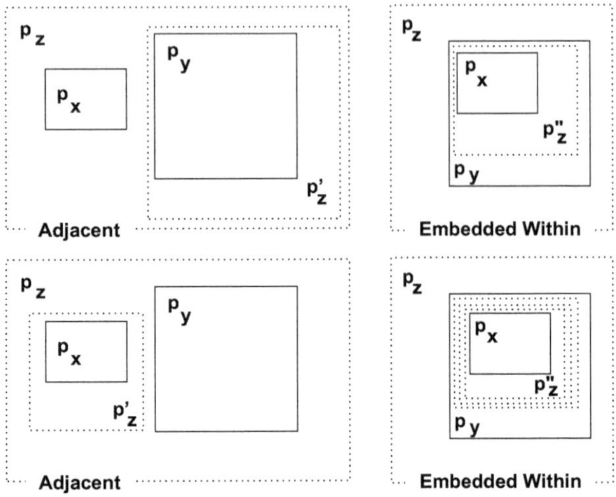

conceptually related. The relation: "being based", shall be made clear in this contribution.

Accordingly two parts, p_x and p_y, (of a same "whole") are either "adjacent", or are "embedded within" one another as loosely indicated in Fig. 12.1.

'Adjacent' parts are direct parts of a same third part, p_z, i.e., p_x and p_y are "embedded within" p_z; or one (p_x) or the other (p_y) or both (p_x and p_y) are parts of a same third part, p_z' "embedded within" p_z; etcetera; as loosely indicated in Fig. 12.2 or one is "embedded within" the other—etc. as loosely indicated in Fig. 12.2.

Parts, whether adjacent or embedded within one another, can share properties. For adjacent parts this sharing seems, in the literature, to be diagrammatically expressed by letting the part rectangles "intersect". Usually properties are not spatial hence 'intersection' seems confusing. We refer to Fig. 12.3.

Instead of depicting parts sharing properties as in the [L]eft side of Fig. 12.3, where dashed rounded edge rectangles stands for 'sharing', we shall (eventually) show parts sharing properties as in the [R]ight side of Fig. 12.3 where •—• connections connect those parts.

[L] 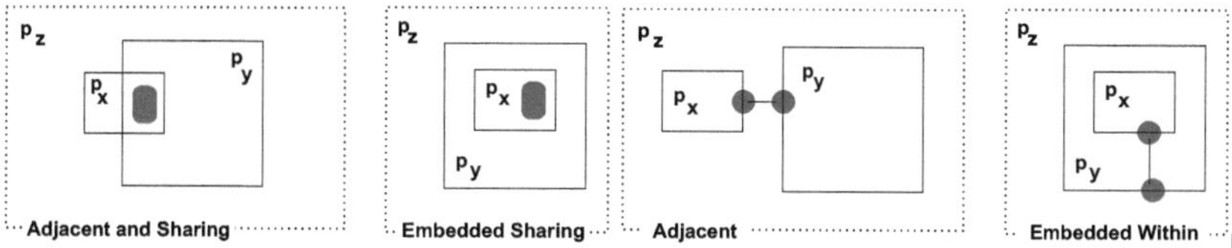

[R]

Fig. 12.3 Two models, [L,R], of parts that share properties

12.1.2 From Domains via Requirements to Software

One reason for our interest in mereology is that we find that concept relevant to the modelling of domains. A derived reason is that we find the modelling of domains relevant to the development of software. Conventionally a first phase of software development is that of requirements engineering. To us domain engineering is (also) a prerequisite for requirements engineering (Bjørner 2008, 2010). Thus to properly *design* $\mathbb{S}$oftware we need to *understand* its or their $\mathbb{R}$equirements; and to properly *prescribe* $\mathbb{R}$equirements one must *understand* its $\mathbb{D}$omain. To *argue* correctness of $\mathbb{S}$oftware with respect to $\mathbb{R}$equirements one must usually *make assumptions* about the $\mathbb{D}$omain: $\mathbb{D}, \mathbb{S} \models \mathbb{R}$. Thus *description* of $\mathbb{D}$omains become an indispensable part of $\mathbb{S}$oftware development.

12.1.3 Domains: Science and Engineering

Domain science is the study and knowledge of domains. Domain engineering is the practice of "walking the bridge" from domain science to domain descriptions: to create domain descriptions on the background of scientific knowledge of domains, the specific domain "at hand", or domains in general; and to study domain descriptions with a view to broaden and deepen scientific results about domain descriptions. This contribution is based on the engineering and study of many descriptions, of air traffic, banking, commerce (the consumer/retailer/wholesaler/producer supply chain), container lines, health care, logistics, pipelines, railway systems, secure [IT] systems, stock exchanges, etcetera.

12.1.4 Contributions of This Contribution

A general contribution is that of providing elements of a domain science. Three specific contributions are those of (i) giving a model that satisfies published formal, axiomatic characterisations of mereology; (ii) showing that to every (such modelled)

mereology there corresponds a CSP (Hoare 2004) program; and, related to (ii), (iii) suggesting complementing syntactic and semantic theories of mereology.

12.1.5 Structure of This Contribution

We briefly overview the structure of this contribution. First, in Sect. 12.2, we loosely characterise how we look at mereologies: "*what they are to us !*". Then, in Sect. 12.3, we give an abstract, model-oriented specification of a class of mereologies in the form of composite parts and composite and atomic subparts and their possible connections. The abstract model as well as the axiom system (Sect. 12.4) focuses on the syntax of mereologies. Following that, in Sect. 12.4 we indicate how the model of Sect. 12.3 satisfies the axiom system of that section. In preparation for Sect. 12.6, Sect. 12.5 presents characterisations of attributes of parts, whether atomic or composite. Finally Sect. 12.6 presents a semantic model of mereologies, one of a wide variety of such possible models. This one emphasize the possibility of considering parts and subparts as processes and hence a mereology as a system of processes. Section 12.7 concludes with some remarks on what we have achieved.

12.2 Our Concept of Mereology

12.2.1 Informal Characterisation

Mereology, to us, is the study and knowledge about how physical and conceptual parts relate and what it means for a part to be related to another part: *being disjoint, being adjacent, being neighbours, being contained properly within, being properly overlapped with,* etcetera. By physical parts we mean such spatial individuals which can be pointed to. Examples: *a road net (consisting of street segments and street intersections); a street segment (between two intersections); a street intersection; a road (of sequentially neighbouring street segments of the same name) a vehicle; and a platoon (of sequentially neigbouring vehicles).*

By a conceptual part we mean an abstraction with no physical extent, which is either present or not. Examples: *a bus timetable (not as a piece or booklet of paper, or as an electronic device, but) as an image in the minds of potential bus passengers; and routes of a pipeline, that is, neighbouring sequences of pipes, valves, pumps, forks and joins, for example referred to in discourse*: *the gas flows through "such-and-such" a route"*. The tricky thing here is that a route may be thought of as being both a concept or being a physical part—in which case one ought give them different names: a planned route and an actual road, for example.

The mereological notion of subpart, that is: *contained within* can be illustrated by examples: *the intersections and street segments are subparts of the road net;*

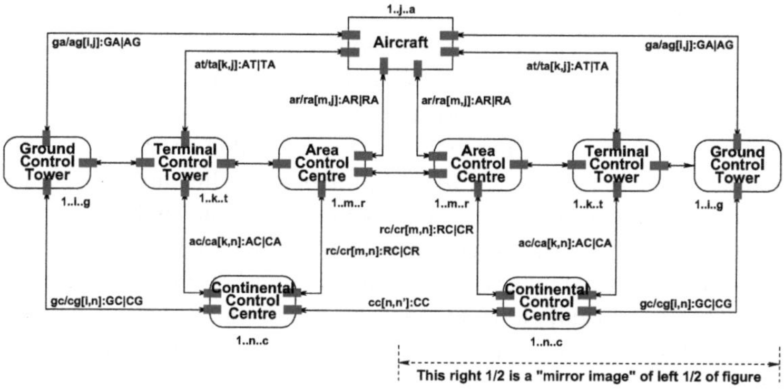

Fig. 12.4 A schematic air traffic system

vehicles are subparts of a platoon; and pipes, valves, pumps, forks and joins are subparts of pipelines. The mereological notion of adjacency can be illustrated by examples. We consider *the various controls of an air traffic system, cf. Fig. 12.4, as well as its aircrafts as adjacent within the air traffic system; the pipes, valves, forks, joins and pumps of a pipeline, cf. Fig. 12.9, as adjacent within the pipeline system; two or more banks of a banking system, cf. Fig. 12.6, as being adjacent.* The mereo-topological notion of neighbouring can be illustrated by examples: *Some adjacent pipes of a pipeline are neighbouring (connected) to other pipes or valves or pumps or forks or joins, etcetera; two immediately adjacent vehicles of a platoon are neighbouring.* The mereological notion of proper overlap can be illustrated by examples some of which are of a general kind: *two routes of a pipelines may overlap; and two conceptual bus timetables may overlap with some, but not all bus line entries being the same*; and some of really reflect adjacency: *two adjacent pipe overlap in their connection, a wall between two rooms overlap each of these rooms — that is, the rooms overlap each other "in the wall".*

12.2.2 Six Examples

We shall, in Sect. 12.3, present a model that is claimed to abstract essential mereological properties of air traffic, buildings and their installations, machine assemblies, financial service industry, the oil industry and oil pipelines, and railway nets.

12.2.2.1 Air Traffic

Figure 12.4 shows nine adjacent (9) boxes and eighteen adjacent (18) lines. Boxes and lines are parts. The line parts "neighbours" the box parts they "connect".

Individually boxes and lines represent adjacent parts of the composite air traffic "whole". The rounded corner boxes denote buildings. The sharp corner box denote an aircraft. Lines denote radio telecommunication. The "overlap" between neigbouring line and box parts are indicated by "connectors". Connectors are shown as small filled, narrow, either horisontal or vertical "filled" rectangle[3] at both ends of the double-headed-arrows lines, overlapping both the line arrows and the boxes. The index ranges shown attached to, i.e., labelling each unit, shall indicate that there are a multiple of the "single" (thus representative) box or line unit shown. These index annotations are what makes the diagram of Fig. 12.4 schematic. Notice that the 'box' parts are fixed installations and that the double-headed arrows designate the ether where radio waves may propagate. We could, for example, assume that each such line is characterised by a combination of location and (possibly encrypted) radio communication frequency. That would allow us to consider all lines for not overlapping. And if they were overlapping, then that must have been a decision of the air traffic system.

12.2.2.2 Buildings

Figure 12.5 shows a building plan—as a composite part.

The building consists of two buildings, A and H. The buildings A and H are neighbours, but shares a common wall. Building A has rooms B, C, D and E, Building H has rooms I, J and K; Rooms L and M are within K. Rooms F and G are within C.

The thick lines labelled N, O, P, Q, R, S, and T models either electric cabling, water supply, air conditioning, or some such "flow" of gases or liquids.

Connection $\kappa\iota o$ provides means of a connection between an environment, shown by dashed lines, and B or J, i.e. "models", for example, a door. Connections κ provides "access" between neighbouring rooms. Note that 'neighbouring' is a transitive relation. Connection $\omega\iota o$ allows electricity (or water, or oil) to be conducted between an environment and a room. Connection ω allows electricity (or water, or oil) to be conducted through a wall. Etcetera.

Thus "the whole" consists of A and B. Immediate subparts of A are B, C, D and E. Immediate subparts of C are G and F. Etcetera.

12.2.2.3 Financial Service Industry

Figure 12.6 is rather rough-sketchy! It shows seven (7) larger boxes [6 of which are shown by dashed lines], six [6] thin lined "distribution" boxes, and twelve (12) double-arrowed lines. Boxes and lines are parts. (We do not described what is meant by "distribution".) Where double-arrowed lines touch upon (dashed) boxes we have

[3] There are 38 such rectangles in Fig. 12.4.

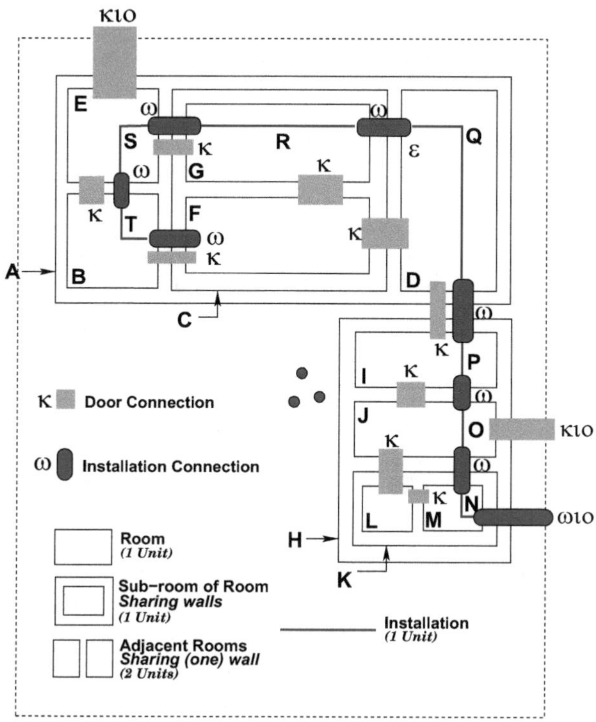

Fig. 12.5 A building plan with installation

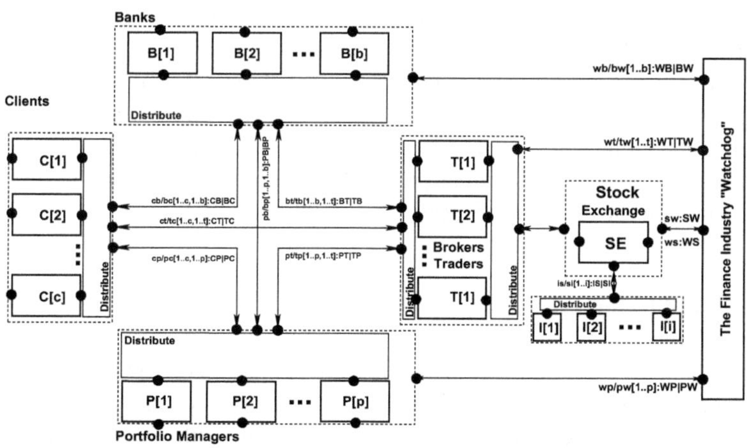

Fig. 12.6 A financial service industry

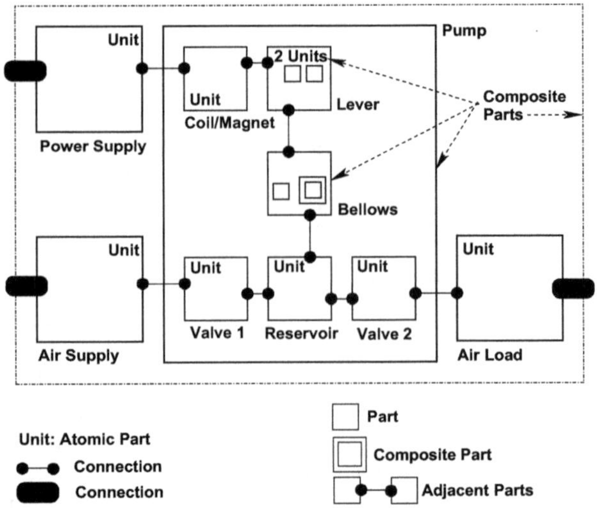

Fig. 12.7 An air pump, i.e., a physical mechanical system

connections. Six (6) of the boxes, the dashed line boxes, are composite parts, five (5) of them consisting of a variable number of atomic parts; five (5) are here shown as having three atomic parts each with bullets "between" them to designate "variability". Clients, not shown, access the outermost (and hence the "innermost" boxes, but the latter is not shown) through connections, shown by bullets, •.

12.2.2.4 Machine Assemblies

Figure 12.7 shows a machine assembly. Square boxes show composite and atomic parts. Black circles or ovals show connections. The full, i.e., the level 0, composite part consists of four immediate parts and three internal and three external connections. The Pump is an assembly of six (6) immediate parts, five (5) internal connections and three (3) external connectors. Etcetera. Some connections afford "transmission" of electrical power. Other connections convey torque. Two connections convey input air, respectively output air.

12.2.2.5 Oil Industry

Figure 12.8 shows a composite part consisting of fourteen (14) composite parts, left-to-right: one oil field, a crude oil pipeline system, two refineries and one, say, gasoline distribution network, two seaports, an ocean (with oil and ethanol tankers and their sea lanes), three (more) seaports, and three, say gasoline and ethanol distribution networks.

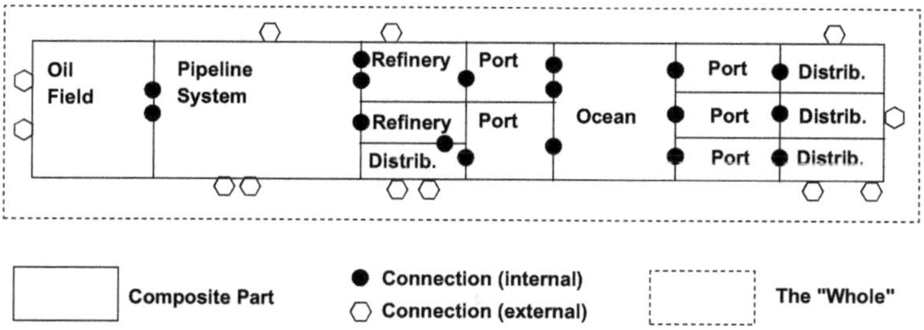

Fig. 12.8 A schematic of an oil industry

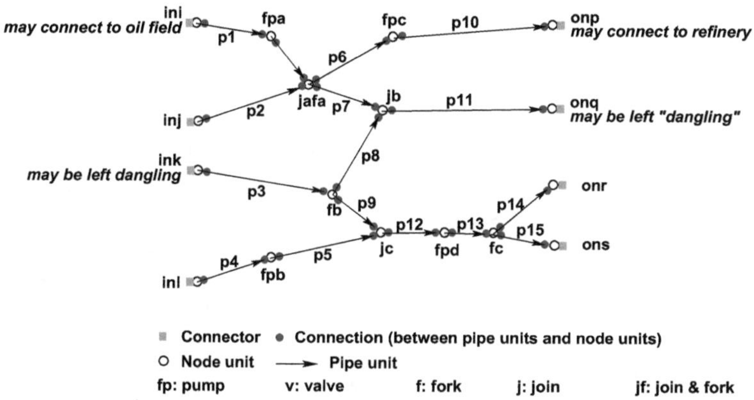

Fig. 12.9 A pipeline system

Between all of the neighbouring composite parts there are connections, and from some of these composite parts there are connections (to an external environment). The crude oil pipeline system composite part will be concretised next.

Figure 12.9 shows a pipeline system. It consists of 32 atomic parts: fifteen (15) pipe units (shown as directed arrows and labelled p1–p15), four (4) input node units (shown as small circles, ∘, and labelled in*i*–in*ℓ*), four (4) flow pump units (shown as small circles, ∘, and labelled fp*a*–fp*d*), five (5) valve units (shown as small circles, ∘, and labelled v*x*–v*w*), three (3) join units (shown as small circles, ∘, and labelled j*b*–jc), two (2) fork units (shown as small circles, ∘, and labelled f*b*–fc), one (1) combined join & fork unit (shown as small circles, ∘, and labelled j*a*f*a*), and four (4) output node units (shown as small circles, ∘, and labelled on*p*–on*s*).

In this example the routes through the pipeline system start with node units and end with node units, alternates between node units and pipe units, and are connected as shown by fully filled-out dark coloured disc connections. Input and output nodes have input, respectively output connections, one each, and shown as lighter coloured connections.

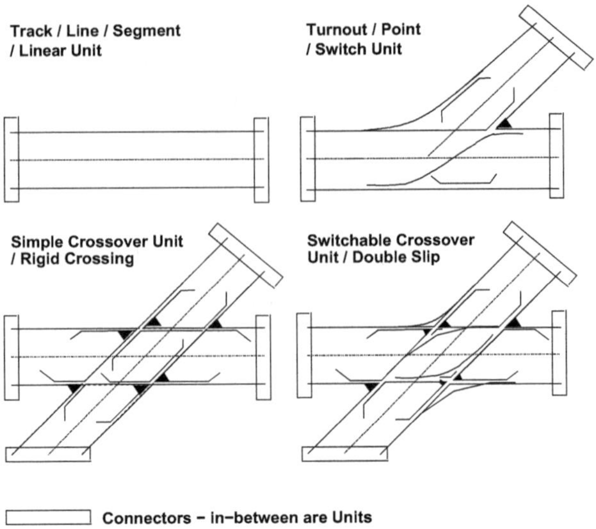

Fig. 12.10 Four example rail units

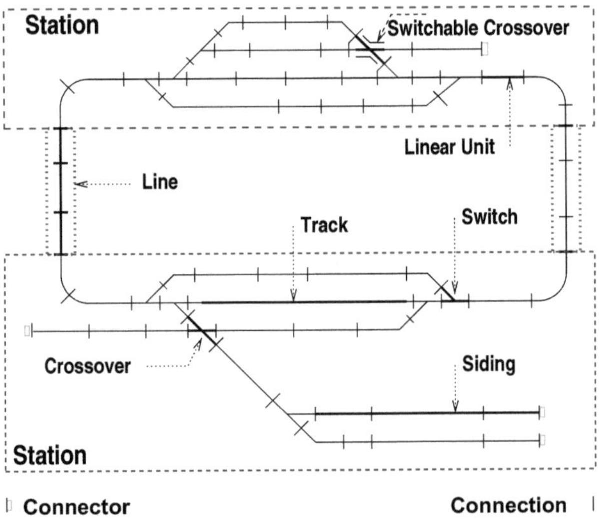

Fig. 12.11 A "model" railway net. An assembly of four assemblies: two stations and two lines; lines here consist of linear rail units; stations of all the kinds of units shown in Fig. 12.10. There are 66 connections and four "dangling" connectors

12.2.2.6 Railway Nets

Figure 12.10 diagrams four rail units, each with two, three or four connectors shown as narrow, somewhat "longish" rectangles. Multiple instances of these rail units can be assembled (i.e., composed) by their connectors as shown on Fig. 12.11 into proper rail nets.

Figure 12.11 diagrams an example of a proper rail net. It is assembled from the kind of units shown in Fig. 12.10. In Fig. 12.11 consider just the four dashed boxes:

The dashed boxes are assembly units. Two designate stations, two designate lines (tracks) between stations. We refer to to the caption four line text of Fig. 12.10 for more "statistics". We could have chosen to show, instead, for each of the four "dangling' connectors, a composition of a connection, a special "end block" rail unit and a connector.

12.2.2.7 Discussion

We have brought these examples only to indicate the issues of a "whole" and atomic and composite parts, adjacency, within, neighbour and overlap relations, and the ideas of attributes and connections. We shall make the notion of 'connection' more precise in the next section. WWW (2007–2010) gives URLs to a number of domain models illustrating a great variety of mereologies.

12.3 An Abstract, Syntactic Model of Mereologies

We distinguish between atomic and composite parts. Atomic parts do not contain separately distinguishable parts. Composite parts contain at least one separately distinguishable part. It is the domain analyser who decides what constitutes "the whole", that is, how parts relate to one another, what constitutes parts, and whether a part is atomic or composite. We refer to the proper parts of a composite part as subparts.

12.3.1 Parts and Subparts

Figure 12.12 illustrates composite and atomic parts. The *slanted sans serif* uppercase identifiers of Fig. 12.12 *A1, A2, A3, A4, A5, A6* and *C1, C2, C3* are meta-linguistic, that is. they stand for the parts they "decorate"; they are not identifiers of "our system".

12.3.1.1 The Model

The formal models of this contribution are expressed in the `RAISE` Specification Language, `RSL` (George et al. 1992, 1995; Bjørner 2006).

1. The "whole" contains a set of parts.
2. A part is either an atomic part or a composite part.
3. One can observe whether a part is atomic or composite.
4. Atomic parts cannot be confused with composite parts.

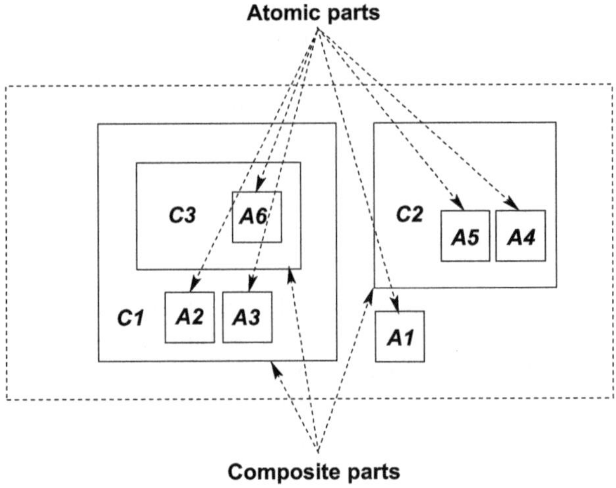

Fig. 12.12 Atomic and composite parts

5. From a composite part one can observe one or more parts.

type
1. W = P-**set**
2. P = A | C
value
3. is_A: P → **Bool**, is_C: P → **Bool**
axiom
4. ∀ a:A,c:C•a≠c, i.e., A∩C={||} ∧ is_A(a)≡∼is_C(a)∧is_C(c)≡∼is_A(c)
value
5. obs_Ps: C → P-**set** **axiom** ∀ c:C • obs_Ps(c)≠{}

The type expression {||} notes the empty type.

Figure 12.12 and the expressions below illustrate the observer function obs_Ps:

- obs_Ps(*C1*) = {*A2, A3, C3*},
- obs_Ps(*C2*) = {*A4, A5*},
- obs_Ps(*C3*) = {*A6*}.

Please note that this example is meta-linguistic. We can define an auxiliary function.

6. From a composite part, c, we can extract all atomic and composite parts
   a. Observable from c or
   b. Extractable from parts observed from c.

value
6. xtr_Ps: C → P-**set**
6. xtr_Ps(c) ≡
6a. **let** ps = obs_Ps(c) **in**
6b. ps ∪ ∪ {obs_Ps(c′)|c′:C • c′ ∈ ps} **end**

ps ∪ ∪ { set-of-sets } expresses that the set ps is union'ed (the first ∪, left-to-right) with the distributed union (the second ∪, left-to-right) of the set of sets.

12.3.2 'Within' and 'Adjacency' Relations

12.3.2.1 'Within'

7. One part, p, is said to be *immediately within*, imm_within(p,p′), another part,
   a. If p′ is a composite part
   b. And p is observable in p′.

value

7. imm_within: $P \times P \xrightarrow{\sim}$ **Bool**
7. imm_within(p,p′) ≡
7a. is_C(p′)
7b. ∧ p ∈ obs_Ps(p′)

12.3.2.2 'Transitive Within'

We can generalise the 'immediate within' property.

8. A part, p, is transitively within a part p′, within(p,p′),
   a. Either if p, is immediately within p′
   b. Or if there exists a (proper) composite part p″ of p′ such that within(p″,p).

value

8. within: $P \times P \xrightarrow{\sim}$ **Bool**
8. within(p,p′) ≡
8a. imm_within(p,p′)
8b. ∨ ∃ p″:C • p″ ∈ obs_Ps(p′) ∧ within(p,p″)

12.3.2.3 'Adjacency'

9. Two parts, p,p′, are said to be *immediately adjacent*, imm_adjacent(p,p′)(c), to one another, in a composite part c, such that p and p′ are distinct and observable in c.

value

9. imm_adjacent: $P \times P \to C \xrightarrow{\sim}$ **Bool**,
9. imm_adjacent(p,p′)(c) ≡ p≠p′ ∧ {p,p′}⊆obs_Ps(c)

12.3.2.4 Transitive 'Adjacency'

We can generalise the immediate 'adjacent' property.

10. Two parts, p,p′, of a composite part, c, are adjacent(p, p′) in c
    a. Either if imm_adjacent(p,p′)(c),
    b. Or if there are two p″ and p‴ of c such that
        i. p″ and p‴ are immediately adjacent parts of c and
        ii. p is equal to p″ or p″ is properly within p and p′ is equal to p‴ or p‴ is properly within p′

value
10. adjacent: P × P → C $\stackrel{\sim}{\to}$ **Bool**
10. adjacent(p,p′)(c) ≡
10a. imm_adjacent(p,p′)(c) ∨
10b. ∃ p″,p‴:P •
10(b)i. imm_adjacent(p″,p‴)(c) ∧
10(b)ii. ((p=p″)∨within(p,p″)(c)) ∧ ((p′=p‴)∨within(p′,p‴)(c))

12.3.3 Unique Identifications

Each physical part can always be uniquely distinguished for example by an abstraction of its properties at a time of origin. In consequence we also endow conceptual parts with unique identifications.

11. In order to refer to specific parts we endow all parts, whether atomic or composite, with **u**nique **id**entifications.
12. We postulate functions which observe these **u**nique **id**entifications, whether as parts in general or as atomic or composite parts in particular.
13. Such that any to parts which are distinct have **u**nique **id**entifications.

type
11. Π
value
12. uid_Π: P → Π
axiom
13. ∀ p,p′:P • p≠p′ ⇒ uid_Π(p)≠uid_Π(p′)

Figure 12.13 illustrates the unique identifications of composite and atomic parts. We exemplify the observer function obs_Π in the expressions below and on Fig. 12.13 on the facing page:

Fig. 12.13 ai_j: atomic part identifiers, ci_k: composite part identifiers

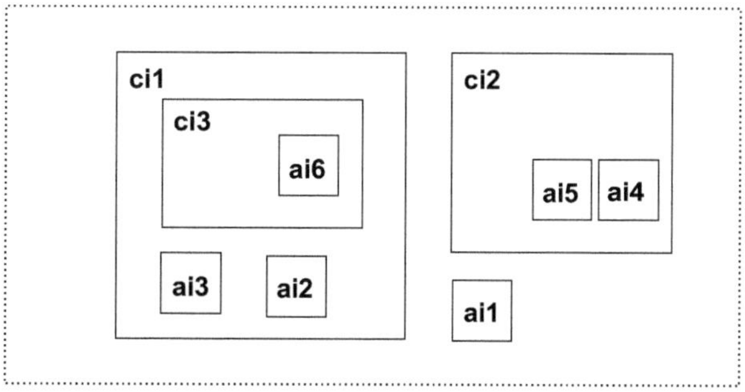

- obs_Π($C1$) = $ci1$, obs_Π($C2$) = $ci2$, etcetera; and
- obs_Π($A1$) = $ai1$, obs_Π($A2$) = $ai2$, etcetera.

Please note that also this example is meta-linguistic.

14. We can define an auxiliary function which extracts all part identifiers of a composite part and parts within it.

value
14. xtr_Πs: C → Π**-set**
14. xtr_Πs(c) ≡ {uid_Π(c)} ∪ ∪ {uid_Π(p)|p:P•p ∈ xtr_Πs(c)}

12.3.4 Attributes

In Sect. 12.5 we shall explain the concept of properties of parts, or, as we shall refer to them, attributes For now we just postulate that

15. Parts have sets of attributes, atr:ATR, (whatever they are!),
16. That we can observe attributes from parts, and hence
17. That two distinct parts may share attributes
18. For which we postulate a membership function ∈.

type
15. ATR
value
16. atr_ATRs: P → ATR-**set**
17. share: P×P → **Bool**
17. share(p,p′) ≡ p≠p′∧∃ atr:ATR•atr ∈ atr_ATRs(p)∧atr ∈ atr_ATRs(p′)
18. ∈: ATR × ATR-**set** → **Bool**

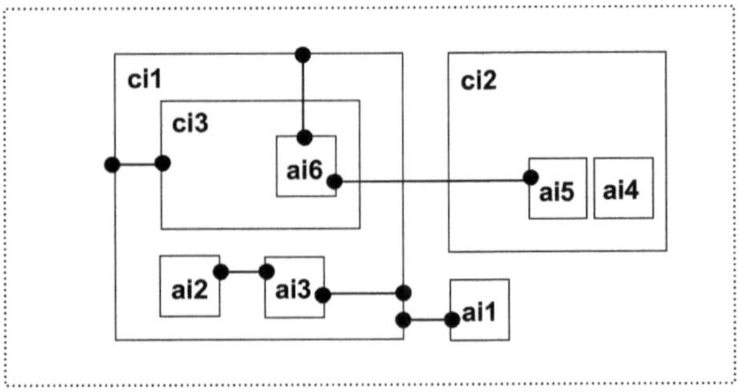

Fig. 12.14 Connectors

12.3.5 Connections

In order to illustrate other than the within and adjacency part relations we introduce the notions of connectors and, hence, connections. Figure 12.14 illustrates connections between parts. A connector is, visually, a •—• line that connects two distinct part boxes.

19. We may refer to the connectors by the two element sets of the unique identifiers of the parts they connect.
For example:

- $\{ci_1, ci_3\}$,
- $\{ai_2, ai_3\}$,
- $\{ai_6, ci_1\}$,
- $\{ai_3, ci_1\}$,
- $\{ai_6, ai_5\}$ and
- $\{ai_1, ci_1\}$.

20. From a part one can observe the unique identities of the other parts to which it is connected.

type
19. K = {| k:Π-**set** • **card** k = 2 |}
value
20. mereo_Ks: P → K-**set**

21. The set of all possible connectors of a part can be calculated.

value
21. xtr_Ks: P → K-**set**
21. xtr_Ks(p) ≡ {{uid_Π(p),π}|π:Π•π ∈ mereo_Πs(p)}

12.3.5.1 Connector Wellformedness

22. For a composite part, s:C,
23. All the observable connectors, ks,
24. Must have their two-sets of part identifiers identify parts of the system.

value
22. wf_Ks: C → **Bool**
22. wf_Ks(c) ≡
23. **let** ks = xtr_Ks(c), πs = mereo_Πs(c) **in**
24. ∀ {π',π''}:Π**-set** • {π',π''}⊆ks ⇒
24. ∃ p′,p″:P • {π',π''}={uid_Π(p′),uid_Π(p″)} **end**

12.3.5.2 Connector and Attribute Sharing Axioms

25. We postulate the following axiom:
    a. If two parts share attributes, then there is a connector between them; and
    b. If there is a connector between two parts, then they share attributes.
26. The function xtr_Ks (Item 21 on the preceding page) can be extended to apply to Wholes.

axiom
25. ∀ w:W•
25. **let** ps = xtr_Ps(w), ks = xtr_Ks(w) **in**
25a. ∀ p,p′:P • p≠p′ ∧ {p,p′}⊆ps ∧ share(p,p′) ⇒
25a. {uid_Π(p),uid_Π(p′)} ∈ ks ∧
25b. ∀ {uid,uid′} ∈ ks ⇒
25b. ∃ p,p′:P • {p,p′}⊆ps ∧ {uid,uid′}={uid_Π(p),uid_Π(p′)} ⇒
25b. share(p,p′) **end**
value
26. xtr_Ks: W → **K-set**
26. xtr_Ks(w) ≡ ∪{xtr_Ks(p)|p:P•p ∈ obs_Ps(p)}

In other words: modelling sharing by means of intersection of attributes or by means of connectors is "equivalent".

12.3.5.3 Sharing

27. When two distinct parts share attributes,
28. Then they are said to be sharing:

27. sharing: P × P → **Bool**
28. sharing(p,p′) ≡ p≠p′∧share(p,p′)

12.3.6 Uniqueness of Parts

There is one property of the model of wholes: W, Item 1, and hence the model of composite and atomic parts and their unique identifiers "spun off" from W (Item 2 [Page 333] to Item 25b [Page 339]). and that is that any two parts as revealed in different, say adjacent parts are indeed unique, where we—simplifying—define uniqueness sôlely by the uniqueness of their identifiers.

12.3.6.1 Uniqueness of Embedded and Adjacent Parts

29. By the definition of the obs_Ps function, as applied obs_Ps(c) to composite parts, c:C, the atomic and composite subparts of c are all distinct and have distinct identifiers (uiids: unique immediate identifiers).

value
29. uiids: C → **Bool**
29. uiids(c) ≡
29. ∀ p,p′:P•p≠p′∧{p,p′}⊆obs_Ps(c)⇒**card**{uid*Π*(p),uid*Π*(p′),uid*Π*(c)}=3

30. We must now specify that that uniqueness is "propagated" to parts that are proper parts of parts of a composite part (uids: unique identifiers).

30. uids: C → **Bool**
30. uids(c) ≡
30. ∀ c′:C•c′ ∈ obs_Ps(c) ⇒ uiids(c′)
30. ∧ **let** ps′=xtr_Ps(c′),ps″=xtr_Ps(c″) **in**
30. ∀ c″:C•c″ ∈ ps′⇒uids(c″)
30. ∧ ∀ p′,p″:P•p′ ∈ ps′∧p″ ∈ ps″⇒uid_*Π*(p′)≠uid_*Π*(p″) **end**

12.4 An Axiom System

Classical axiom systems for mereology focus on just one sort of "things", namely 𝒫arts. Leśniewski had in mind, when setting up his mereology to have it supplant set theory. So parts could be composite and consisting of other, the sub-parts—some of which would be atomic; just as sets could consist of elements which were sets—some of which would be empty.

12.4.1 Parts and Attributes

In our axiom system for mereology we shall avail ourselves of two sorts: $\mathscr{P}$arts, and $\mathscr{A}$ttributes.[4]

- **type** $\mathscr{P}, \mathscr{A}$

$\mathscr{A}$ttributes are associated with $\mathscr{P}$arts. We do not say very much about attributes: We think of attributes of parts to form possibly empty sets. So we postulate a primitive predicate, $\in$, relating $\mathscr{P}$arts and $\mathscr{A}$ttributes.

- $\in: \mathscr{A} \times \mathscr{P} \rightarrow$ **Bool**.

12.4.2 The Axioms

The axiom system to be developed in this section is a variant of that in Casati and Varzi (1999). We introduce the following relations between parts[5]:

$$\begin{aligned}
\text{part_of:}\ & \mathbb{P} : \mathscr{P} \times \mathscr{P} \rightarrow \mathbf{Bool} \\
\text{proper_part_of:}\ & \mathbb{PP} : \mathscr{P} \times \mathscr{P} \rightarrow \mathbf{Bool} \\
\text{overlap:}\ & \mathbb{O} : \mathscr{P} \times \mathscr{P} \rightarrow \mathbf{Bool} \\
\text{underlap:}\ & \mathbb{U} : \mathscr{P} \times \mathscr{P} \rightarrow \mathbf{Bool} \\
\text{over_crossing:}\ & \mathbb{OX} : \mathscr{P} \times \mathscr{P} \rightarrow \mathbf{Bool} \\
\text{under_crossing:}\ & \mathbb{UX} : \mathscr{P} \times \mathscr{P} \rightarrow \mathbf{Bool} \\
\text{proper_overlap:}\ & \mathbb{PO} : \mathscr{P} \times \mathscr{P} \rightarrow \mathbf{Bool} \\
\text{proper_underlap:}\ & \mathbb{PU} : \mathscr{P} \times \mathscr{P} \rightarrow \mathbf{Bool}
\end{aligned}$$

Let $\mathbb{P}$ denote part-hood; p_x is part of p_y, is then expressed as $\mathbb{P}(p_x, p_y)$. Equation (12.1) Part p_x is part of itself (reflexivity). Equation (12.2) If a part p_x is part p_y and, vice versa, part p_y is part of p_x, then $p_x = p_y$ (antisymmetry). Equation (12.3) if a part p_x is part of p_y and part p_y is part of p_z, then p_x is part of p_z (transitivity).

$$\forall p_x : \mathscr{P} \bullet \mathbb{P}(p_x, p_x) \tag{12.1}$$

$$\forall p_x, p_y : \mathscr{P} \bullet (\mathbb{P}(p_x, p_y) \wedge \mathbb{P}(p_y, p_x)) \Rightarrow p_x = p_y \tag{12.2}$$

$$\forall p_x, p_y, p_z : \mathscr{P} \bullet (\mathbb{P}(p_x, p_y) \wedge \mathbb{P}(p_y, p_z)) \Rightarrow \mathbb{P}(p_z, p_z) \tag{12.3}$$

[4] Identifiers P and A stand for model-oriented types (parts and atomic parts), whereas identifiers $\mathscr{P}$ and $\mathscr{A}$ stand for property-oriented types (parts and attributes).

[5] Our notation now is not RSL but a conventional first-order predicate logic notation.

Let $\mathbb{PP}$ denote *proper part-hood.* p_x is a proper part of p_y is then expressed as $\mathbb{PP}(p_x, p_y)$. $\mathbb{PP}$ can be defined in terms of $\mathbb{P}$. $\mathbb{PP}(p_x, p_y)$ holds if p_x is part of p_y, but p_y is not part of p_x.

$$\mathbb{PP}(p_x, p_y) \triangleq \mathbb{P}(p_x, p_y) \wedge \neg\mathbb{P}(p_y, p_x) \tag{12.4}$$

Overlap, $\mathbb{O}$, expresses a relation between parts. Two parts are said to overlap if they have "something" in common. In classical mereology that 'something' is parts. To us parts are spatial entities and these cannot "overlap". Instead they can 'share' attributes.

$$\mathbb{O}(p_x, p_y) \triangleq \exists a : \mathscr{A} \bullet a \in p_x \wedge a \in p_y \tag{12.5}$$

Underlap, $\mathbb{U}$, expresses a relation between parts. Two parts are said to underlap if there exists a part p_z of which p_x is a part and of which p_y is a part.

$$\mathbb{U}(p_x, p_y) \triangleq \exists p_z : \mathscr{P} \bullet \mathbb{P}(p_x, p_z) \wedge \mathbb{P}(p_y, p_z) \tag{12.6}$$

Think of the underlap p_z as an "umbrella" which both p_x and p_y are "under".

Over-cross, $\mathbb{OX}$, p_x and p_y are said to over-cross if p_x and p_y overlap and p_x is not part of p_y.

$$\mathbb{OX}(p_x, p_y) \triangleq \mathbb{O}(p_x, p_y) \wedge \neg\mathbb{P}(p_x, p_y) \tag{12.7}$$

Under-cross, $\mathbb{UX}$, p_x and p_y are said to under cross if p_x and p_y underlap and p_y is not part of p_x.

$$\mathbb{UX}(p_x, p_y) \triangleq \mathbb{U}(p_x, p_z) \wedge \neg\mathbb{P}(p_y, p_x) \tag{12.8}$$

Proper Overlap, $\mathbb{PO}$, expresses a relation between parts. p_x and p_y are said to properly overlap if p_x and p_y over-cross and if p_y and p_x over-cross.

$$\mathbb{PO}(p_x, p_y) \triangleq \mathbb{OX}(p_x, p_y) \wedge \mathbb{OX}(p_y, p_x) \tag{12.9}$$

Proper Underlap, $\mathbb{PU}$, p_x and p_y are said to properly underlap if p_x and p_y under-cross and p_x and p_y under-cross.

$$\mathbb{PU}(p_x, p_y) \triangleq \mathbb{UX}(p_x, p_y) \wedge \mathbb{UX}(p_y, p_x) \tag{12.10}$$

12.4.3 Satisfaction

We shall sketch a proof that the *model* of the previous section, Sect. 12.3, *satisfies*—is a model for—the *axioms* of this section. To that end we first define the notions of *interpretation, satisfiability, validity* and *model.*

Interpretation: By an interpretation of a predicate we mean an assignment of a truth value to the predicate where the assignment may entail an assignment of values, in general, to the terms of the predicate.

Satisfiability: By the satisfiability of a predicate we mean that the predicate is true for some interpretation.

Valid: By the validity of a predicate we mean that the predicate is true for all interpretations.

Model: By a model of a predicate we mean an interpretation for which the predicate holds.

12.4.3.1 A Proof Sketch

We assign

31. P as the meaning of $\mathscr{P}$
32. ATR as the meaning of $\mathscr{A}$,
33. imm_within as the meaning of $\mathbb{P}$,
34. within as the meaning of $\mathbb{PP}$,
35. $\in_{(\text{of type:}ATR \times ATR\textbf{-set} \rightarrow \textbf{Bool})}$ as the meaning of $\in_{(\text{of type:}\mathscr{A} \times \mathscr{P} \rightarrow \textbf{Bool})}$ and
36. sharing as the meaning of $\mathbb{O}$.

With the above assignments one can prove that the other axiom-operators $\mathbb{U}$, $\mathbb{PO}$, $\mathbb{PU}$, $\mathbb{OX}$ and $\mathbb{UX}$ can be modelled by means of $\in_{(\text{of type:}ATR \times ATR\textbf{-set} \rightarrow \textbf{Bool})}$ imm_within, within and sharing.

12.5 An Analysis of Properties of Parts

So far we have not said much about *"the nature"* of parts other than composite parts having one or more subparts and parts having attributes. In preparation also for the next section, Sect. 12.6 we now take a closer look at the concept of 'attributes'. We consider three kinds of attributes: their unique identifications [uid_Π]—which we have already considered; their connections, i.e., their mereology [mereo_P]—which we also considered; and their "other" attributes which we shall refer to as properties. [prop_P]

12.5.1 *Mereological Properties*

12.5.1.1 An Example

Road nets, n:N, consists of a set of street intersections (hubs), h:H, uniquely identified by hi's (in HI), and a set of street segments (links), l:L, uniquely identified by li's (in LI). such that from a street segment one can observe a two element set of street intersection identifiers, and from a street intersection one can observe a set of street segment identifiers. Constraints between values of link and hub identifiers must be satisfied. The two element set of street intersection identifiers express that the street segment is connected to exactly two existing and distinct street intersections, and the zero, one or more element set of street segment identifiers express that the street intersection is connected to zero, one or more existing and distinct street segments. An axiom expresses these constraints. We call the hub identifiers of hubs and links, the link identifiers of links and hubs, and their fulfilment of the axiom the connection **mereo**logy.

```
type
    N, H, L, HI, LI
value
    obs_Hs: N→H-set, obs_Ls: N→L-set
    uid_HI: H→HI, uid_LI: L→LI
    mereo_HIs: L→HI-set axiom ∀ l:L•card mereo_HIs(l)=2
    mereo_LIs: H→LI-set
axiom
  ∀ n:N•
      let hs=obs_Hs(n),ls=obs_Ls(n) in
      ∀ h:H•h ∈ hs⇒∀ li:LI•li ∈ mereo_LIs(h)⇒∃ l:L•uid_LI(l)=li
      ∧ ∀ l:L•l ∈ ls⇒∃ h,h′:H•{h,h′}⊆hs∧mereo_HIs(l)={uid_HI(h),uid_HI(h′)}
      end                                                              •
```

12.5.1.2 Unique Identifier and Mereology Types

In general we allow for any embedded (within) part to be connected to any other embedded part of a composite part or across adjacent composite parts. Thus we must, in general, allow for a family of part types P1, P2, ..., Pn, for a corresponding family of part identifier types Π1, Π2, ..., Πn, and for corresponding observer **u**nique **id**entification and **mer**eology functions:

```
type
    P = P1 | P2 | ... | Pn
    Π = Π1 | Π2 | ... | Πn
value
    uid_Πj: Pj → Πj for 1≤j≤n
    mereo_Πs: P → Π-set
```

Example: Our example relates to the abstract model of Sect. 12.3.

37. With each part we associate a unique identifier, π.
38. And with each part we associate a set, $\{\pi_1, \pi_2, \ldots, \pi_n\}, n \leq 0$ of zero, one ore more other unique identifiers, different from π.
39. Thus with each part we can associate a set of zero, one or more connections, viz.: $\{\pi, \pi_j\}$ for $0 \leq j \leq n$.

type
37. Π
value
37. uid_Π: P $\rightarrow$ Π
38. mereo_Πs: P $\rightarrow$ Π**-set**
axiom
38. $\forall$ p:P•uid_Π(p) $\notin$ mereo_Πs(p)
value
39. xtr_Ks: P $\rightarrow$ K**-set**
39. xtr_Ks(p) $\equiv$
39. **let** (π,πs)=(uid_Π,mereo_Πs)(p) **in**
39. $\{\{\pi',\pi''\}|\pi',\pi'':\Pi\bullet\pi'=\pi\wedge\pi'' \in \pi s\}$ **end**

•

12.5.2 Properties

By the properties of a part we mean such properties additional to those of unique identification and mereology. Perhaps this is a cryptic characterisation. Parts, whether atomic or composite, are there for a purpose. The unique identifications and mereologies of parts are there to refer to and structure (i.e., relate) the parts. So they are there to facilitate the purpose. The properties of parts help towards giving these parts "their final meaning". (We shall support his claim ("their final meaning") in Sect. 12.6.) Let us illustrate the concept of properties.

Examples: (i) Typical properties of street segments are: length, cartographic location, surface material, surface condition, traffic state—whether open in one, the other, both or closed in all directions. (ii) Typical properties of street intersections are: design,[6] location, surface material, surface condition, traffic state—open or closed between any two pairs of in/out street segments. (iii) Typical properties of road nets are: name, owner, public/private, free/tool road, area, etcetera. •

40. Parts are characterised (also) by a set of one or more distinctly named and not necessarily distinctly typed property values.

---

[6] For example, a simple 'carrefour', or a (circular) roundabout, or a free-way interchange, or a cloverleaf, or a stack, or a clover-stack, or a turbine, or a roundabout, or a trumpet, or a directional, or a full Y, or a hybrid interchange.

a. Property names are further undefined tokens (i.e., simple quantities).
b. Property types are either sorts or are concrete types such as integers, reals, truth values, enumerated simple tokens, or are structured (sets, Cartesians, lists, maps) or are functional types.
c. From a part
   i. One can observe its sets of property names
   ii. And its set (i.e., enumerable map) of distinctly named and typed property values.
d. Given an property name of a part one can observe the value of that part for that property name.
e. For practical reasons we suggest **prop**erty named **prop**erty value observer function—where we further take the liberty of using the **prop**erty type name in lieu of the **prop**erty name.

type
40. Props = PropNam $\overrightarrow{m}$ PropVAL
40a. PropNam
40b. PropVAL
value
40(c)i. obs_Props: P → Props
40(c)ii. xtr_PropNams: P → PropNam-**set**
40(c)ii. xtr_PropNams(p) ≡ **dom** obs_Props(p)
40d. xtr_PropVAL: P → PropNam $\stackrel{\sim}{\rightarrow}$ PropVAL
40d. xtr_PropVAL(p)(pn) ≡ (obs_Props(p))(pn)
40d. **pre**: pn ∈ xtr_PropNams(p)

Here we leave PropNames and PropVALues undefined.
Example:

type
NAME, OWNER, LEN, DESIGN, PP == public | private, ...
LΣ, HΣ, LΩ, HΩ
value
obs_Props: N → {| [″name″↦nm,″owner″↦ow,″public/private″↦pp,...]
| nm:NAME, ow:OWNER, ..., pp:PP |}
obs_Props: L → {| [″length″↦len,...,″state″↦lσ,″state space″↦lω:LΩ]
| **len**:LEN,...,lσ:LΣ,lω:LΩ |}
obs_Props: H → {| [″design″↦des, ...,″state″↦hσ,″state space″↦hω]
| des:DESIGN,...,hσ:HΣ,hω:HΩ |}
prop_NAME: N → NAME
prop_OWNER: N → OWNER
prop_LEN: L → LEN
prop_LΣ: L → LΣ, obs_LΩ: L → LΩ
prop_DESIGN: H → DESIGN
prop_HΣ: H → HΣ, obs_HΩ: H → HΩ
...

We trust that the reader can decipher this example. •

12.5.3 Attributes

There are (thus) three kinds of part attributes:

- **u**nique **id**entifier "observers" (uid_),
- **mereo**logy "observers (**mereo_**), and
- **prop**erty "observers" (**prop_...**, obs_Props)

We refer to Sect. 12.3.4, and to Items 15–16.

type
15.′ ATR = $\Pi \times \Pi$**-set** × Props
value
16.′ atr_ATR: P → ATR
axiom
∀ p:P • **let** (π,πs,props) = atr_ATR(p) **in** $\pi \notin \pi$s **end**

In preparation for redefining the share function of Item 17 we must first introduce a modification to property values.

41. A property value, pv:PropVal, is
either a simple property value (as was hitherto assumed), or is a unique part identifier.

type
40. Props = PropNam $\overrightarrow{m}$ PropVAL_or_Π
41. PropVAL_or_Π :: mk_Simp:PropVAL | mk_Π:Π

42. The idea a property name pn, of a part p', designating a Π-valued property value π is
    a. That π refers to a part p'
    b. One of whose property names must be pn
    c. And whose corresponding property value must be a proper, i.e., simple property value, v,
    d. Which is then the property value in p' for pn.

value
42. get_VAL: P × PropName → W → PropVAL
42. get_VAL(p,pn)(w) ≡
44. **let** pv = (obs_Props(p))(pn) **in**
42. **case** pv **of**
42. mk_Simp(v) → v,
42a. mk_Π(π) →
42a. **let** p′:P•p′ ∈ xtr_Ps(w)∧uid_Π(p′)=π **in**
42c. (obs_Props(p′))(pn) **end**
42. **end end**

42c. **pre**: pn ∈ obs_PropNams(p)
42b. ∧ pn ∈ obs_PropNams(p′)
42c. ∧ is_PropVAL((obs_Props(p′))(pn))

The three bottom lines above, Items 42b–42c, imply the general constraint now formulated.

43. We now express a constraint on our modelling of attributes.
    a. Let the attributes of a part p be $(\pi, \pi s, \mathsf{props})$.
    b. If a property name pn in props has (associates to) a Π value, say π'
    c. Then π' must be in πs.
    d. And there must exist another part, p', distinct from p, with unique identifier π', such that
    e. It has some property named pn with a simple property value.

value
43. wf_ATR: ATR → W → **Bool**
43a. wf_ATR(π,πs,props)(w) ≡
43a. π ∉ πs ∧
43b. ∀ π′:Π • π′ ∈ **rng** props ⇒
43c. **let** pn:PropNam•props(pn)=π′ **in**
43c. pi′∈ πs
43d. ∧ ∃ p′:P•p′∈ xtr_Ps(w)∧uid_Π(p′)=π′ ⇒
43e. pn ∈ obs_PropNams(obs_Props(p′))
43e. ∧ ∃ mk_SimpVAL(v):VAL•(obs_Props(p′))(pn)=mk_SimpVAL(v) **end**

44. Two distinct parts share attributes
    a. If the unique part identifier of one of the parts is in the mereology of the other part, or
    b. If a property value of one of the parts refers to a property of the other part.

value
44. share: P × P → **Bool**
44. share(p,p′) ≡
44. p ≠ p′ ∧
44. **let** (π,πs,props) = atr_ATR(p),(π′,πs′,props′) = atr_ATR(p′),
44. pns = xtr_PropNams(p), pns′ = xtr_PropNams(p′) **in**
44a. π ∈ πs′ ∨ π′ ∈ πs ∨
44b. ∃ pn:PropNam•pn ∈ pns ∩ pns′ ⇒
44b. **let** vop = props(pn), vop′ = props′(pn) **in**
44b. **case** (vop,vop′) **of**
44b. (mk_Π(π″),mk_Simp(v)) → π″=π′,
44b. (mk_Simp(v),mk_Π(π″)) → π=π″,
44b. _ → **false**
44. **end end end**

Comment: v is a shared attribute.

12.5.4 Discussion

We have now witnessed four kinds of observer function:

- The above three kinds of mereology and property 'observers' and the
- Part (and subpart) **obs**_ervers.

These observer functions are postulated. They cannot be defined. They "just exist" by the force of our ability to observe and decide upon their values when applied by us, the domain observers.

Parts are either composite or atomic. Analytic functions are postulated. They help us decide whether a part is composite or atomic, and, from composite parts their immediate subparts.

Both atomic and composite parts have all three kinds of attributes: unique identification, mereology (connections), and properties. Analytic functions help us observe, from a part, its unique identification, its mereology, and its properties.

Some attribute values may be static, that is, constant, others may be inert dynamic, that is, can be changed. It is exactly the inert dynamic attributes which are the basis for the next sections semantic model of parts as processes.

In the above model (of this and Sect. 12.3) we have not modelled distinctions between static and dynamic properties. You may think, instead of such a model, that an **always** temporal operator, $\Box$, being applied to appropriate predicates.

12.6 A Semantic CSP Model of Mereology

The model of Sect. 12.3 can be said to be an abstract model-oriented definition of the syntax of mereology. Similarly the axiom system of Sect. 12.4 can be said to be an abstract property-oriented definition of the syntax of mereology. With the analysis of attributes of parts, Sect. 12.5, we have begun a semantic analysis of mereology. We now bring that semantic analysis a step further.

12.6.1 A Semantic Model of a Class of Mereologies

We show that to every mereology there corresponds a program of cooperating sequential processes CSP (Hoare 2004).

Some of the attributes of parts may be static, i.e., constants, others may be dynamic, i.e., variable. The latter form the state of parts. Actions change part states. Processes, P, Q, are sets of sequences of actions and sets of processes. Processes may communicate values (of type M) and then do so via channels, ch. Schematic process P and Q definitions

```
    type
[1]     A, B, M
    channel
[2]     ch:M
    value
[3]     P: A → out ch process,          Q: B → in ch process
[4]     P(a) ≡ ... ;                    Q(b) ≡ ... ;
[5]           ch!f(a);                        let v = ch? in
[6]           ... ;                           ... ;
[7]           P(g(a))                         Q(h(v,b)) end
```

[1] A, B and M are sets of values. [3] P and Q are names of processes. [4] Process P initially accepts arguments a of type A, may offer outputs, ch!f(a), of type M on channel ch and otherwise continues "infinitely" [7] (hence type **process** in line [3]). Similarly Q initially accepts arguments of type B, may accept input, ch? (of type M) also on channel ch and otherwise continues "infinitely" [7]. [5] The output offering, ch!f(a), of process P may be accepted, ch?, by Q. The "..." accounts for our saying "may". We leave the d and h functions undefined.

12.6.1.1 Parts ≃ Processes

The model of mereology presented in Sect. 12.3 (Pages 333–340) focused on (i) parts and (ii) connectors. To parts we associate CSP processes. Part processes are indexed by the unique part identifiers. The connectors form the mereological attributes of the model.

12.6.1.2 Connectors ≃ Channels

The CSP channels are indexed by the two-set (hence distinct) part identifier connectors. From a whole we can extract (xtr_Ks, Item 26) all connectors. They become indexes into an array of channels. Each of the connector channel index identifiers indexes exactly two part processes. Let w:W be the whole under analysis.

```
value
   w:W
   ps:P-set = ∪{xtr_Ps(c)|c:C•c ∈ w} ∪ {a|a:A•a ∈ w}
   ks:K-set = xtr_Ks(w)
type
   K = Π-set axiom ∀ k:K•card k=2
   ChMap = Π →m K-set
value
   cm:ChMap = [ uid_Π(p)↦xtr_Ks(p)|p:P•p ∈ ps ]
```

channel
ch[k|k:K•k ∈ ks] MSG

We leave channel messages. m:MSG, undefined.

12.6.1.3 Process Definitions

The whole, w:W, where W=P-**set**, is semantically seen as the distributed parallel composition of part processes one for each part, p_i in w where w = $\{p_1, p_2, \ldots, p_m\}$.

value
system: W → **process**
system(w) ≡ ‖ {part_process(uid_Π(p))(p)|p:P•p ∈ w}

A part process is either a composite (part) process or an atomic (part) process.

value
part_process: Π → P → **process**
part_process(π)(p) ≡ **assert:** π = uid_Π(p)
  is_C(p) → composite_process(π)(p),
  is_A(p) → atomic_process(π)(p)

A composite process, c, is the parallel composition of the core composite process, $\mathscr{M}_{\mathscr{C}}$, with the distributed parallel composition of part processes, one for each part observed from c.

value
composite_process: π:Π → c:C → **in,out** {ch(k)|k:K•k ∈ cm(π)} **process**
composite_process(π)(c) ≡ **assert:** π = uid_Π(c)
  $\mathscr{M}_{\mathscr{C}}$(π)(c)(atr_ATR(c)) ‖
  ‖ {part_process(uid_Π(p))(p)|p:P•ps ∈ obs_Ps(p)}

ATR and atr_ATR are defined in Items 15.′ and 16.′ (Page 347).

The core composite process, $\mathscr{M}_{\mathscr{C}}$ (of a composite part c), is an in[de]finite cyclic process which evolves around the attributes, atr_ATR(c), of c. $\mathscr{M}_{\mathscr{C}}$ is based on a postulated, i.e., an undefined attribute update action C$\mathscr{F}$.

$\mathscr{M}_{\mathscr{C}}$: π:Π → C → ATR → **in,out** {ch(k)|k:K•k ∈ cm(π)} **process**
$\mathscr{M}_{\mathscr{C}}$(π)(c)(c_attrs) ≡ $\mathscr{M}_{\mathscr{C}}$(π)(c)(C$\mathscr{F}$(π)(c)(c_attrs))
  assert: π = uid_Π(c) ∧ atr_ATR(c) ≡ c_attrs

C$\mathscr{F}$ potentially communicates with all those part processes (of the whole, w) with which c, the part on which $\mathscr{M}_{\mathscr{C}}$(π)(c)(atr_ATR(c)) is based, shares attribites, that is, has connectors.

$\mathcal{CF}$: $\pi{:}\Pi \to$ c:C $\to$ ATR $\to$ **in**,**out** {ch[em(i)]|i:KI•i $\in$ cm(π)} ATR

Atomic processes are just that: They evolve sôlely around a core atomic process $\mathcal{M}_{\mathcal{A}}$(a)(atr_ATR(a)).

atomic_process: $\pi{:}\Pi \to$ A $\to$ **in**,**out** {ch[cm(k)]|:K•k $\in$ cm(π)} **process**
atomic_process(π)(a) $\equiv$ $\mathcal{M}_{\mathcal{A}}$(a)(atr_ATR(a)) **assert:** π = uid_Π(a)

The core atomic process, $\mathcal{M}_{\mathcal{A}}(\pi)$(a)(atr_ATR(a)), is based on a postulated, i.e., an undefined attribute update action A$\mathcal{F}$.

$\mathcal{M}_{\mathcal{A}}$: $\pi{:}\Pi \to$ A $\to$ ATR $\to$ **in**,**out** {ch[cm(k)]|k:K•k $\in$ cm(π)} **process**
$\mathcal{M}_{\mathcal{A}}(\pi)$(a)(a_attrs) $\equiv$ $\mathcal{M}_{\mathcal{A}}(\pi)$(a)($A\mathcal{F}$(a)(a_attrs))
    assert: π = uid_Π(a) $\wedge$ atr_ATR(a) $\equiv$ a_attrs

A$\mathcal{F}$ potentially communicates with all those part processes (of the whole, w) with which a, the part on which $\mathcal{M}_{\mathcal{A}}(\pi)$(a)(atr_ATR(a))is based, shares attribites, that is, has connectors.

A$\mathcal{F}$: $\pi{:}\Pi \to$ A $\to$ ATR $\to$ **in**,**out** {ch[em(k)]|k:K • k $\in$ cm(π)} ATR

The meaning processes $\mathcal{M}_{\mathcal{C}}$ and $\mathcal{M}_{\mathcal{A}}$ are generic. Their sôle purpose is to provide a never ending recursion. "In-between" they "make use" of *C*omposite, respectively *A*tomic specific $\mathcal{F}$unctions here symbolised by $C\mathcal{F}$, respectively $A\mathcal{F}$.

Both $C\mathcal{F}$ and $A\mathcal{F}$ are expected to contain input/output clauses referencing the channels of their signatures; these clauses enable the sharing of attributes. We illustrate this "sharing" by the schematised function $\mathcal{F}$ standing for either $C\mathcal{F}$ or $A\mathcal{F}$.

The $\mathcal{F}$ action non-deterministically internal choice chooses between either [1,2,3] accepting input from another part process, then optionally offering a reply to that other process, and finally delivering an updated state; or [4,5] offering an output to another part process, and then delivering an updated state; or [6] doing own work resulting in an updated state.

value
    $\mathcal{F}$: p:P $\to$ ATR $\to$ **in**,**out** {ch[em(k)]|k:K • k $\in$ cm(uid_Π(p))} ATR
    $\mathcal{F}$(p)(π,πs,props) $\equiv$ **assert:** uid_Π(p)=π
[1]   ⌷ {**let** av = ch[em({π,j})] ? **in**
[2]   ... ; [optional] ch[em({π,j})] ! in_reply(props)(av);
[3]   (π,πs,in_update_ATR(props)(j,av)) **end** | {π,j}:K•{π,j} $\in$ πs}
[4] $\sqcap$ ⌷ { ... ; ch[em({π,j})] ! out_reply(props);
[5]   (π,πs,out_update_ATR(props)(j)) | {π,j}:K•{π,j} $\in$ πs}
[6] $\sqcap$ (π,πs,own_work(props))
    assert: π = uid_Π(p)

    in_reply: Props $\to$ $\Pi \times$ VAL $\to$ VAL
    in_update_ATR: Props $\to$ $\Pi \times$ VAL $\to$ Props

out_reply: Props → VAL
out_update_ATR: Props →Π → Props
own_work: Props → Props

We leave VAL undefined.

12.6.2 Discussion

Parts, subparts and their relations, that is, mereology, reflect *syntactic properties* of wholes. The interpretations of parts as processes and mereology as channels reflect *semantic properties* of wholes. What we have shown in this section is that to every mereology there corresponds a *normal form* `CSP` "program schema".

12.7 Concluding Remarks

A first basic idea of this paper has been to to take axiom systems of mereology and render them mathematical in the sense of showing that a mathematical model satisfies the axiom system A second basic idea of this paper has then been to extend this model-oriented treatment to not just covering syntactic aspects of mereology but also to cover, albeit schematically, normative, schematic models of semantic aspects of mereology.

12.7.1 Relation to Other Work

The present contribution has been conceived in the following model-oriented context.

My first awareness of the concept of 'mereology' was from listening to many presentations by Douglas T. Ross (1929–2007) at IFIP working group WG3.2 meetings over the years 1980–1999. In Douglas T. Ross and John E. Ward (1968) reports on the 1958–1967 MIT project for *computer-aided design (CAD) for numerically controlled production.*[7] Pages 13–17 of Ross and Ward (1968) reflects on issues bordering to and behind the concerns of mereology. Ross' thinking is clearly seen in the following text: *"... our consideration of fundamentals begins not with design or problem-solving or programming or even mathematics, but with philosophy (in the old-fashioned meaning of the word) – we begin by establishing a "world-view". We have repeatedly emphasized that there is no*

[7]Doug is said to have coined the term and the abbreviation CAD (Ross 1961).

Douglas T. Ross 1927–2007.
Courtesy MIT Museum

way to bound or delimit the potential areas of application of our system, and that we must be prepared to cope with any conceivable problem. Whether the system will assist in any way in the solution of a given problem is quite another matter, ..., but in order to have a firm and uniform foundation, we must have a uniform philosophical basis upon which to approach any given problem. This "world-view" must provide a working framework and methodology in terms of which any aspect of our awareness of the world may be viewed. It must be capable of expressing the utmost in reality, giving expression to unending layers of ever-finer and more concrete detail, but at the same time abstract chimerical visions bordering on unreality must fall within the same scheme. "Above all, the world-view itself must be concrete and workable, for it will form the basis for all involvement of the computer in the problem-solving process, as well as establishing a viewpoint for approaching the unknown human component of the problem-solving team." Yes, indeed, the philosophical disciplines of ontology, epistemology and mereology, amongst others, ought be standard curricula items in the computer science and software engineering studies, or better: domain engineers cum software system designers ought be imbued by the wisdom of those disciplines as was Doug. *"... in the summer of 1960 we coined the word plex to serve as a generic term for these philosophical ruminations. "Plex" derives from the word plexus, "An interwoven combination of parts in a structure", (Webster). ... The purpose of a* ***'modeling plex'*** *is to represent completely and in its entirety a "thing", whether it is concrete or abstract, physical or conceptual. A 'modeling plex' is a trinity with three primary aspects, all of which must be present. If any one is missing a complete representation or modeling is impossible. The three aspects of plex are* ***data, structure,*** *and* ***algorithm.*** *... "* which *"... is concerned with the behavioral characteristics of the plex model– the* *interpretive rules for making meaningful the data and structural aspects of the plex, for assembling specific instances of the plex, and for interrelating the plex with other plexes and operators on plexes. Specification of the algorithmic aspect removes the ambiguity of meaning and interpretation of the data structure and provides a complete representation of the thing being modeled."* In the terminology of the current paper a plex is a part (whether composite or atomic), the data are the properties (of that part), the structure is the mereology (of that part) and the algorithm is the process (for that part). Thus Ross was, perhaps, a first instigator (around 1960) of object-orientedness. A first, "top of the iceberg" account of the mereology-ideas that Doug had then can be found in the much later (1976) three page note (Ross 1976). Doug not only 'invented' CAD but was also the father of AED (Algol Extended for Design), the Automatically Programmed Tool (APT) language, SADT (Structured Analysis and Design Technique) and helped develop SADT into the IDEF0 method for the Air Force's Integrated Computer-Aided Manufacturing (ICAM) program's IDEF suite of analysis and design methods. Douglas T. Ross went on for many years thereafter, to deepen and expand his ideas of relations between mereology and the programming language concept of type at the IFIP WG2.3 working group meetings. He did so in the, to some, enigmatic, but always fascinating style you find on Page 63 of Ross (1976).

In Henry S. Leonard and Henry Nelson Goodman (1940): *A Calculus of Individuals and Its Uses* present the American Pragmatist version of Leśniewski's mereology. It is based on a single primitive: *discrete.* The idea the calculus of individuals is, as in Leśniewski's mereology, to avoid having to deal with the empty sets while relying on explicit reference to classes (or parts). The treatment of Leonard and Goodman (1940) is axiomatic.

R. Casati and A. Varzi (1999): *Parts and Places: the structures of spatial representation* has been the major source for this paper's understanding of mereology. Although our motivation was not the spatial or topological mereology, Smith (1996), and although the present paper does not utilize any of these concepts' axiomatision in Casati and Varzi (1999) and Smith (1996) it is best to say that it has benefitted much from these publications. The treatments of these papers are axiomatic.

Domain descriptions, besides mereological notions, also depend, in their successful form. on FCA: Formal Concept Analysis. Here a main inspiration has been drawn, since the mid 1990s from B. Ganter and R. Wille's *Formal Concept Analysis—Mathematical Foundations* (Ganter and Wille 1999). *The approach takes as input a matrix specifying a set of objects and the properties thereof, called attributes, and finds both all the "natural" clusters of attributes and all the "natural" clusters of objects in the input data, where a "natural" object cluster is the set of all objects that share a common subset of attributes, and a "natural" property cluster is the set of all attributes shared by one of the natural object clusters. Natural property clusters correspond one-for-one with natural object clusters, and a concept is a pair containing both a natural property cluster and its corresponding natural object cluster. The family of these concepts obeys the mathematical axioms defining a lattice, a Galois connection).* Thus the choice of adjacent and embedded ('within') parts and their connections is determined after serious formal concept analysis. In Bjørner and Eir (2008) we present a 'concept analysis' approach to domain description, where the present paper presents the mereological approach.

The present paper is based on Bjørner (2009) of which it is an extensive revision and extension.

12.7.2 What Has Been Achieved?

We have given a model-oriented specification of mereology. We have indicated that the model satisfies a widely known axiom system for mereology. We have suggested that (perhaps most) work on mereology amounts to syntactic studies. So we have suggested one of a large number of possible, schematic semantics of mereology. And we have shown that to every mereology there corresponds a set of communicating sequential `CSP` processes.

12.7.3 Future Work

There are four kinds of 'future works': (i) studies that give us further insight into the syntactic mereology operators: `overlap`, `underlap`, `over-crossing`, `under-crossing`, `proper overlap` and `proper underlap`; (ii) studies that explore further semantic models of mereology, we, for example, need to characterise the class of `CSP` programs for which there corresponds a mereology; (iii) refinements of the normative, schematics `CSP` (Sect. 12.6) models of mereology; and (iv) an extensive editing and publication of Doug Ross' surviving notes.

Acknowledgements I thank Dr. Claudio Calosi and Dr. Pierluigi Graziani, University of Urbino, Italy, for inviting this paper for a Springer Verlag *Synthese Library* volume on *Mereology and the Sciences*. I further thank Patricia M. Ross for permission to dedicate this paper to the memory of her husband of many years.

References

Bjørner, D. (2006). *Software engineering, Vol. 1: Abstraction and modelling; Vol. 2: Specification of systems and languages; Vol. 3: Domains, requirements and software design* (Texts in theoretical computer science, the EATCS series). Berlin/New York: Springer.

Bjørner, D. (2008). From domains to requirements. In P. Degano, Rocco De Nicola, & J. Meseguer (Eds.), *Montanari festschrift* (Lecture notes in computer science, pp. 1–30). Heidelberg: Springer.

Bjørner, D. (2009). On mereologies in computing science. In B. Roscoe (Ed.), *Festschrift for Tony Hoare* (History of computing london). UK: Springer.
Bjørner, D. (2010). The role of domain engineering in software development. Why current requirements engineering seems flawed!. In *Perspectives of systems informatics* (Volume 5947 of lecture notes in computer science, pp. 2–34). Heidelberg: Springer.
Bjørner, D., & Eir, A. (2010). Compositionality: Ontology and mereology of domains. Some clarifying observations in the context of software engineering in July 2008. In M. Steffen, D. Dams, & U. Hannemann (Eds.), *Festschrift for Prof. Willem Paul de Roever concurrency, compositionality, and correctness* (Volume 5930 of lecture notes in computer science, pp. 22–59). Heidelberg: Springer.
Casati, R., & Varzi, A. (1999). *Parts and places: The structures of spatial representation.* Cambridge: MIT.
Ganter, B., & Wille, R. (1999). *Formal concept analysis—mathematical foundations.* Springer. ISBN:3540627715, 300 pages, Amazon price: US $44.95.
George, C. W., Haff, P., Havelund, K., Haxthausen, A. E., Milne, R., Nielsen, C. B., Prehn, S., & Wagner, K. R. (1992). *The RAISE specification language* (The BCS practitioner series). Hemel Hampstead: Prentice-Hall.
George, C. W., Haxthausen, A. E., Hughes, S., Milne, R., Prehn, S., & Pedersen, J. S. (1995). *The RAISE development method* (The BCS practitioner series). Hemel Hampstead: Prentice-Hall.
Hoare, C. (1985). *Communicating sequential processes* (C.A.R. Hoare series in computer science). Prentice-Hall International. Published electronically: http://www.\discretionary-using\discretionary-csp.\discretionary-com/\discretionary-csp\discretionary-book.pdf (2004).
Lejewski, C. (1983). A note on Leśniewksi's axiom system for the ereological notion of ingredient or element. *Topoi, 2*(1), 63–71.
Leonard, H. S., & Goodman, N. (1940). The calculus of individuals and its uses. *Journal of Symbolic Logic, 5*, 45–44.
Ross, D. T. (1961). Computer-aided design. *Communications of the ACM, 4*(5), 41–63.
Ross, D. T. (1976). Toward foundations for the understanding of type. In *Proceedings of the 1976 conference on data: Abstraction, definition and structure*, Salt Lake City (pp. 63–65). New York: ACM.
Ross, D. T., & Ward, J. E. Investigations in computer-aided design for numerically controlled production. Final technical report ESL-FR-351, May 1968. 1 December 1959–3 May 1967. Electronic Systems Laboratory Electrical Engineering Department, MIT, Cambridge.
Smith, B. (1996). Mereotopology: A theory of parts and boundaries. *Data and Knowledge Engineering, 20*, 287–303.
WWW. Domain Descriptions:

1. *A Container Line Industry Domain:* http://www2.imm.dtu.dk/~db/container-paper.pdf
2. *What is Logistics:* http://www2.imm.dtu.dk/~db/logistics.pdf
3. *The "Market": Consumers, Retailers, Wholesalers, Producers* http://www2.imm.dtu.dk/~db/themarket.pdf
4. *MITS: Models of IT Security. Security Rules & Regulations:* http://www2.imm.dtu.dk/~db/it-security.pdf
5. *A Domain Model of Oil Pipelines:* http://www2.imm.dtu.dk/~db/pipeline.pdf
6. *A Railway Systems Domain:* http://euler.fd.cvut.cz/railwaydomain/PDF/tb.pdf
7. *Transport Systems:* http://www2.imm.dtu.dk/~db/comet/comet1.pdf
8. *The Tokyo Stock Exchange* http://www2.imm.dtu.dk/~db/todai/tse-1.pdf and http://www2.imm.dtu.dk/~db/todai/tse-2.pdf
9. *On Development of Web-based Software. A Divertimento of Ideas and Suggestions:* http://www2.imm.dtu.dk/~db/wfdftp.pdf

R&D Experiments, Dines Bjørner, DTU Compute, Technical University of Denmark, 2007–2010.

Appendix: Formal Theories of Parthood

Achille C. Varzi

This Appendix gives a brief overview of the main formal theories of parthood, or mereologies, to be found in the literature.[1] The focus is on classical theories, so the survey is not meant to be exhaustive. Moreover, it does not cover the many philosophical issues relating to the endorsement of the theories themselves, concerning which the reader is referred to the Selected Bibliography at the end of the volume. In particular, we shall be working under the following simplifying assumptions[2]:

— *Absoluteness*: Parthood is a two-place relation; it does not hold relative to time, space, spacetime regions, sortals, worlds, or anything else.[3]
— *Monism*: There is a single relation of parthood that applies to every entity independently of its ontological category.[4]
— *Precision*: Parthood is not a source of vagueness: there is always a fact of the matter as to whether the parthood relation obtains between any given pair of things.[5]

For definiteness, all theories will be formulated in a standard first-order language with identity, supplied with a distinguished binary predicate constant, 'P', to be interpreted as the parthood relation. The underlying logic will be the classical predicate calculus.

[1]The exposition follows Varzi (2014). For a thorough survey, see Simons (1987).

[2]The labels and formulations of these assumptions are from Sider (2007).

[3]For the view that parthood should be a three-place relation relativized to time, see e.g. Thomson (1983). For the view that it should be a four-place relation, see Gilmore (2009).

[4]For misgivings about *Absoluteness* and related worries, see e.g. Mellor (2006) and McDaniel (2009).

[5]For mereologies that allow for indeterminate or "fuzzy" parthood relations, see e.g. Smith (2005) and Polkowski (2011).

C. Calosi and P. Graziani (eds.), *Mereology and the Sciences*, Synthese Library 371,
DOI 10.1007/978-3-319-05356-1,

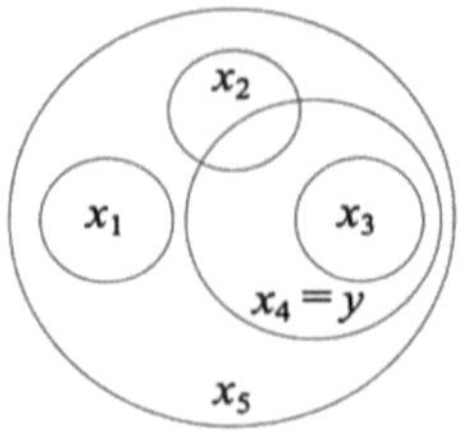

	Ux_ny	Ox_ny	PPx_ny	EQx_ny	PEx_ny
$n = 1$	+	–	–	–	–
$n = 2$	+	+	–	–	–
$n = 3$	+	+	+	–	–
$n = 4$	+	+	–	+	–
$n = 5$	+	+	–	–	+

Fig. 1 Basic patterns of mereological relations (*shaded cells* indicate parthood)

Core Principles

As a minimal requirement on 'P', it is customary to assume that it stands for a partial order—a reflexive, transitive, and antisymmetric relation[6]:

(P.1) Pxx *Reflexivity*
(P.2) $(Pxy \wedge Pyz) \rightarrow Pxz$ *Transitivity*
(P.3) $(Pxy \wedge Pyx) \rightarrow x = y$ *Antisymmetry*

Together, these three axioms are meant to fix the intended meaning of the parthood predicate. They form the "core" of any standard mereological theory, and the theory that comprises just them is called *Ground Mereology*, or M for short.[7] A number of additional mereological predicates may then be introduced by definition (Fig. 1):

(1) $EQxy =_{df} Pxy \wedge Pyx$ *equality*
(2) $PPxy =_{df} Pxy \wedge \neg x = y$ *proper parthood*[8]
(3) $PExy =_{df} Pyx \wedge \neg x = y$ *proper extension*
(4) $Oxy =_{df} \exists z(Pzx \wedge Pzy)$ *overlap*
(5) $Uxy =_{df} \exists z(Pxz \wedge Pyz)$ *underlap*

Fig. 1 Given (P.1)–(P.3), it follows immediately that *EQ* is an equivalence relation. Moreover, *PP* and *PE* are irreflexive, asymmetric, and transitive whereas *O* and *U* are reflexive and symmetric, but not transtive. Since the following is a also theorem of M,

(6) $Pxy \leftrightarrow (PPxy \vee x = y)$

'*PP*' could have been used as a primitive instead of 'P'. Similarly for '*PE*'. Sometimes 'P' is also defined in terms of 'O' via the biconditional

---

[6]Unless otherwise specified, all formulas are to be understood as universally closed.

[7]For a survey of the motivations that may lead to the development of non-standard mereologies in which P is weaker than a partial order, see Varzi (2014: §2.1).

[8]In the literature, proper parthood is sometimes defined as asymmetric parthood: $PPxy =_{df} Pxy \wedge \neg Pyx$. Given *Antisymmetry*, this definition is equivalent to (2). Without *Antisymmetry*, however, the two definitions would come apart. (Similarly for '*PE*'.) See Cotnoir (2010).

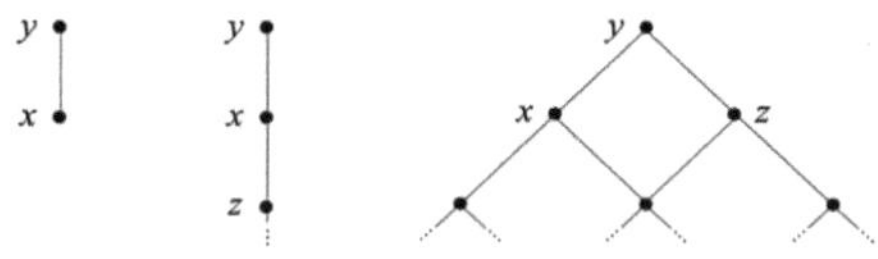

Fig. 2 Three unsupplemented models (connecting lines going upwards indicate proper parthood)

(7) $Pxy \leftrightarrow \forall z(Ozx \rightarrow Ozy)$.

However, (7) is not provable in M and calls for stronger axioms (specifically, the axioms of theory EM defined below). Since those stronger axioms reflect substantive philosophical theses, 'P' and 'PP' (or 'PE') are the best options to start with. Here we stick to 'P'.

Decomposition Principles

M is standardly viewed as embodying the common core of any mereological theory. Yet not every partial order qualifies as parthood, and establishing what further requirements should be added to (P.1)–(P.3) is precisely the question a good mereological theory is meant to answer.

One way to extend M is by means of *decomposition principles*, i.e., principles concerning the part structure of a given whole. Here, one fundamental intuition is that no whole can have a single proper part. There are several ways in which this intuition can be captured, beginning with the following:

(P.4$_a$) $PPxy \rightarrow \exists z(PPzy \wedge \neg z = x)$ *(Weak) Company*
(P.4$_b$) $PPxy \rightarrow \exists z(PPzy \wedge \neg Pzx)$ *Strong Company*
(P.4) $PPxy \rightarrow \exists z(PPzy \wedge \neg Ozx)$ *(Weak) Supplementation*[9]

(P.4$_a$) is the literal rendering of the idea in question, but it is too weak: it rules out certain implausible finitary models (Fig. 2, left) but not, for example, models with infinitely descending chains in which the additional parts do not leave any mereological "remainder" (Fig. 2, center). (P.4$_b$) is stronger, but it still admits of models in which a whole can be decomposed into several proper parts all of which overlap one another (Fig. 2, right). In such cases it is unclear what would be left of the whole upon the removal of any of its proper parts (along with all proper parts thereof). It is only (P.4) that appears to capture the full spirit of the above-mentioned intuition: every proper part must be "supplemented" by another part—a proper part that is completely disjoint (i.e., does not overlap) the first. (P.4) entails both (P.4$_a$)

[9]In the literature, this principle is sometimes formulated using 'P' in place of 'PP' in the consequent. In M the two formulations are equivalent.

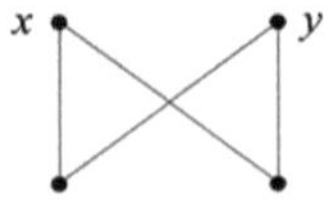

Fig. 3 A weakly supplemented model violating strong supplementation

and ($P.4_b$) and rules out each of the models in Fig. 2. The extension of M obtained by adding this principle to (P.1)–(P.3) is called *Minimal Mereology*, or MM.[10]

There is another, stronger way of expressing the supplementation intuition. It corresponds to the following axiom, which differs from (P.4) in the antecedent:

(P.5) $\neg Pyx \rightarrow \exists z(Pzy \wedge \neg Ozx)$ *Strong Supplementation*

In M this principle entails (P.4). The converse, however, does not hold, as shown by the model in Fig. 3. The stronger mereological theory obtained by adding (P.5) to the three core principles of M is called *Extensional Mereology*, EM.

The extensional character of EM may not be manifest in (P.5) itself, but it becomes clearer in view of the following theorem:

(8) $\exists zPPzx \rightarrow (\forall z(PPzx \rightarrow PPzy) \rightarrow Pxy)$

from which it follows that sameness of mereological composition is both necessary and sufficient for identity:

(9) $\exists zPPzx \rightarrow (x = y \leftrightarrow \forall z(PPzx \leftrightarrow PPzy))$.

Thus, EM is "extensional" precisely insofar as it rules out *any* model of the sort depicted in Fig. 3, where distinct objects decompose into the same proper parts.

There is yet a further way of capturing the supplementation intuition. It corresponds to the following axioms, which differs from (P.5) in the consequent:

(P.6) $\neg Pyx \rightarrow \exists z\forall w(Pwz \leftrightarrow (Pwy \wedge \neg Owx))$. *Complementation*[11]

Informally, (P.6) states that whenever an object fails to include another among its parts, there is something that amounts exactly to the *difference* or relative *complement* between the first object and the second. Once again, it is easily checked that in M this principle entails (P.5)—thus, a fortiori, (P.4)—whereas the converse does not hold (Fig. 4). It should be noted, however, that (P.6) goes beyond the original supplementation intuition. For while it guarantees that a whole cannot have a single proper part, it also pronounces on the specific mereological makeup of the supplementary part. In particular, it requires the relative complement to exist regardless of its internal structure. If, for example, y is a wine glass and x the stem of the glass, (P.6) entails the existence of something composed exactly of the base and the bowl—a spatially disconnected entity. Whether there exist entities of this

[10]Strictly speaking, in MM (P.3) is redundant, as it follows from (P.4) along with (P.1) and (P.2). For ease of reference, however, we shall continue to treat (P.3) as an axiom.

[11]In the literature, (P.6) is also known as the *Remainder Principle*.

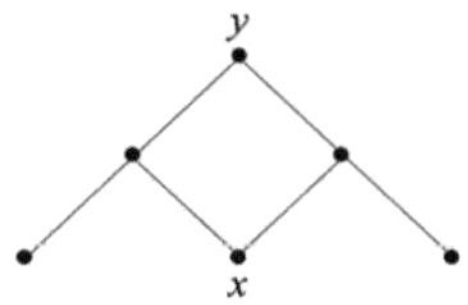

Fig. 4 A strongly supplemented model violating complementation

sort, and more generally whether the remainder between a whole and any one of its proper parts adds up to a *bona fide* entity of its own, is really a question about mereological composition, over and above the conditions on decomposition set by (P.4) and (P.5).

Before turning to issues regarding composition, a different sort of decomposition principles is worth mentioning. Let a mereological atom be any entity with no proper parts:

(10) $Ax =_{\text{df}} \neg\exists yPPyx.$ *atom*

Obviously, all the theories considered so far are compatible with the existence of such things. But one may want to demand more than mere compatibility, just as one may want to preclude it. Thus, one may want to require that everything is ultimately composed of atoms, or else that everything is made up of "atomless gunk"[12] that divides forever into smaller and smaller parts. These two options are usually formulated as follows:

(P.7) $\exists y(Ay \wedge Pyx)$ *Atomicity*
(P.8) $\exists yPPyx$ *Atomlessness*

These postulates are mutually inconsistent, but taken in isolation they can consistently be added to any mereological theory mentioned so far to yield either an *atomistic* variant or an *atomless* variant, respectively.

Atomistic mereologies admit significant semplifications in the axiomatics. For example, Atomistic EM can be simplified by merging *Strong Supplementation* (P.5) and *Atomicity* (P.7) into a single axiom:

(P.5′) $\neg Pxy \rightarrow \exists z(Az \wedge Pzx \wedge \neg Pzy)$ *Atomistic Supplementation*

and the the extensionality thesis (9) can be put more perspicuously as follows:

(9′) $x = y \leftrightarrow \forall z(Az \rightarrow (Pzx \leftrightarrow Pzy))$

This is especially significant if one considers that (P.7) does not quite say that everything is *made up* of atoms; it merely says that everything *has* atomic parts, which is consistent with the possibility of infinitely descending chains of decomposition that never bottom out (Fig. 5). Whether stronger versions of (P.7)

[12]The phrase is from Lewis (1991: 20).

Fig. 5 An infinitely descending atomistic model

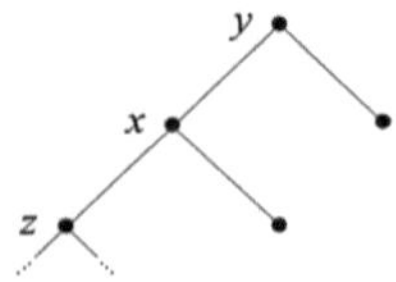

can be formulated that rule out such dubious patterns is, at the moment, a question that has not been fully explored.[13]

Concerning atomless mereologies, one may similarly remark that (P.8) is by itself rather weak. For one thing, the unsupplemented model in Fig. 2, middle, qualifies as atomless. To the extent that such models run afoul of the intended notion of a gunky world, this means that (P.8) calls for teories at least as strong as MM, in which case the relevant axiomatization may again be simplified by merging (P.4) and (P.8) into a single axiom:

(P.4′) $Pxy \to \exists z(PPzy \wedge (Ozx \to x = y))$ *Atomless Supplementation*

Moreover, infinite divisibility is loose talk. Given (P.8), gunk may have as few as denumerably many parts; but can it have more? Is there an upper bound on the cardinality on the number of pieces of gunk? Should it be allowed that for *every* cardinal number there may be more than that many pieces of gunk? (P.8) is silent on these questions, yet these are certainly aspects of atomless mereology that deserve further scrutiny.

Composition Principles

The other main way of extending M is via *composition principles*, i.e. principles governing the behavior of Pin the bottom-up direction: from the parts to the wholes that they compose. We have already seen that the *Complementation* axiom (P.6) is, in a way, a principle of this sort. Another such principle would be the dual of *Atomlessness*, to the effect that everything might be "worldless junk"[14] that composes forever into greater and greater wholes:

(P.9) $\exists yPPxy$ *Ascent*

Both (P.6) and (P.9) are consistent with any of the theories considered so far. They are, however, fairly strong principles, which reflect specific views on the overall mereological structure of the universe. More generally, it is customary to consider ways of extending M by means of composition principles that specify the *conditions* under which one or more things qualify as parts of a larger whole.[15]

---

[13]See Cotnoir (2013) for some work in this direction.

[14]The phrase is from Schaffer (2010: 64).

[15]This is a version of the so-called "Special Composition Question". See van Inwagen (1990: Chap. 2).

Fig. 6 An a-sum that is not a b- or c-sum, and a c-sum that is not an a- or b-sum[18]

The most basic principles of this sort have the following form, to the effect that for any pair of suitably related entities, i.e., any two entities satisfying a given provision ξ, there is something of which both are part—an underlapper:

(P.10) $\xi xy \rightarrow Uxz$ *ξ-Bound*

Such principles are quite weak. For example, regardless of how exactly ξ is construed, (P.10) is trivially satisfied in any model that includes a universal entity of which everything is part.

A stronger sort of requirement is that any pair of suitably related entities have a *minimal* underlapper, something composed of their parts and nothing else. There are at least three ways of formulating such a requirement, corresponding to three different ways of characterizing the relevant notion of a minimal underlapper, also known as a mereological *sum* of the two entities in question[16]:

(11_a) $S_a zxy =_{df} \quad Pxz \wedge Pyz \wedge \forall w((Pxw \wedge Pyw) \rightarrow Pzw)$ *a-sum*[17]
(11_b) $S_b zxy =_{df} \quad Pxz \wedge Pyz \wedge \forall w(Pwz \rightarrow (Oxw \vee Oyw))$ *b-sum*
(11_c) $S_c zxy =_{df} \quad \forall w(Owz \leftrightarrow (Oxw \vee Oyw))$ *c-sum*

In M these three notions are pairwise distinct (Fig. 6), though they may coincide in the presence of further axioms. For instance, given *Strong Supplementation*, (11_b) and (11_c) are equivalent (though stronger that (11_a)), whereas in the presence of *Complementation* all three notions coincide so long as there is a universal entity: in that case, each sum of any two things is just the complement of the difference between the complement of one minus the other. (Such is the strength of (P.6)—a genuine cross between decomposition and composition principles.)

For each i $\in$ {a, b, c}, we can then extend M by adding a corresponding axiom as follows, where again ξ specifies a suitable binary condition:

($P.11_i$) $\xi xy \rightarrow \exists z S_i zxy$ *ξ-Sum_i*

The non-equivalence of these axioms is immediately verified by taking ξ to be satisfied by all pairs of objects and considering the models in Fig. 6. But the axioms may also differ when ξ is more restrictive. For instance, with ξ expressing overlap,

---

[16]The first notion may be found in Eberle (1967) and Bostock (1979), the second in Tarski (1935) and Lewis (1991), the third in Simons (1987) and Casati and Varzi (1999).

[17]Given *Reflexivity* and *Transitivity*, the definiens in (11_a) is equivalent to $\forall w(Pzw \leftrightarrow (Pxw \wedge Pyw))$.

[18]The non-extensional model of Fig. 3 also depicts a case in which x and y have a c-sum, in fact two c-sums (themselves), though no a- or b-sum. This runs contrary the intended meaning of 'sum', suggesting that (11_c) is best suited to theories at least as strong as EM. See Hovda (2009) for discussion.

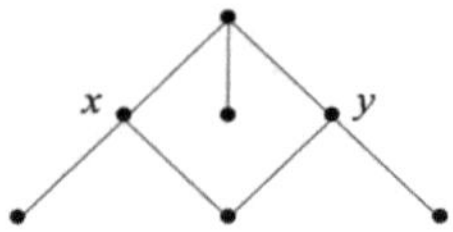

Fig. 7 A model of ξ-sum$_a$ violating ξ-sum$_b$ and ξ-sum$_c$

the model in Fig. 6, right, still satisfies (P.11$_c$), but not (P.11$_a$) or (P.11$_b$), whereas the model in Fig. 7 satisfies (P.11$_a$), but not (P.11$_b$) or (P.11$_c$). In EM, however, (P.11$_b$) or (P.11$_c$) are equivalent, since the corresponding notions of sum coincide.

The intuitive force of each (P.11$_i$) is in fact best appreciated in the context of EM, for in that case the relevant sums must be unique. If we introduce a corresponding binary operator (using 'ι' for the definite descriptor),

(12$_i$) $x +_i y =_{df} \quad \iota z F_i zxy$ *i-sum*

then is then easy to see that EM warrants all the "Boolean" properties one might expect. For instance, as long as the arguments satisfy the relevant condition ξ,[19] each operator is idempotent, commutative, and associative:

(13) $x = x +_i x$
(14) $x +_i y = y +_i x$
(15) $x +_i (y +_i z) = (x +_i y) +_i z$

and well-behaved with respect to parthood:

(16) $Px(x +_i y)$
(17) $Pxy \rightarrow Px(y +_i z)$
(18) $P(x +_i y)z \rightarrow Pxz$
(19) $Pxy \leftrightarrow x +_i y = y$

Each (P.11$_i$) is still fairly weak, for it governs only finitary composition. We get even stronger composition principles by requiring a minimal underlapper to exist for *any* set of objects satisfying a given condition, including infinite sets (whose sums—or *fusions*—cannot be generated by means of the binary operators defined above). There is, of course, a technical obstacle to formulating such principles in their full generality without resorting to explicit quantification over sets, since a standard first-order language does not have the resources to specify all sets, but only a denumerable number (in any given domain).[20] However, one can achieve a sufficient degree of generality by relying on axiom *schemas* where the relevant sets are identified through open formulas. Thus, let 'φ' be any formula in the language,

[19] If the condition is *not* satisfied, the sum may not exist, in which case the standard treatment of descriptive terms implies that the corresponding instances of the theorems that follow are false. In classical logic, (13)–(19) should therefore be taken to hold conditionally on the assumption that the relevant variables range over ξ-related entities.

[20] To overcome this limitation, some early theories such as those of Tarski (1929) and Leonard and Goodman (1940) resort to explicit quantification over sets. Others, such as Lewis (1991), resort to the machinery of plural quantification.

and let 'ψ' expresses the condition in question. Infinitary variants of the three notions of sum in (11_a)–(11_c) can be defined as follows, respectively[21]:

(20_a) $F_a z\varphi w =_{df} \forall w(\varphi w \to Pwz) \wedge \forall v(\forall w(\varphi w \to Pwv) \to Pzv)$ *a-fusion*
(20_b) $F_b z\varphi w =_{df} \forall w(\varphi w \to Pwz) \wedge \forall v(Pvz \to \exists w(\varphi w \wedge Owv))$ *b-fusion*
(20_c) $F_c z\varphi w =_{df} \forall v(Ovz \leftrightarrow \exists w(\varphi w \wedge Owv))$ *c-fusion*

('$F_i z\varphi w$' may be read as 'z is an i-fusion of the φ-ers'.) For each such notion, we may then introduce a corresponding principle of infinitary fusion through the following axiom schema, which asserts the existence of an i-fusion ($i \in \{a, b, c\}$) for every non-empty set of objects satisfying ψ:

($P.12_i$) $(\exists w\varphi w \wedge \forall w(\varphi w \to \psi w)) \to \exists z F_i z\varphi w$ *ψ-Fusion$_i$*

It can be checked that each ($P.12_i$) includes the corresponding finitary principle ($P.11_i$) as a special case, taking 'φw' to be the formula '$w = x \vee w = y$' and 'ψw' the condition '$(w = x \to \xi wy) \wedge (w = y \to \xi xw)$'. Thus, again, these principles are pairwise distinct in M, though it turns out that in the presence of *Strong Supplementation* ($P.12_b$) and ($P.12_c$) are equivalent.

Finally, the strongest versions of all these composition principles are obtained by asserting them as axiom schemas holding for *every* condition ψ, i.e., effectively, by foregoing any reference to ψ altogether. Formally this amounts in each case to dropping the second conjunct of the antecedent of ($P.12_i$), i.e., to asserting the schema expressed by the relevant consequent for any non-empty set of objects specifiable in the language:

($P.13_i$) $\exists w\varphi w \to \exists z F_i z\varphi w$ *Unrestricted Composition$_i$*

Once again, the relative strength of these principles varies for each $i \in \{a, b, c\}$. In particular, it is noteworthy that adding ($P.13_b$) to MM would suffice to warrant the equivalence of *Weak* and *Strong Supplementation*, (P.4) and (P.5), whereas adding ($P.13_c$) would not (Fig. 4 would still count as a counteremodel). Given (P.5), however, the two composition principles are equivalent, which means that the theory obtained by adding every instance of ($P.13_b$) to MM[22] is the same theory obtained by adding every instance of ($P.13_c$) to EM. This theory is known in the literature as *General Extensional Mereology*, or GEM. The same theory can be obtained by extending MM with ($P.13_a$), provided the following axiom is also added[23]:

(P.14) $(F_a z\varphi w \wedge Pxz) \to \exists w(\varphi w \wedge Owx)$ *Filtration*

[21](20_a)–(20_c) are to be read on the assumption that the variables 'z' and 'v' do not occur free in φ. Similar restrictions will apply below.

[22]Indeed, (P.2) and (P.4) would suffice.

[23]From Hovda (2009).

Classical Mereology

GEM is a powerful theory, and it was meant to be so by its nominalistic forerunners, who were thinking of mereology as a fundamental alternative to set theory.[24] Indeed, GEM has such a distinguished pedigree that it has earned the title of *Classical Mereology*. It is also a decidable theory, whereas for example M, MM, EM, and many extensions thereof are not.[25] To see just how powerful GEM is, consider the following operator, where 'F' is any of the 'F_i's defined above (which GEM forces to coincide):

(21) $\sigma x \varphi x =_{\text{df}} \iota z F z \varphi x$ *general fusion*

In terms of this operator—the fusion of all φ-ers—GEM can be further simplified, for example by merging (P.5) and (P.13$_c$) into a single axiom schema:

(P.13) $\exists x \varphi x \rightarrow \exists z(z = \sigma x \varphi x)$ *Unique Unrestricted Fusion*

and we can introduce the following definitions:

(22) $x + y =_{\text{df}} \sigma z(Pzx \vee Pzy)$ *sum*[26]
(23) $x \times y =_{\text{df}} \sigma z(Pzx \wedge Pzy)$ *product*
(24) $x - y =_{\text{df}} \sigma z(Pzx \wedge \neg Ozy)$ *difference*
(25) $\sim x =_{\text{df}} \sigma z \neg Ozx$ *complement*
(26) $\text{U} =_{\text{df}} \sigma z Pzz$ *universe*

The full strength of GEM can then be appreciated by considering that its models are closed under each of these notions, subject to the satisfiability of the relevant conditions. More exactly: the condition '$\neg Oz\text{U}$' is unsatisfiable, so U cannot have a complement. Likewise products are defined only for overlappers and differences only for pairs that leave a remainder. In all other cases, however, (22)–(26) yield perfectly well-behaved operators. Since such operators are the natural mereological analogues of the familiar set-theoretic operators, with 'σ' in place of set abstraction, it follows that the parthood relation axiomatized by GEM has essentially the same properties as the inclusion relation in standard set theory, modulo the absence of a null entity corresponding to the empty set. Indeed, P is virtually isomorphic to the inclusion relation restricted to the set of all non-empty subsets of a given set, which is to say a complete Boolean algebra with the zero element removed. We say 'virtually' because, strictly speaking, this is only true of stronger versions of GEM in which infinitary sums are defined using explicit quantification over sets.[27] For set-free formulations that, like those considered here, strictly adhere to a standard

[24]See the classical works of Lésniewski (1927–1931) and Leonard and Goodman (1940).

[25]For a comprehensive picture of decidability in mereology, see Tasi (2013b).

[26]In GEM, this definition is equivalent to (12$_i$), for each i $\in$ {a, b, c}.

[27]As such, the result goes back to Tarski (1935: n. 4).

first-order language with a denumerable supply of open formulas, the isomorphism does not quite hold. However, this is only a minor limitation, and we can still characterize the exact algebraic strength of GEM as follows: any model of this theory is isomorphic to a Boolean *subalgebra* of a complete Boolean algebra with the zero element removed (a subalgebra that is not necessarily complete if Zermelo-Frankel set theory with the axiom of Choice is consistent).[28]

In this connection, two further points are worth stressing. First, the existence of a "null entity" which is part of everything—the analogue of the empty set—is not in principle incompatible with GEM. However, it is easy to see that the only models of GEM with such an entity are trivial one-element models, owing to *Weak Supplementation*. It is for this reasons that the principles of *Unrestricted Composition* in (P.13$_i$) are stated as conditionals warranting the existence of a fusion for any given *non-empty* set of φ-ers. Dropping such a proviso would have disastrous effects, for then the existence of a null entity—*the* null entity—would be guaranteed by taking 'φw' to be the condition '$\forall xPwx$'. The only way around the disaster would be to revisit the non-basic vocabulary by carefully distinguishing trivial cases of parthood and overlap (involving the ubiquitous null entity) and non-trivial, genuine ones, as in

(27) $GPxy =_{\text{df}} \quad Pxy \wedge \exists z \neg Pxz$ *genuine parthood*
(28) $GOxy =_{\text{df}} \quad \exists z(GPzx \wedge GPzy)$ *genuine overlap*

and by reformulating all non-core axioms accordingly.[29] In this way, one can actually arrive at a variant of GEM that inherits all the strength of a complete Boolean algebra. Nonetheless, the philosophical import of such a theory would remain dubious.

Second, note that GEM is fully committed to the existence of U, a "universal entity" of which everything is part. This is not by itself a problem, barring any philosophical concerns about the gerrymendered nature of such an entity. It is, however, not without consequences. In particular, while GEM admits of models in which everything is composed of atoms as well as "gunky" models in which everything divides forever, the necessary existence of U deprives GEM of any "junky" model in which everything composes forever. Thus, while GEM admits of both atomistic and atomless extensions, adding the *Ascent* principle (P.9) would immediately result in an inconsistent theory.

[28]See Pontow and Schubert (2006), Theorem 34, for details and proof.

[29]This strategy is not uncommon in the mathematically oriented literature; see again Pontow and Schubert (2006) for a comprehensive treatment.

Summary of GEM

For ease of reference, we conclude by summarizing the main axiomatizations of GEM mentioned above, with some rewriting of bound variables and dropping all redundancies[30]:

(I)	$(Pxy \wedge Pyz) \rightarrow Pxz$	*Transitivity*	(P.2)
	$PPxy \rightarrow \exists z(PPzy \wedge \neg Ozx)$	*Weak Supplementation*	(P.4)
	$\exists x\varphi x \rightarrow \exists z F_{\mathrm{a}} z\varphi x$	*Unrestricted Composition*$_{\mathrm{a}}$	(P.13$_{\mathrm{a}}$)
	$(F_{\mathrm{a}} z\varphi x \wedge Pyz) \rightarrow \exists x(\varphi x \wedge Oxy)$	*Filtration*	(P.14)
(II)	$(Pxy \wedge Pyz) \rightarrow Pxz$	*Transitivity*	(P.2)
	$PPxy \rightarrow \exists z(PPzy \wedge \neg Ozx)$	*Weak Supplementation*	(P.4)
	$\exists x\varphi x \rightarrow \exists z F_{\mathrm{b}} z\varphi x$	*Unrestricted Composition*$_{\mathrm{b}}$	(P.13$_{\mathrm{b}}$)
(III)	Pxx	*Reflexivity*	(P.1)
	$(Pxy \wedge Pyz) \rightarrow Pxz$	*Transitivity*	(P.2)
	$Pxy \wedge Pyx) \rightarrow x = y$	*Antisymmetry*	(P.3)
	$\neg Pyx \rightarrow \exists z(Pzy \wedge \neg Ozx)$	*Strong Supplementation*	(P.5)
	$\exists x\varphi x \rightarrow \exists z F_{\mathrm{c}} z\varphi x$	*Unrestricted Composition*$_{\mathrm{c}}$	(P.13$_{\mathrm{c}}$)
(IV)	$(Pxy \wedge Pyz) \rightarrow Pxz$	*Transitivity*	(P.3)
	$\exists x\varphi x \rightarrow \exists z(z = \sigma x\varphi x)$	*Unique Unrestricted Fusion*	(P.13)

[30]See also Simons (1987) and Hovda (2009) for additional axiom sets. The elegant axiomatization in (IV) is essentially due to Tarski (1929), though the axioms are explicitly given only in the 1956 English translation.

Selected Bibliography

Claudio Calosi and Pierluigi Graziani

Aiello, M., Pratt-Hartmann, I., & van Benthem, J. (Eds.). (2007). *Handbook of spatial logics*. Berlin: Springer.
Armstrong, D. (1978). *Universals and scientific realism*. Cambridge: Cambridge University Press.
Bostock, D. (1979). *Logic and arithmetic, Vol. 2: Rational and irrational numbers*. Oxford: Clarendon.
Burkhardt, H., & Dufour, C. A. (1991). Part/whole I: History. In H. Burkhardt & B. Smith (Eds.), *Handbook of metaphysics and ontology* (pp. 663–673). Munich: Philosophia.
Burkhardt, H., Seibt, J., & Imaguire, G. (Eds.). (2014). *Handbook of mereology*. Munich: Philosophia.
Casati, R., & Varzi, A. (1999). *Parts and places*. Cambridge: MIT.
Chisolm, R. M. (1973). Parts as essentials to their wholes. *Review of Metaphysics, 26*, 581–603.
Clarke, B. L. (1981). A calculus of individuals based on connection. *Notre Dame Journal of Formal Logic, 22*, 204–218.
Cotnoir, A. J. (2010). Anti-symmetry and non-extensional mereology. *Philosophical Quarterly, 60*, 396–405.
Cotnoir, A. J. (2013). Beyond atomism. *Thought*, 2, 67–72.
Cruse, D. A. (1979). On the transitivity of the part-whole relation. *Journal of Linguistics, 15*, 29–38.
Eberle, R. A. (1967). Some complete calculi of individuals. *Notre Dame Journal of Formal Logic, 8*, 267–278.
Eberle, R. A. (1970). *Nominalistic systems*. Dordrecht: Reidel.
Forrest, P. (1996). How innocent is mereology? *Analysis, 56*, 127–131.
Forrest, P. (2002). Nonclassical mereology and its application to sets. *Notre Dame Journal of Formal Logic, 43*, 79–94.
Gilmore, C. (2009). Why parthood might be a four place relation and how it behaves if it is. In L. Hoonefelder, E. Runggaldier, & B. Schick (Eds.), *Unity and time in metaphysics* (pp. 83–133). Berlin: De Gruyter.
Goodman, N. (1951). *The structure of appearance*. Cambridge: Harvard University Press.
Hempel, C. G. (1953). Reflections on Nelson Goodman's the structure of appearance. *Philosophical Review, 62*, 108–116.
Hodges, W., & Lewis, D. K. (1968). Finitude and infinitude in the atomic calculus of individuals. *Nous, 2*, 405–410.

C. Calosi (✉) • P. Graziani
Department of Basic Sciences and Foundations, Urbino University, Urbino, Italy
e-mail: claudio.calosi@uniurb.it; pierluigi.graziani@uniurb.it

C. Calosi and P. Graziani (eds.), *Mereology and the Sciences*, Synthese Library 371,
DOI 10.1007/978-3-319-05356-1,

Hoffman, J., & Rosenkrantz, G. (1999). Mereology. In R. Audi (Ed.), *The Cambridge dictionary of philosophy* (2nd ed., pp. 557–558). Cambridge: Cambridge University Press.
Hovda, P. (2009). What is classical mereology. *Journal of Philosophical Logic, 38*(1), 55–82.
Iris, M. A., Litowitz, B. E., & Evens, M. (1988). Problems of the part-whole relation. In M. Evens (Ed.), *Relations models of the lexicon* (pp. 261–288). Cambridge: Cambridge University Press.
Johansson, I. (2004). On the transitivity of parthood relations. In H. Hochberg & K. Mulligan (Eds.), *Relations and predicates* (pp. 161–181). Lancaster/Frankfurt: Ontos.
Johansson, I. (2006). Formal mereology and ordinary language: Reply to Varzi. *Applied Ontology, 1*(2), 157–161.
Leonard, H. S., & Goodman, N. (1940). The calculus of individuals and its uses. *Journal of Symbolic Logic, 5*, 45–55.
Leśniewski, S. (1927–1931). O podstawach matematyki. *Przegląd Filozoficzny, 30*, 164–206; *31*, 261–291; *32*, 60–101; *33*, 77–105; *34*, 142–170. Eng. trans. by Barnett, D. I. (1992). On the foundations of mathematics. In S. J. Surma et al. (Eds.), *S. Leśniewski, collected works* (Vol. 1, pp. 174–382). Dordrecht: Kluwer.
Lewis, D. K. (1991). *Parts of classes*. Oxford: Blackwell.
McDaniel, K. (2009). Structure-making. *Australasian Journal of Philosophy, 87*, 251–274.
Mellor, D. H. (2006). Wholes and parts: The limits of composition. *South African Journal of Philosophy, 25*, 138–145.
Moltmann, F. (1997). *Parts and wholes in semantics*. Oxford: Oxford University Press.
Niebergall, K.-G. (2011). Mereology. In L. Horsten & R. Pettigrew (Eds.), *The continuum companion to philosophical logic* (pp. 271–298). New York: Continuum.
Polkowski, L. (2011). *Approximate reasoning by parts. An introduction to rough mereology*. Berlin: Springer.
Polkowski, L., & Skowron, A. (1994). Rough mereology. In Z. W. Ras & M. Zemankova (Eds.), *Proceedings of the 8th international symposium on methodologies for intelligent systems (ISMIS 94)*, Charlotte (pp. 85–94). Berlin: Springer.
Pontow, C. (2004). A note on the axiomatics of theories in parthood. *Data & Knowledge Engineering, 50*, 195–213.
Pontow, C. & Schubert, R. (2006). A mathematical analysis of theories of parthood. *Data & Knowledge Engineering, 59*, 107–138.
Rescher, N. (1955). Axioms for the part relation. *Philosophical Studies, 6*, 8–11.
Schaffer, J. (2010). Monism: The priority of the whole. *Philosophical Review, 119*, 31–76.
Sider, T. (2007). Parthood. *Philosophical Review, 116*, 51–91.
Simons, P. (1987). *Parts: A study in ontology*. Oxford: Clarendon.
Simons, P. (1991a). Free part-whole theory. In K. Lambert (Ed.), *Philosophical applications of free logic* (pp. 285–306). Oxford: Oxford University Press.
Simons, P. (1991b). Part/whole II: Mereology since 1900. In H. Burkhardt & B. Smith (Eds.), *Handbook of metaphysics and ontology* (pp. 209–210). Munich: Philosophia.
Simons, P. (2003). The universe. *Ratio, 16*, 237–250.
Sällström, P. (Ed.). (1983–1986). *Parts & wholes: An inventory of present thinking about parts and wholes* (Vol. 4). Stockholm: Forskningsrdsnmnden.
Smith, B. (1982). Annotated bibliography of writings on part-whole relations since Brentano. In B. Smith (Ed.), *Parts and moments. Studies in logic and formal ontology* (pp. 481–552). Munich: Philosophia.
Smith, N. J. J. (2005). A plea for things that are not quite all there. *Journal of Philosophy, 102*, 381–421.
Tarski, A. (1929). Les fondements de la géométrie des corps. Ksiega Pamiatkowa Pierwszkego Polskiego Zjazdu Matematycznego. Suppl. to *Annales de la Société Polonaise de Mathématique, 7*, 29–33; Eng. trans. by Woodger, J. H. (1956). Foundations of the geometry of solids. In A. Tarski (Ed.), *Logics, semantics, metamathematics. Papers from 1923 to 1938* (pp. 24–29). Oxford: Clarendon.
Tarski, A. (1935). Zur Grundlegung der Booleschen Algebra. I. *Fundamenta Mathematicae, 24*, 177–198; Eng. trans. by Woodger, J. H. (1956). On the foundations of the Boolean algebra, In

A. Tarski (Ed.), *Logics, semantics, metamathematics, papers from 1923 to 1938* (pp. 320–341). Oxford: Clarendon.
Thomson, J. J. (1983). Parthood and identity across time. *Journal of Philosophy, 80*, 201–220.
Tsai H.-c. (2009). Decidability of mereological theories. *Logic and Logical Philosophy, 18*, 45–63.
Tsai H.-c. (2011). More on the decidability of mereological theories. *Logic and Logical Philosophy, 20*, 251–265.
Tsai H.-c. (2013a). Decidability of general extensional mereology. *Studia Logica, 101*(3), 619–636.
Tsai H.-c. (2013b). A comprehensive picture of the decidability of mereological theories. *Studia Logica, 101*(5), 987–1012.
Uzquiano, G. (2006) The price of universality. *Philosophical Studies, 129*, 137–169.
van Inwagen, P. (1990). *Material beings*. Ithaca: Cornell University Press.
Varzi, A.C. (2000). Mereological commitments. *Dialectica, 54*, 283–305.
Varzi, A.C. (2006). A note on transitivity of parthood. *Applied Ontology, 1*, 141–146.
Varzi, A.C. (2008). The extensionality of parthood and composition. *The Philosophical Quarterly, 58*, 108–133.
Varzi, A.C. (2009b) Universalism entails extensionalism. *Analysis, 69*(4), 599–604.
Varzi, A. C. (2014). Mereology. In E. N. Zalta (Ed.), The Stanford encyclopedia of philosophy (Spring 2014 ed.). http://plato.stanford.edu/archives/spr2014/entries/mereology.
Whitehead, A. N. (1929). *Process and reality: An essay in cosmology*. New York: McMillian.
Winston, M., Chaffin, R., & Herrmann, D. (1987). A taxonomy of part-whole relations. *Cognitive Science, 11*, 417–444.
Yi, B. U. (1999). Is mereology ontologically innocent? *Philosophical Studies, 93*, 141–160.

Index

C. Calosi and P. Graziani (eds.), *Mereology and the Sciences*, Synthese Library 371, DOI 10.1007/978-3-319-05356-1, © Springer International Publishing Switzerland 2014

MIX
Papier aus verantwortungsvollen Quellen
Paper from responsible sources
FSC® C105338

If you have any concerns about our products,
you can contact us on
ProductSafety@springernature.com

In case Publisher is established outside the EU,
the EU authorized representative is:
Springer Nature Customer Service Center GmbH
Europaplatz 3, 69115 Heidelberg, Germany

Printed by Libri Plureos GmbH
in Hamburg, Germany